Encyclopaedia of Mathematical Sciences

Volume 9

Editor-in-Chief: R.V. Gamkrelidze

G. M. Khenkin (Ed.)

Several Complex Variables III

Geometric Function Theory

Springer-Verlag
Berlin Heidelberg NewYork
London Paris Tokyo

Title of the Russian edition:
Itogi nauki i tekhniki, Sovremennye problemy matematiki,
Fundamental'nye napravleniya, Vol. 9,
Kompleksnyĭ analiz – mnogie peremennye 3
Publisher VINITI, Moscow 1986

Mathematics Subject Classification (1980):
32A15, 32CXX, 32F05, 32F15, 32HXX, 53C15, 53C40

ISBN-13: 978-3-642-64785-7 e-ISBN-13: 978-3-642-61308-1
DOI. 10.1007/978-3-642-61308-1

Library of Congress Cataloging-in-Publication Data
Kompleksnyĭ analiz – mnogie peremennye 3. English.
Several complex variables III/G.M. Khenkin (ed.).
p. cm. – (Encyclopaedia of mathematical sciences; v. 9)
Translation of: Kompleksnyĭ analiz – mnogie peremennye 3.
Bibliography: p. Includes indexes.
ISBN-13: 978-3-642-64785-7
1. Functions of several complex variables. I. Khenkin, G.M.
II. Title. III. Title: Several complex variables 3. IV. Series.
QA331.K738313 1989 515.9'4 – dc19 88-4670 CIP

Softcover reprint of the hardcover 1st edition 1989

Typesetting, printing and binding:
Universitätsdruckerei H. Stürtz AG, D-8700 Würzburg
2141/3140-543210 – Printed on acid-free paper

List of Editors, Contributors and Translators

Editor-in-Chief

R.V. Gamkrelidze, Academy of Sciences of the USSR, Steklov Mathematical Institute, ul. Vavilova 42, 117966 Moscow, Institute for Scientific Information (VINITI), Baltiiskaya ul. 14, 125219 Moscow, USSR

Consulting Editor

G.M. Khenkin, Academy of Sciences of the USSR, Central Economic and Mathematical Institute, ul. Krasikova 32, 117418 Moscow, USSR

Contributors

I.M. Dektyarev, Vladimir State University, 600028 Vladimir, USSR

O.M. Khudaverdyan, Erevan State University, 375049 Erevan, USSR

V.Ya. Lin, Academy of Sciences of the USSR, Central Economic and Mathematical Institute, ul. Krasikova 32, 117418 Moscow, USSR

S.I. Pinchuk, Bashkir State University, 450074 Ufa, USSR

E.A. Poletskiĭ, Central Research Institute of Information and Technical and Economic Research, ul. Verkhnyaya Pervomajskaya, d. 47, korp. 11, 105264 Moscow, USSR

L.I. Ronkin, Low Temperature Physics Institute of the Academy of Sciences of the Ukrainian SSR, Prosp. Lenina 47, 310164 Kharkov, USSR

A.A. Roslyĭ, All-union Research Institute of Drilling Techniques, Leninskij Prospekt 6, 117957 Moscow, USSR

A.S. Schwarz, Moscow Physical Engineering Institute, Kashirskoe Sh. 31, Moscow M-409, USSR

B.V. Shabat †

A.E. Tumanov, Research Institute of Computing Complexes, 117485 Moscow, USSR

M.G. Zaĭdenberg, Orel State Pedagogical Institute, ul. Komsomolskaya, 302015 Orel, USSR

Translator

J. Peetre, University of Stockholm, Department of Mathematics, Box 6701, S-113 85 Stockholm, Sweden

Contents

I. Entire Functions

L.I. Ronkin

Translated from the Russian
by J. Peetre

Contents

Introduction

We consider the basic problems, notions and facts in the theory of entire functions of several variables, i.e. functions $f(z)$ holomorphic in the entire space $\mathbb{C}^n$ (i.e. $f \in H(\mathbb{C}^n)$). Such functions constitute an independent object of study and are often encountered in applications of complex analysis to other branches of mathematics. For instance, entire functions of several variables are widely used in the theory of linear p.d.e. [8], [32], in the theory of convolution equations [31], in the theory of distributions [12], [51], in probability theory [26]. There are also numerous applications of entire functions of several variables in various branches of physics.

In the theory of entire functions of $n>1$ variables, as in the case $n=1$, a central theme deals with questions of growth of functions and the distribution of their zeros. However, there are significant differences between the cases of one and several variables. In the first place there is the fact that for $n>1$ the zero set of an entire function is not discrete and therefore one has no analogue of a tool such as the canonical Weierstrass product, which is fundamental in the case $n=1$. Second, for $n>1$ there exist several different natural ways of exhausting the space $\mathbb{C}^n$, which for $n=1$ give rise to one and the same way of exhausting the plane with disks. Such exhaustions are necessary for the description of the growth of entire functions and the distribution of their zero sets. These circumstances have the effect that even in problems induced by corresponding problems in the case of one variable there often arise qualitatively new phenomena, for the description of which new methods are needed and also new notions, which in many cases have no direct counterpart for $n=1$, or no counterpart at all.

In accordance with what was said above, the material set forth in this part is directly or indirectly connected with problems of growth or distribution of zeros. In § 1 we describe the most frequently used characteristics of growth: orders, types and indicators. In § 2 we present the basic laws of the distribution of zeros of entire functions of several variables. In particular, several functions characterizing the distribution of zeros are given, the properties of the restrictions of the zero set of an entire function to complex lines are studied and estimates are found for the ($(2n-2)$-dimensional) volume of the zero sets. In § 3 we study the connection between the growth of an entire function and the "size" of its zero set. The main emphasis is here on the construction of an entire function which, given the zero set, admits minimal growth. In § 4 we set forth results concerning the holomorphic continuation to $\mathbb{C}^n$ of functions defined on analytic subsets of $\mathbb{C}^n$. Here the classes of functions in which we seek the solution of our problem are given in terms of growth estimates. The following Section is devoted to entire functions of exponential type. Such functions are very often encountered in applications. Here the main attention is given to the connections between these functions and Laplace and Fourier transforms and to the study of their discrete sets of uniqueness. In the last Section a number of supplementary results are set forth and also some results pertaining to functions with a completely regular growth and to quasipolynomials.

In our report we will not consider a number of other important directions in the theory of entire functions which have been developed intensively in the last 10–15 years. Among them are: 1) the study of the properties of the "level surfaces" of functions $f \in H(\mathbb{C}^2)$ as Riemann surfaces (Nisino, Saito, Yamaguchi and others); 2) the study of entire arithmetic functions, i.e. entire functions which on points $z \in \mathbb{Z}^n$ take algebraic values (Gramain, Pisot, Waldschmidt, Avanissian and Gay and others); 3) The study of the growth and the zero distribution of entire characteristic functions of probability laws (I.V. Ostrovskiĭ, I.P. Kamynin, Schopf and others); 4) the study of ideals in algebras of entire functions and of invariant subspaces in modules of entire functions (D.N. Gurevich, Taylor, Kelleher, Berenstein, Yger etc.); 5) the study of analytic sets of arbitrary dimension defined by systems of entire functions (Stoll, Gruman, Sibony and P.-M. Wong and others).

Likewise, we have in this report refrained from considering questions close in nature to corresponding questions for entire functions but referring to functions holomorphic in a cone, in radial tube domains or on an analytic set in $\mathbb{C}^n$.

Let us further remark that in an attempt to cut down the size of the references we have when there is a whole series of papers by one and the same author as a rule only listed one of these papers. In addition, if a result is included in a monograph then we refer to the monograph and not to the original paper. In many cases, instead of a complete reference we have restricted ourselves to giving just the name of the original of the result or investigation. In view ot this it is clear that from the references given it will not be possible to infer the time of the first appearance of the results set forth here.

§ 1. Characteristics of Growth

1.1. Orders and Types. The Class 𝔄. The measurement of the growth of (arbitrary) functions is similar to any measurement process. First one chooses the scale of measurement – in the case at hand some family of sufficiently simple functions. Thereafter one picks from this family a function whose asymptotic behavior, in some specially preassigned sense, is closest in behavior to the function whose growth we are measuring. Very often one takes for such a family (or scale) the family of power functions. Fundamental here are the notions of *order* ρ and *type* σ with respect to the order ρ, which are defined in the following manner for the *function* $f: \mathbb{R}_+ \to \mathbb{R}_+$:

$$\rho = \rho(f) = \inf\{a: \exists\, C_a < \infty, f(t) < C_a + t^a, \forall\, t > 0\},$$

$$\sigma = \sigma(f; \rho) = \inf\{a: \exists\, C_a < \infty, f(t) < C_a + a\,t^{\rho}, \forall\, t > 0\}.$$

These definitions are equivalent to the following:

$$\rho=\varlimsup_{t\to\infty}\frac{\ln^+ f(t)}{\ln t},\qquad \sigma=\varlimsup_{t\to\infty} t^{-\rho} f(t).$$

In order to define the order and type of an entire function $f(z)$, $z\in\mathbb{C}^n$, we pass at a preliminary step to the auxiliary *function* (A.A. Gol'dberg [37])

$$M_{f,G}(R)=\max_{z/R\in G}|f(z)|,$$

where G is a logarithmically convex complete polycircular domain[1] (a Reinhardt domain). Set now $\rho_f=\rho(\ln M_{f,g})$ and $\sigma_{f,G}=\sigma(\ln M_{f,G};\rho)$. Let us remark that the order ρ_f does not depend on the choice of the domain G, whereas the type $\sigma_{f,G}$ does depend on G. Frequently one takes for G one of the following three domains: $B_1=\{z\in\mathbb{C}^n\colon |z|<1\}$, $U^n=\{z\in\mathbb{C}^n\colon |z_1|<1,\dots,|z_n|<1\}$, $\{z\in\mathbb{C}^n\colon |z_1|+\dots+|z_n|<1\}$.

The order and the type are connected with the Taylor coefficients a_k, $k\in\mathbb{Z}_+^n$, of the function $f(z)$ by the following relations [37]:

$$\rho_f=\varlimsup_{\|k\|\to\infty}\frac{\|k\|\ln\|k\|}{-\ln|a_k|}\qquad(\|k\|=k_1+\dots+k_n),$$

$$(e\rho\sigma_{f,G})^{1/\rho}=\varlimsup_{\|k\|\to\infty}\{\|k\|^{1/\rho}[a_k d_k(G)]^{1/\|k\|}\},$$

where

$$d_k(G)=\sup\{|z_1|^{k_1}\dots|z_n|^{k_n}\colon z\in G\}.$$

For a more subtle description of the growth of the function $f(z)$ one has to pass, instead of the function $M_{f,G}(R)$, to the *function* of n variables

$$M(r;f)=M(r_1,\dots,r_n;f)=\max\{|f(z)|\colon |z_1|=r_1,\dots,|z_n|=r_n\}.$$

This function, apparently, grows steadily in each variable, whereas the function $\ln M(r;f)$, as was shown already by Valiron, is convex with respect to $\ln r_1,\dots,\ln r_n$ [37]. These properties, i.e. the monotonicity and the convexity in $\ln r_1,\dots,\ln r_n$, are shared by some other functions which arise in a natural way in the study of the zero sets of holomorphic functions. It is therefore natural to collect the functions $\Phi\colon\mathbb{R}_+^n\to\mathbb{R}$ with these properties into a special *class*, denoted by $\mathfrak{A}$ [37], and to study their growth, while as the characteristics of the growth of entire functions, similarly as was done with order and type, one declares the corresponding characteristics of growth of the functions $\ln M(r;f)$ considered as functions of the class $\mathfrak{A}$. Let us remark that the class $\mathfrak{A}$ can be defined in an equivalent manner as

[1] Let us recall that a domain $G\in\mathbb{C}^n$ is called complete polycircular if $z^0\in G$ implies that the polydisk $\{z\colon |z_1|\le|z_1^0|,\dots,|z_n|\le|z_n^0|\}$ belongs to G. Such a domain is said to be logarithmically convex if the set $|G|=\{(t,\dots,t_n\}\colon t_1=\ln|z_1|,\dots,t_n=\ln|z_n|,z\in g\}$ is convex.

$$\mathfrak{A}=\{\Phi(r)\colon \Phi(|z_1|, \ldots, |z_n|)\in \mathrm{PSH}(\mathbb{C}^n)\},$$

where PSH(G) is the set of all plurisubharmonic functions in G. For functions $\Phi\in\mathfrak{A}$ one defines in the obvious way the order $\sigma(\Phi)$ and the type $\sigma_{|G|}(\Phi)$:

$$\rho(\Phi)=\overline{\lim_{|r|\to\infty}}\frac{\ln\Phi(r)}{\ln|r|},$$

$$\sigma_{|G|}(\Phi)=\sigma_{|G|}(\Phi, \rho)=\overline{\lim_{r\to\infty}}\frac{1}{R^\rho}\sup\{\Phi(r)\colon r/R\in G\}.$$

One gets more subtle characteristics of growth if one compares the growth of functions $\Phi\in\mathfrak{A}$ with the growth of functions of the form

$$a_1 r_1^{b_1}+\ldots+a_n r_n^{b_n}$$

or if the growth of the functions

$$\Phi(a_1 t^{b_1}, \ldots, a_n t^{b_n})$$

for various values of the parameters a_i and b_i is compared with the growth of t, $t\to\infty$. The characteristics which arise in this way – the hypersurfaces of *conjugate orders and types*, order functions and type functions – have given rise to a variety of investigations (cf., in particular, [37], [30], [13], [44]). We will not dwell upon this in any detail. Let us just remark that the characteristic properties found for these notions, both for the class $\mathfrak{A}$ itself as well as for the functions $N_f(r)$ introduced in the following Section, are connected with the distribution of zeros of entire functions [37], [45].

1.2. Growth in a Distinguished Variable. The Class $\mathfrak{B}$. The growth of functions $\Phi\in\mathfrak{A}$ in one of the variables, for example, the variable r_n, the other ones taking on fixed values, is characterized by the *order in this variable*

$$\rho_n(\Phi)=\rho(\Phi('r, t))\quad ('r=(r_1, \ldots, r_{n-1}),$$

which quantity does not depend on $'r\in\operatorname{int}\mathbb{R}_+^{n-1}$, and the type

$$\sigma_n('r; \Phi, \rho)=\sigma(\Phi('r, t); \rho),$$

which belongs to the class $\mathfrak{A}$. Let us notice that $\rho(\Phi)\leqq\sum_{j=1}^n \rho_j(\Phi)$ (É. Borel, Valiron, cf. [37]).

For entire functions $f(z)$ one has an even more subtle phenomenon than the independence of $\rho(\Phi('r, t))$ on $'r$. Roughly speaking, this has to do with the circumstance that the growth of the function $f(z)$ in the variable z_1 only little depends on the choice of $'z=(z_2, \ldots, z_n)$ [2]. Thanks to this it is possible

---

[2] It is clear that in place of the growth in the variable z_1 one can speak of the growth of the function on complex lines belonging to any family of parallel lines (the growth in a given "complex direction") or lines issuing from a fixed point.

to apply in many multivariate problems the well developed machinery of the theory of entire functions of one variable. There are many investigations devoted to this phenomenon: Cire, Lelong, M.S. Stavskiĭ, B.I. Lokshin, S.Yu. Favorov, Kiselman, L.I. Ronkin and others (cf. [25], [37]). In these investigations various characteristics of growth were used and the above "small" dependence of the growth in z_1 from the choice of $'z$ was characterized in several ways. Here we give just one result, which has a definitive character and pertains to the case when the growth in the distinguished variable is characterized by the order. Before giving the exact formulation of this result, let us remark that the *function* $\ln M_f(r_1; 'z)$, where

$$M_f(r_1; 'z) = \max_{|z_1|=r_1} |f(z_1, 'z)|,$$

has the property that $\ln M_f(|z_1|; 'z) \in \mathrm{PSH}(\mathbb{C}^n)$. The set of all such functions $\Phi(t; 'z)(\Phi(|z_1|, 'z) \in \mathrm{PSH}(\mathbb{C}^n))$ will be denoted by $\mathfrak{B}$ (the *class* $\mathfrak{B}$). Let us remark that if $\Phi(r_1, 'z) \in \mathfrak{B}$ then

$$M_\Phi(r) = \sup\{\Phi(r_1, 'z) : |z_1| = r_1, \ldots, |z_{n-1}| = r_{n-1}\}$$

belongs to the class $\mathfrak{A}$ and its characteristics of growth are obtain in a natural way from the characteristics of growth of the function $\Phi(r_1, 'z)$.

Theorem 1. 1) *If* $\Phi(t, 'z) \in \mathfrak{B}$ *then the* set $E(\Phi)$ *of those* $'z \in \mathbb{C}^{n-1}$ *for which* $\rho(\Phi(t, 'z)) < \rho_1(M_\Phi(r))$ *is* $\mathbb{C}^{n-1}$*-polar*, [25].

2) *If* E *is a closed* $\mathbb{C}^{n-1}$*-polar set then there exists an entire function* $f(z)$ *such that* $E \subset E_f = E(\ln M_f(r_1; 'z))$, [29].

Corollary. *The order of growth of the restriction of a function* $f \in H(\mathbb{C}^n)$ *to complex lines* (the points of a projective space $\mathbb{P}^{n-1}$) *coincides everywhere on* $\mathbb{P}^{n-1}$ *with the order of growth* ρ_f *of the function* $f(z)$, *excepting possibly a* $\mathbb{C}^{n-1}$*-polar set.*

Let us remark that the order ρ_f is not changed if we make a nondegenerate linear change of coordinates.

1.3. Indicators of Growth. In the previous Sections the characteristics of growth were defined not directly for the entire function but with the aid of an intermediate passage to functions depending on the moduli of the coordinates (all or perhaps only one). In case when such an intermediate passage is not done we obtain for functions $f(z)$ of finite order certain characteristics of growth known as indicators. The most common of these is the so-called *regularized radial indicator* ([25], [37]) for the order $\rho > 0$:

$$\mathscr{L}_f^*(z) = \lim_{\varepsilon \to 0} \sup_{|\zeta - z| < \varepsilon} \mathscr{L}_f(\zeta),$$

where $\mathscr{L}_f(z) = \overline{\lim}_{t \to \infty} t^{-\rho} \ln |f(tz)|$. As follows from the recent (positive) solution by Bedford and Taylor to Lelong's problem (cf. Vol. 8, Chap. II), the set $\{z: \mathscr{L}_f(z) < \mathscr{L}_f^*(z)\}$ must be $\mathbb{C}^n$-polar. The obvious properties of the indicator $\mathscr{L}_f^*$ are:

1) $\mathscr{L}_f^* \in \mathrm{PSH}(\mathbb{C}^n)$;

2) $\mathscr{L}_f^*(tz) = t^\rho \mathscr{L}_f^*(z)$, $\forall t > 0$ (positively homogeneous).

Theorem 2. *If $\rho > 0$ and $\varphi(z) \in \mathrm{PSH}(\mathbb{C}^n)$ with $\varphi(tz) = t^\rho \varphi(z)$, $\forall t > 0$ then there exists a function $f(z) \in H(\mathbb{C}^n)$ of finite order ρ such that $\mathscr{L}_f^*(z) = \varphi(z)$.*

In the special case $\rho = 1$ this theorem was obtained by Kiselman [22] and in the general case by Martineau [37].

For entire functions $f(z, w)$, $z \in \mathbb{C}^n$, $w \in \mathbb{C}$, of finite order with respect to the variable w, one also considers the regularized *indicator with respect to the distinguished variable* [1]:

$$h_f^*(z, w) = \lim_{\varepsilon \to 0} \sup_{|\zeta - z| < \varepsilon, |\lambda - w| < \varepsilon} h_f(\zeta, \lambda),$$

where

$$h_f(\zeta, \lambda) = \overline{\lim_{t \to \infty}}\, t^{-\rho} \ln |f(\zeta, t\lambda)|.$$

The characteristic properties of this indicator are:

1) $h_f^*(z, w) \in \mathrm{PSH}(\mathbb{C}^{n+1})$; 2) $h_f^*(z, tw) = t^\rho h_f^*(z, w)$.

For entire functions of exponential type (i.e. of order $\rho_f = 1$ and of finite type) other types of indicators are studied. We will return to them in § 4. Let us further remark that in many investigations (F.I. Geche, P.Z. Agranovich and L.I. Ronkin) some of the facts mentioned here are obtained for so-called proximate orders.

§ 2. Distribution of Zeros

In the one dimensional case the distribution of zeros of an entire function $f(z)$ in a neighborhood of the point at infinity is characterized by the growth of the so-called *counting function* $n_f(t)$, which equals the numbers of zeros of $f(z)$ in the disk $|z| < t$ counted with their multiplicities. One also considers the logarithmic average of $n_f(t)$, that is, the function

$$N_f(t) = \int_0^t (n_f(s) - n_f(0))\, d \ln s + n_f(0) \cdot \ln t.$$

In the multivariate case the corresponding functions may be defined in several ways.

2.1. The Functions $n_f(t)$ and $N_f(t)$. Let $f \in H(\mathbb{C}^n)$ and put $\Lambda_f = \{z \in \mathbb{C}^n : f(z) = 0\}$, denoting by Λ_f^* the set of regular points of Λ_f. Consider D_f, the divisor of $f(z)$, that is, the pair consisting on the set Λ_f (the support of the divisor) and the *function* γ_f: $\Lambda_f \to \mathbb{N}$ which equals at a point $z \in \Lambda_f$ the multiplicity of the zeros of $f(z)$ at that point. Let us remark that $\gamma_f(z)$ is constant on each connected component of Λ_f^*.

Let us denote by $\mu_f(t)$ the $(2n-2)$-dimensional volume of the divisor D_f in the ball $B_t=\{z\in\mathbb{C}^n: |z|<t\}$, that is,

$$\mu_f(t)=\int_{B_t\cap\Lambda_f}\gamma_f\,dV_{2n-2}, \quad \text{(the } \textit{function } \mu_f(t)\text{)},$$

where

$$dV_{2n-2}=\frac{1}{(n-1)!}(d\,d^c|z|^2)^{n-1}.$$

Theorem 3 (Lelong [25], [37], [49], [14]).

$$\mu_f(t)=\frac{1}{2\pi}\int_{B_t}\Delta\ln|f|\,dV_{2n-2}=\frac{2}{\pi(n-1)!}\int_{B_t}d\,d^c\ln|f|\wedge(d\,d^c|z|^2)^{n-1}.$$

Here

$$\Delta=4\sum_i\frac{\partial^2}{\partial z_i\,\partial\bar z_i},$$

where derivation is taken in the sense of distributions.

Let $n_f(t)$ be the projective volume of the closure in $\mathbb{P}^{n-1}$ of the set obtained by projecting the intersection $D_f\cap B_t$. As the volume form in the projective space $\mathbb{P}^{n-1}$ equals $(d\,d^c\ln|z|^2)^{n-1}$ (cf., for example, [37]) we get

$$n_f(t)=\frac{1}{(n-1)!}\int_{B_t\cap\Lambda_f}\gamma_f(z)(d\,d^c\ln|z|^2)^{n-1}+\gamma_f(0).$$

This definition of the *function* $n_f(t)$ (cf., for example, [37]) may be replaced by an equivalent one.

Theorem 4.

$$n_f(t)=\frac{1}{\sigma_{2n-1}}\int_{\partial B_1}n_f(t;z)\,d\sigma,$$

where $n_f(t;z)$ is the "counting" function of the function $\varphi_z(w)=f(wz)$, $w\in\mathbb{C}$, and $d\sigma$ the element of the $(2n-1)$-dimensional volume of the sphere ∂B_1,

$$s_{2n-1}=\int_{\partial B_1}d\sigma.$$

The following relation connects the quantities $n_f(t)$ and $\mu_f(t)$.

Theorem 5 (Kneser, Lelong [25], [37]) [3]

$$n_f(t)=\frac{1}{V_{2n-2}}\frac{\mu_f(t)}{t^{2n-2}},$$

where V_{2n-2} is the volume of the ball B_1 in $\mathbb{C}^{n-1}$.

---

[3] In Kneser's paper it is assumed that $\gamma_f(0)\neq 0$. Let us also remark that Lelong establishes a similar result for analytic sets of arbitrary dimension and, moreover, for closed positive currents.

As the function $n_f(t)$, apparently, is steadily increasing and, moreover, $\lim_{t\to 0} n_f(t)=\gamma_f(0)$, it follows that the volume of the divisor D in the ball B_t, that is, $\mu_f(t)$ can be estimated from below by the quantity[4] $\gamma_f(0)\, V_{2n-2}\, t^{2n-2}$. Thus, if the support of the divisor contains the center of this ball then the quantity $\mu_f(t)$ attains its minimum when $f(z)$ is a linear function.

Assuming, in order to simplify the notation, that $f(0)\neq 0$, let us set

$$N_f(t)=\int_0^t n_f(s)\, d\ln s.$$

The function $N_f(t)$ can also be expressed in terms of the *functions*

$$N_f(t;z)=N_{\varphi_z}(t);$$

namely, we have

$$N_f(t)=\frac{1}{\sigma_{2n-1}}\int_{\partial B_1} N_f(t;z)\, d\sigma.$$

From this representation, using the usual Jensen's formula, it follows readily that

$$N_f(t)=\frac{1}{\sigma_{2n-1}}\int_{\partial B_1} \ln|f(tz)|\, d\sigma-\ln|f(0)|.$$

This identity is likewise called Jensen's formula.

With the aid of the functions $n_f(t;z)$ and $N_f(t;z)$ one measures the size of the divisor D_f on complex lines $\{\zeta\in\mathbb{C}^n\colon \zeta=wz,\ w\in\mathbb{C}\}$. It is interesting to note that there exists a uniform estimate of these quantities using the functions $N_f(t)$ and $n_f(t)$, which characterize the size of the divisor globally. Indeed, one has the following

Theorem 6 ([37]). $\forall\,\delta>1,\ \exists\, C_\delta<\infty$,

$$N_f(t;z)\leqq C_\delta\, N_f(t\delta), \quad \forall\, t>0, \quad z\in\partial B_1,$$

and

$$\forall\,\delta>1, \quad \forall\, t_0>0, \quad \exists\, C_{\delta,t_0}<\infty,$$

$$n_f(t;z)\leqq C_{\delta,t_0}\ln\frac{t}{t_0}\, n_f(\delta t), \quad \forall\, t>t_0, \quad z\in\partial B_1.$$

The existence of estimates in t for the functions $N_f(t;z)$ or $n_f(t;z)$ for z in some large set always entails estimates also for the functions $n_f(t)$ and $N_f(t)$ ([37], [16], [10]). Thus, in particular, if the type of the function $N_f(t;z)$ in the variable t (for the order ρ) is finite for $z\in E$, where the set E is not $\mathbb{C}^n$-polar, then the type of $N_f(t)$ must be finite too ([10], [37]).

[4] If $n=2$ this result was established simultaneously with Lelong also by Rutishauser.

2.2. The Functions $n_f(r_1, \ldots, r_n)$ and $N_f(r_1, \ldots, r_n)$. For a more exact characterization of the divisor D_f one uses the functions in n variables $n_f(r)$ and $N_f(r)$, where $r=(r_1, \ldots, r_n)$, to be defined below.

We denote by $n_f(r)$ the n-dimensional projective volume of the closure in $\mathbb{P}^{n-1}$ of the intersection of D_f with the cone $K_r=\{z: |z_1|/r_1=\ldots=|z_n|/r_n\}$ and the ball $B_{|r|}$. One has the formula

$$n_f(r)=\frac{|r|^n}{(2\pi)^n r_1 \ldots r_n} \int_0^{\pi} n_f(1; r_1 e^{i\varphi_1}, \ldots, r_n e^{i\varphi_n})\, d\varphi_1 \ldots d\varphi_n,$$

which may be taken as an equivalent definition. The corresponding function $N_f(r)$ is defined as follows:

$$N_f(r)=\int_0^{|r|} n_f(r/|r|\cdot t)\frac{dt}{t},$$

or, equivalently, by the formula

$$N_f(r)=\frac{1}{(2\pi)^n}\frac{|r|^n}{r_1 \ldots r_n}\int_0^{2\pi} \ldots \int_0^{2\pi} N_f(1; r_1 e^{i\varphi_1}, \ldots, r_n e^{i\varphi_n})\, d\varphi_1 \ldots d\varphi_n,$$

from which it follows, in particular, that the function $N_f(r)$ belongs to the class $\mathfrak{A}$ introduced in Sec. 1.1. Consequently, for the functions $N_f(r)$ the hypersurfaces of conjugate orders and types, the order-functions and the type functions mentioned in § 1 are defined and have the corresponding properties.

Concluding this Section let us state an inequality connecting $N_f(t)$ and $N_f(r)$. Indeed, for $\forall \delta>1$, $\exists C_\delta<\infty$, such that

$$N_f(t/|r|\cdot \min_{1\leqq i\leqq n} r_i)\leqq N_f(tr/|r|)\leqq C_\delta N_f(\delta t), \quad \forall t>0, \quad r\in \operatorname{int}\mathbb{R}^n_+.$$

2.3. Distribution of the Points of the Divisor D_f in a Distinguished Direction. Let us set $'z=(z_2, \ldots, z_n)$ writing $\psi_{'z}(z_1)=f(z_1, 'z)$. Using the function $\psi_{'z}(z_1)$ we construct in the standard way its counting function $n_f('z; t)$ and the corresponding average. Set

$$n_f('r; t)=\frac{1}{(2\pi)^{n-1}}\int_0^{2\pi} \ldots \int_0^{2\pi} n_f(r_2 e^{i\varphi_2}, \ldots, r_n e^{i\varphi_n}; t)\, d\varphi_2 \ldots d\varphi_n,$$

(the *function* $n_f('r; t)$) and

$$N_f('r; t)=\frac{1}{(2\pi)^{n-1}}\int_0^{2\pi} \ldots \int_0^{2\pi} N_f(r_2 e^{i\varphi_2}, \ldots, r_n e^{i\varphi_n}; t)\, d\varphi_2 \ldots d\varphi_n$$

(the *function* $N_f('r; t)$).

The function $n_f('r; t)$ admits the standard interpretation as the average value of the $(n-1)$-dimensional volume of the projection onto $T^{n-1}('r)=\{'z: |z_2|=r_2, \ldots, |z_n|=r_n\}$ of the intersection $D_f\cap\{z: |z_1|\leqq t\}$. The growth in the variable t of the functions introduced, apparently, characterizes the distribution of the points of the divisor D_f for $z_1 \to \infty$. It is not hard to see that

$$N_f(r) - N_f('r; r_1) = N_f(0, 'r)$$

(here we assume, of course, that $f(0, 'z) \not\equiv 0$). From this identity it is clear that the study of the growth in t of the function $N_f('r; t)$ may be reduced to the study in r_1 of the function $N_f(r)$, which belongs to the above class $\mathfrak{A}$. At the same time, as follows from Jensen's formula for $n=1$, we see that the function $N_f('z; t)$ up to an inessential term (for $t \to \infty$) coincides with the function

$$\Phi('z, t) = \frac{1}{2\pi} \int_0^{2\pi} \ln |f(t e^{i\varphi}, 'z)| \, d\varphi,$$

belonging to the class $\mathfrak{B}$ [1]. Consequently, the asymptotic properties (as $t \to \infty$) of the functions $N_f('z; t)$, and therefore, with some reserve, also of the functions $n_f('z; t)$ are the same as for functions in class $\mathfrak{B}$. Let us in particular remark that if $n_f('z; t) = 0$, $\forall t > 0$, $'z \in E$, where E is not $\mathbb{C}^n$-polar, then $n_f('z; t) = 0$, $\forall t > 0$, $'z \in \mathbb{C}^n \setminus E_1$, where E_1 either is an analytic set of codimension 1 or $E_1 = \emptyset$ ([25], [39], [42]). For $n=2$ the converse holds true too:

Theorem 7 ([2]). *For every compact set $K \subset \mathbb{C}$ of zero capacity there exists an entire function $f(z_1, z_2)$ such that $n_f(z_2; t) = 0$, $\forall t > 0$, $\forall z_2 \in K$, and at the same time* $\sup_{0 < t < \infty} n_f(z_2; t) > 0$, $\forall z_2 \notin K$.

Between the functions $N_f('z; t)$ and $N_f('r; t)$ there exist relations similar to those mentioned above between $N_f(t; z)$ and $N_f(t)$ [32].

Concluding this Section, let us remark that along with the divisor D_f of a function $f \in H(\mathbb{C}^n)$ we may consider the notion of divisor as an independent object, that is, as a pair (Λ, γ), comprising an analytic set Λ of pure codimension 1 and a function $\gamma: \Lambda^* \to \mathbb{N}$, which is constant on each connected component of the set Λ^* of regular points of Λ. Each such divisor coincides (in an obvious sense) with the divisor D_F of some function $F \in H(\mathbb{C}^n)$. It is clear that all the characteristics of the divisor D_F introduced above using the function F can be translated into characteristics of the divisor D, replacing in the corresponding denotations f by D, for example, writing $N_D(t)$ for $N_f(t)$.

§ 3. Analogues of the Canonical Product of Weierstrass

A fundamental tool in the theory of entire functions of one variable is certainly the representation of the functions with the aid of Weierstrass's canonical product. The special character of the multivariate case does not allow to give an immediate construction of this notion for functions of several variables. One can, however, use the fact that among all functions with a given zero set the canonical product in some sense has minimal growth.

In fact, as follows from Jensen's formula, for every function $f \in H(\mathbb{C})$ holds the formula $n_f(R) < \ln M_f(eR) - \ln|f(0)|$. Consequently $\rho_f \geqq \rho(n_f) = \rho(N_f)$. If now f is a canonical product then $\rho_f = \rho(n_f)$. Therefore, because of this and some other pecularities of the canonical product, guaranteeing the uniqueness of its construction from a given zero set, the following definition is natural.

The *function* $f(z) \in H(\mathbb{C}^n)$, $f(0) \neq 0$, is called *canonical* if

1) $\rho_f = \rho(n_f)$ and, provided $\rho_f = \rho \in \mathbb{N}$ and $\int^\infty t^{-\rho-1} n_f(t)\,dt < \infty$, in addition $\sigma(\ln M_f; \rho) = 0$.

2) In a neighborhood of the origin the Taylor development of $\ln f(z)$ has the form

$$\ln f(z) = \sum_{\|k\| = q+1}^{\infty} a_k z^k,$$

where q is the smallest integer such that the integral $\int^\infty t^{-q-2} n_f(t)\,dt$ is convergent.

It is not hard to see that if f and φ are canonical and $D_f = D_\varphi$ then $f(z) = \varphi(z)$. Let us also remark that it follows from the results mentioned in § 2 on the distribution of zeros of the restriction of an entire function to complex lines passing through the origin (cf. [37], [10]) that for almost all such lines the restriction of a canonical function is an ordinary Weierstrass canonical product. From this, in turn, it is not hard to see that for canonical functions holds the estimate

$$\ln M_f(R) \leqq C_{q,n} R^q \left\{ \int_0^R t^{-q-1} n_f(t)\,dt + R \int_R^\infty t^{-q-2} n_f(t)\,dt \right\},$$

where q is the number entering in the definition of canonical function.

The problem of representing an entire functions in terms of canonical functions or, equivalently, the construction of a canonical function with a given divisor was solved with different methods by Lelong and Stoll (cf. [25], [49] and likewise [37]).

Theorem 8. *For $F \in H(\mathbb{C}^n)$, $F(0) \neq 0$ and put $\rho(n_F) = \rho < \infty$. Then there exist a canonical function $f_F(z)$ and a function $g_F(z) \in H(\mathbb{C}^n)$ such that $F(z) = f_F(z)\, e^{g_F(z)}$.*

Lelong gives the following representation for $\ln|f_F|$:

$$\ln|f_F(z)| = \frac{\pi}{(n-1)\,\sigma_{2n}} \int_{A_F} \gamma_F(\zeta)\, A_q(z, \zeta)\, dV_{2n-2}, \quad \forall z \in \mathbb{C}^n, \tag{1}$$

where

$$A_q(z, \zeta) = -\frac{1}{|z-\zeta|^{2n-2}} + \frac{1}{|\zeta|^{2n-2}} - D_1(z, \zeta) - \dots - D_q(z, \zeta),$$

with

$$D_l(z, \zeta) = \frac{1}{l!} \left\{ \frac{\partial^l}{\partial t^l} \left(-\frac{1}{|t\zeta - z|^{2n-2}} \right) \right\} \Bigg|_{t=0}, \quad l = 1, 2, \dots.$$

The integral representation obtained by Stoll has the following form

$$\ln f_F(z) = \int_{A_F} \gamma_F(\zeta)\, B_q(z, \zeta)\, d V_{2n-2}, \tag{2}$$

where $B_q(z, \zeta)$ is a certain kernel guaranteeing that the integral in (2) and therefore the identity (2) itself do exist, at least in balls B_t, satisfying the condition $B_t \cap A_F = \emptyset$.

Let us remark that when $n=1$ and f_F is the canonical product of Weierstrass then

$$\ln f_F(z) = \sum_{l=1}^{\infty} \ln G_q(z/a_l), \tag{3}$$

where $G_q(z/a_l)$ is the Weierstrass prime factor and a_l are the roots of the function under consideration. The right hand side may be viewed as a zero-dimensional integral over the set of roots of F, that is, over the set A_F. Therefore the representations (1) and (2) may be considered as analogues of the formula (3).

It is readily seen that the intersection of the cone K_r (cf. § 2) with D_F is a uniqueness set of the divisor D_F. Therefore there arises the question of representing $\ln f_F(z)$ in the form

$$\ln f_F(z) = \int_{K_r \cap A_F} \gamma_F(\zeta)\, D(z, \zeta)\, d\omega_{n-1}.$$

This problem is solved in [37]. We do not give the concrete form of the kernel $D(z, \zeta)$. Let us just point out that the representation obtained, as well as the representation (2), holds only in a neighborhood of the origin.

In the multivariate case the minimality of the growth of the function with a given divisor may be treated in various ways. Thus, besides the treatment just given, minimality of growth may be interpreted as minimality in a distinguished variable. Such an approach to the problem of constructing analogues of the canonical product was given by L.I. Ronkin [37] and A. Sadullaev [43]. In the case when the growth of the entire function is characterized by the hypersurface of conjugate orders the minimality of the growth with a given divisor was studied by Schopf.

The problem of constructing an entire function with minimal growth with a given divisor has a sense also when the function $N_D(t)$ corresponding to that divisor does not have finite order ([48], [24]). However, in this case one can, of course, not have the same exactness of the estimates as in the case of finite order and, correspondingly, it is not possible to pick the canonical function in a unique way. Using the methods of $\bar{\partial}$-estimates Skoda obtained the following result.

Theorem 9. *Let $F \in H(\mathbb{C}^n)$ and assume that $n_F(t) \leqq C e^{\psi(t)}$, where $\psi(t)$ is a function with $\psi(|z|) \in \mathrm{PSH}(\mathbb{C}^n)$. Then for any $\alpha > 0$ and $\varepsilon > 0$ there exists a function $f \in H(\mathbb{C}^n)$ such that $D_F = D_f$ and*

$$\ln |f(z)| \leqq C(\varepsilon, \alpha)(1 + |z|)^{\alpha} e^{\psi(|z|(1+\varepsilon))}, \quad \forall z \in \mathbb{C}^n.$$

Skoda proved also other results close in nature to Theorem 9, in particular a theorem on the existence of entire functions $f_1, \ldots, f_{n+1}$ giving a purely p-dimensional analytic set X and having in one sense or other a minimal growth or a growth close to minimal growth.

Theorem 10 ([46]). *Let X be an analytic set of pure dimension p, $0 \leqq p \leqq n-1$. Denote by $\sigma(t)$ the $2p$-dimensional Euclidean volume of $X \cap B_t$ and put $\nu(t) = \pi^{-p} p!\, t^{-2p} \sigma(t)$. Then for each $\varepsilon > 0$ there exist entire functions $f_1(z), \ldots, f_{n+1}(z)$ such that $X = \{z : f_1(z) = \ldots = f_{n+1}(z) = 0\}$ and for $|z| \to \infty$ holds one* (but an arbitrary one, taking account of the restriction in 4)) *of the estimates*

1) $\ln |f_j(z)| \leqq C(\varepsilon) |z|^2 \sigma(|z| + \varepsilon)$, $j = 1, \ldots, n+1$;
2) $\ln |f_j(z)| \leqq C(\varepsilon) \ln^2 |z|\, \sigma(|z| + \varepsilon |z|)$, $j = 1, \ldots, n+1$;
3) *given, besides ε, a number $d > 0$*

$$\ln |f_j(z)| \leqq C(\varepsilon, d)(1 + |z|)^d \int_1^{1+|z|} t^{-d-1} \nu(t + \varepsilon t)\, dt, \quad j = 1, \ldots, n+1;$$

4) if $0 \notin X$ and $\int_0^\infty t^{-2} \nu(t + \varepsilon t)\, dt < \infty$

$$\ln |f_j(z)| \leqq C(\varepsilon) \left\{ \int_0^{|z|+\varepsilon} t^{-1} \nu(t)\, dt + (|z| + \varepsilon) \int_{|z|+\varepsilon}^{\infty} t^{-2} \nu(t)\, dt \right\}, \quad j = 1, \ldots, n+1.$$

The case of an arbitrary growth is likewise considered in Kujala [49] who gives conditions which the divisor D must satisfy in order that there exists, given *a priori* an increasing function $\lambda \colon \mathbb{R}_+^n \to \mathbb{R}_+$, an entire function $f(z)$ such that

1) $D_f = D$;
2) $\exists B > 0$, $\overline{\lim}_{R \to \infty} \ln M_f(R) / \lambda(BR) < \infty$.

Also the results in [39], [42] are related to the construction of canonical products. Here it is shown that if $\sup_t n_f(z, t) < \infty$, $\forall z \in E$, where E is in a suitable sense large set, then the function $f(z)$ must have the form $P(z)\, e^{g(z)}$ where $g \in H(\mathbb{C}^n)$ and $P(z) = P('z, z_1)$ is a pseudopolynomial with respect to z_1. In the case when one instead of $n_f('z; t)$ considers $n_f(z)$ one has as before the representation $f(z) = P(z)\, e^{g(z)}$ but now $P(z)$ is a polynomial in $z_1, \ldots, z_n$.

§ 4. Interpolation

The questions of interpolation constitute one of the traditional topics of the theory of entire functions. The classical interpolation problem consists of constructing a holomorphic function $f(z)$, $z \in \mathbb{C}$, taking at given points λ_j (the nodes of the interpolation) given values. One considers also interpolation problems of a different character and, notably, the problem of multiple interpolation, where at the nodes of the interpolation λ_j not only the values

$f(\lambda_j)$ of the sought functions but also the values of the derivatives $f'(\lambda_j), \ldots, f^{(q_j)}(\lambda_j)$ are given. The solution of these problems usually belongs to some class of functions. In the case when it is question of entire functions these classes are as a rule defined by growth estimates for the function at infinity. In this connection the properties of the set of nodes of interpolation, a discrete set, is usually characterized by the properties of the entire function of minimal growth (in a suitable sense) which vanishes on this set. Let us remark that interpolation problems have numerous applications within function theory itself as well as in its applications.

In the multivariate situation the zero set of a holomorphic function and an arbitrary discrete set in $\mathbb{C}^n$ are just special cases of analytic sets (respectively of dimension $n-1$ and 0). Therefore it is natural to treat the interpolation problem in $\mathbb{C}^n$ as a problem of holomorphic continuation from a given analytic set Λ in $\mathbb{C}^n$ (or in a region $G \subset \mathbb{C}^n$). As, generally speaking, Λ is not discrete, the given function on Λ (i.e., the one to be continued) cannot be arbitrary. It has to belong to the *space* $H(\Lambda)$ of functions $f(z)$ analytic on Λ. Let us recall that a function $f(z)$ defined on Λ is called analytic on Λ if $\forall z_0 \in \Lambda$ there exist a ball $B_{r_0}(z_0)$ and a function $f_{z_0}(z) \in H(B_{r_0}(z_0))$ such that $f_{z_0}(z)=f(z)$, $\forall z \in B_{z_0}(z_0) \cap \Lambda$. For such functions the problem of the holomorphic continuation to $\mathbb{C}^n$ is solvable, i.e. $\forall f \in H(\Lambda)$, $\exists F \in H(\mathbb{C}^n)$, $F(z)=f(z)$, $\forall z \in \Lambda$ (Cartan's Theorem A). In additional restrictions on the growth in one sense or other of the continuation $F(z)$ the problem gets considerably more complicated. The study of this problem began rather late and is far from being complete. We may distinguish two aspects of it. On the one hand, it is desirable to have the existence of the holomorphic continuation in question for a as large as possible class ot sets Λ. On the other hand, in given restrictions on the growth of the function $f \in H(\Lambda)$ it is natural to try to find the function $F(z)$ (the continuation of $f(z)$) of minimally possible growth. It is clear that when the class of sets Λ is extended the possibility of getting good estimates for $F(z)$ shrinks. The most general, but perhaps not the only way of constructing the continuation $F(z)$ is to first construct local continuations of $f(z)$, satisfying the required restrictions, and then to "paste together" these local continuations to a global one. This again leads to proving that the corresponding cohomology group, which has to be defined taking account of the given growth conditions, is trivial.

Let us further remark that, as in the one dimensional case, the interpolation problems in $\mathbb{C}^n$ have various applications, in particular in the study of systems of differential equations with constant coefficients, about which some details will be given in the following section, in the study of convolution equations, and in many questions of harmonic analysis.

4.1. Holomorphic Continuation from Algebraic Varieties. Recall that an analytic set $\Lambda \subset \mathbb{C}^n$ is called an algebraic variety if it can be presented in the form $\Lambda = \{z: P_1(z)=\ldots=P_l(z)=0\}$ where $P_1, \ldots, P_l$ are polynomials in

$z_1, \ldots, z_n$. In the following the set of all polynomials in $z \in \mathbb{C}^n$ will be denoted by $\mathscr{P}_n$ or simply by $\mathscr{P}$.

The first papers on holomorphic continuation from algebraic varieties were the papers by Ehrenpreis and V.P. Palamodov (cf. [32], [8]). In these papers a certain special multiple interpolation problem is considered, which serves as a tool in the study of systems of linear partial differential equations with constant coefficients. Let us elaborate this in some detail.

Every solution of a linear ordinary differential equation with constant coefficients has the form $y(t)=\sum_i P_i(t)\, e^{\lambda_i t}$, where the λ_i are the distinct roots of the corresponding characteristic equation and $P_i(t)$ has degree one unit less than the multiplicity of the root λ_i. Thus, the solution may be written in the form

$$y=\sum_j \sum_{k=0}^{q_j-1} C_{k,j}\left(\frac{\partial^k}{\partial \lambda^k}\, e^{\lambda t}\right)\Bigg|_{\lambda=\lambda_j}. \tag{4}$$

The sum entering here may be treated as an integral with respect to a measure concentrated at the points λ_j and, consequently, the multidimensional analogue of (4) ought to be of the form

$$Y=\sum_j \int_{\Lambda_j} D_j\, e^{\langle t, z\rangle}\, d\mu^{(j)}, \tag{5}$$

where the vector function $Y=Y(t)$ is a solution of the system under consideration, D_j are matrices whose elements are differential operators, Λ_j are analytic sets and $\mu^{(j)}$ are vector valued measures on these sets. It turned out that the proof of the representation (5) leads to the construction of a holomorphic continuation from $\cup \Lambda_j$ satisfying a series of special constraints. This problem was solved by the authors mentioned. We give here a partial, somewhat simplified version of their result. Let us introduce the definitions and notions required for this purpose.

We denote by $\mathscr{H}$ the *space* of holomorphic functions $f(z)$, $z=x+iy \in \mathbb{C}^n$ satisfying with some $C=C(f)>0$, $q=q(f)>-\infty$ and $a=a(f)\geqq 0$ the condition

$$|f(z)| \leqq C(1+|x|)^q\, e^{a|y|}, \quad \forall z \in \mathbb{C}^n. \tag{6}$$

We denote by $I=I(P)$, where $P=(P_1, \ldots, P_m)\in \mathscr{P}^m$, the ideal in $\mathscr{H}$ generated by the polynomials $P_1, \ldots, P_m$. In [8], [32] it is shown that for any I there exists a finite collection of analytic sets

$$\Lambda_j \subset \Lambda_I = \{z : f(z)=0,\ \forall f \in I\}, \quad j=1, \ldots, N=N(I)<\infty$$

and linear differential operators D_j with constant coefficients such that

$$f \in I \text{ iff } D_j f|_{\Lambda_j}=0, \quad j=1, \ldots, N.$$

Let us point out that it is not assumed that $\Lambda_j \neq \Lambda_i$ for $i \neq j$ and that the operators D_j are not determined uniquely by I. We denote further by $\mathscr{H}(I, d)$ the *space* of vector functions $F=(F_1, \ldots, F_N)\in H^N(\Lambda_I)$ satisfying the conditions:

a) $\exists c'=c'(f)>0,\ q'=q'(f)>-\infty,\ a'=a'(f)\geqq 0,$

$$|F(z)|\leqq c'(1+|x|)^{q'}e^{a'|y|},\quad \forall z\in\Lambda_I; \tag{7}$$

b) for any point $z_0\in\Lambda_I$ there exists a function $\varphi_{z_0}(z)$ holomorphic in a neighborhood ω_{z_0} such that $D_j\varphi_{z_0}(z)=F_j(z)$, $\forall z\in\omega_{z_0}\cap\Lambda_j$, $j=1,2,\ldots,N$.

Theorem 11 ([8], [32]). *Let $F\in\mathscr{H}(I,d)$. Then there exists a function $f\in\mathscr{H}$ such that $D_jf|\Lambda_j=F_j$, $\forall j=1,\ldots,N$. For any $\kappa>0$, prescribed in advance, it is possible to choose f such that any estimate for F of the form (7) forces f to satisfy (6) with $q\leqq q'+\gamma$, $a\leqq a'+\kappa$, where the constant γ depends only on the ideal I and the operators D_j, $j=1,\ldots,N$.*

Let us note that if $I=\operatorname{rad}I$ then one can take $N=1$, $\Lambda_1=\Lambda_I$ and D_1 equal to the identity operator. Consequently, from Theorem 11 follows

Corollary. *Let Λ be an arbitrary algebraic set in $\mathbb{C}^n$ and let the function $F\in H(\Lambda)$ satisfy condition (7) on Λ. Then for any preassigned $\kappa>0$ there exists a function $f(z)$ satisfying (6) such that $a(f)-a'(F)\leqq\kappa$, $q(f)-q'(F)\leqq\gamma$ where $\gamma=\gamma(\Lambda)$.*

In the case of holomorphic continuation from an arbitrary algebraic variety Λ let us further remark that if $f\in H(\Lambda)$ and $|f(z)|\leqq C_1|z|^a+C_2$ then $f(z)$ may be continued to $\mathbb{C}^n$ as a polynomial of degree $\leqq a+\gamma$, where $\gamma=\gamma(\Lambda)$ [7].

In the case when the form of the algebraic set is imposed additional restrictions of one kind or other, one has many other results which do not follow (at least, not immediately) from the results of Ehrenpreis and V.P. Palamodov.

The simplest algebraic varieties are complex surfaces (subspaces) of some dimension or other. In this case Hörmander proved the following result, having many applications.

Theorem 12 [17]. *Let $\varphi\in\mathrm{PSH}(\mathbb{C}^n)$ and assume that $\sup\{|\varphi(z')-\varphi(z'')|: |z'-z''|<1\}=C<\infty$. Let further Λ be a complex linear subspace of $\mathbb{C}^n$ of codimension k, $d\sigma$ being the volume element on Λ induced by the metric of $\mathbb{C}^n$. If $f\in H(\Lambda)$ and $\int_\Lambda|f|^2e^{-\varphi}d\sigma=\kappa<\infty$, then $\exists F\in H(\mathbb{C}^n)$ such that $F|\Lambda=f$ and*

$$\int_{\mathbb{C}^n}|F|^2e^{-\varphi}(1+|z|^2)^{-3k}\,dV_{2n}\leqq\text{const}\cdot\kappa.$$

A complex surface of (complex) codimension $n-1$ is a special case of an algebraic variety of the form $\Lambda_P=\{z: P(z)=0\}$, $P\in\mathscr{P}$. For such manifolds holds

Theorem 13 ([40]). *Let $f\in H(\Lambda_P)$ and assume that $\ln|f(z)|\leqq u(z)$, $\forall z\in\Lambda_P$, where $u(z)\in\mathrm{PSH}(\mathbb{C}^n)$. Then $\forall\varepsilon>0\ \exists F\in H(\mathbb{C}^n)$ such that $F|\Lambda_P=f$ and $\ln|F(z)|\leqq C_\varepsilon+N_{\varepsilon,P}\ln(1+|z|)+\sup\{u(z): |z-\zeta|<\varepsilon\}$.*

For functions $f\in H(\Lambda)$, where Λ is an arbitrary analytic set, it is possible to introduce in a natural way the notion of *order* $\rho=\rho(f,\Lambda)$ and *type*

$\sigma = \sigma(f;\Lambda,\rho)$. Indeed, we set

$$\rho = \varlimsup_{t\to\infty} \frac{1}{\ln t}\, \ln\ln M_{f,\Lambda}(t),$$

where $M_{f,\Lambda}(t)=\sup\{|f(z)|\colon z\in\Lambda,|z|\leq t\}$, and

$$\sigma = \varlimsup_{t\to\infty} t^{-\rho}\ln M_{f,\Lambda}(t).$$

The question of introducing for $f\in H(\Lambda)$ the notion of radial indicator is, however, somewhat harder. This is connected with the fact that a function $f\in H(\Lambda)$ cannot be defined even for just only one point for almost all complex lines passing through the origin. It turns out that it is natural to define the *indicator* $L(z;f,\Lambda)$ of a function $f\in H(\Lambda)$ as follows:

$$L(z;f,\Lambda)=\lim_{\varepsilon\to 0}\ \varlimsup_{t\to\infty}\sup\{\ln|f(tz')|\colon|z'-z|<\varepsilon,\,tz'\in\Lambda\},$$

with the convention $\sup_{\emptyset}\{\,:\,\} \overset{\text{def}}{=} -\infty$.

Theorem 14 ([40]). *Let $u\in\mathrm{PSH}(\mathbb{C}^n)\cap C(\mathbb{C}^n)$ and assume that $u(tz)=t^\rho u(z)$, $\forall t>0$, $z\in\mathbb{C}^n$. Let further $f\in H(\Lambda_P)$ with $L(z;f,\Lambda_P)\leq u(z)$, $\forall z\in\mathbb{C}^n$. Then there exists an entire function $F(z)$ such that $\mathscr{L}_F(z)\leq u(z)$, $\forall z\in\mathbb{C}^n$.*

From this theorem it follows, in particular, that each function $f\in H(\Lambda)$ which is of finite type σ on Λ for some $\rho>0$ can be continued to an entire function without the type being enlarged.

Results similar to Theorem 13 and Theorem 14 have also been established (L.I. Ronkin, A.M. Russakovskiĭ) for a set Λ_I in the case when the generators $P_1,\ldots,P_m$ of I satisfy the conditions: 1) for each $j=2,\ldots,n$ the set $\{z\colon Q_1(z)=\ldots=Q_j(z)=0\}$, where the Q_i are the homogeneous constituents of highest degree of P_i, has pure codimension j; 2) on each irreducible component of the set $\{z\colon P_1(z)=\ldots=P_j(z)=0\}$, $j=1,\ldots,m$, the maximum of the rank of the matrix $\{\partial P_i/\partial z_j\}$ equals j. Moreover, similar theorems have also been obtained in the case when Λ is the zero set of a pseudopolynomial, i.e. $\Lambda=\{(z,w)\colon z\in\mathbb{C}^n,\,w\in\mathbb{C},\,\sum_{j=0}^{q} f_j(z)w^{q-j}\}$, with $f_j\in H(\mathbb{C}^n)$, $j=0,1,\ldots,q$, and then the usual order has in the estimates to be replaced by the proximate order ([40]).

4.2. Continuation from Analytic Sets of General Character. In the past years some progress has been witnessed in the construction of holomorphic continuation from analytic sets of a more general character than algebraic varieties (Jennane, Demailly, Nisimura, Ioshioka, Berenstein and Taylor, Berndtsson, and others). Let us list some of the results obtained in this direction.

Assume that the function $p(z)\in\mathrm{PSH}(\mathbb{C}^n)$ satisfies the conditions:
1) $p(z)\geq 0$;

2) $\ln(1+|z|)=O(p(z))$;

3) $\exists c_1, c_2, c_3, c_4$ such that $|z-\zeta|\leqq\exp\{-c_1 p(z)-c_2\} \Rightarrow p(\zeta)\leqq c_3 p(z)+c_4$.

Let us denote by $H(\mathbb{C}^n; p(z))$ and $H(\Lambda; p(z))$ the sets (*spaces*) of functions in $H(\mathbb{C}^n)$ and $H(\Lambda)$ satisfying, respectively, the conditions $|f(z)|\leqq A e^{Bp(z)}$, $\forall z\in\mathbb{C}^n$, and $|f(z)|\leqq A e^{Bp(z)}$, $\forall z\in\Lambda$, $A=A(f)$, $B=B(f)$.

Let $\Phi=(\Phi_1, \ldots, \Phi_N)\in H^N(\mathbb{C}^n; p(z))$ and assume that the rank of the Jacobi matrix J_Φ of the map Φ equals k at each point $z\in\Lambda_\Phi=\{z: \Phi(z)=0\}$. Let us denote by $\Delta_{\Phi,k}(z)$ the sum of the moduli of all $(k\times k)$-minors of J_Φ.

Theorem 15 ([4]). *If there exist $\varepsilon>0$ and $C>0$ such that $|\Delta_{\Phi,k}(z)|\geqq \varepsilon\cdot\exp\{-C p(z)\}$, $\forall z\in\Lambda_\Phi$, then for each function $f\in H(\Lambda_\Phi; p(z))$ there exists $F\in H(\mathbb{C}^n; p(z))$ such that $F(z)=f(z)$, $\forall z\in\Lambda_\Phi$.*

In the case $N=k=n$ and, correspondingly, a discrete set Λ_Φ the sufficient conditions of Theorem 15 are, in some auxiliary conditions on the weight $p(z)$ and the map Φ, also necessary. Namely, let us assume that $|z-\zeta| \leqq 1 \Rightarrow p(\zeta)\leqq c_1 p(z)+c_2$ and that the *map* Φ is *slowly decreasing*, that is, there exist positive numbers ε, c_1, c_2, c_3 such that

1) all components Ω_j of the set $\{z: |\Phi(z)|\geqq\varepsilon e^{-c_1 p(z)}\}$ are bounded;

2) $p(\zeta)\leqq c_1 p(z)+c_2$ as soon as the points z and ζ lie in the same component Ω_j.

Theorem 16 ([3]). *Assume that the map $\Phi\in H^n(\mathbb{C}^n; p(z))$ is slowly decreasing and let its Jacobian* $\det J_\Phi(z)$ *be different from zero at each point $z\in\Lambda_\Phi$. Then for each function $f\in H(\Lambda_\Phi; p(z))$ to be continued holomorphically to a function $F\in H(\mathbb{C}^n; p(z))$ it is necessary and sufficient that for some $\varepsilon>0$ and $c>0$ there holds the inequality* $|\det J_\Phi(z)|\geqq\varepsilon e^{-cp(z)}$, $\forall z\in\Lambda_\Phi$.

Let us remark that the problem of continuation from a discrete set, satisfying certain special, but purely geometric conditions, has been considered after [3] in works by Kiselman and V.N. Logvinenko.

In the general case continuation from a set Λ_Φ can in a specified way be reduced to the case when Λ_Φ is discrete. Namely, in [3] there are given conditions on the map $\Phi\in H^k(\mathbb{C}^n; p(z))$ and the family L of analytic k-dimensional subspaces l which are sufficient for the solvability of the problem of holomorphic continuation from $l\cap\Lambda_\Phi$ (with growth conditions) for each $l\in L$ to entail the solvability of the corresponding problem for the set Λ_Φ itself. These conditions are rather cumbersome and we just remark that if $p(z)=p_1(|z|)$ and $p_1(2|z|)=O(p_1(|z|))$ and, in addition, $k=1$ then they are fulfilled if L is a family of complex lines passing through the origin and the restriction Φ_l of Φ, $\Phi(0)\neq 0$, to each line l is slowly decreasing on l.

In [3] one considers also the multiple interpolation problem in the case when the initial data are not functions in $H(\Lambda_\Phi; p(z))$ but elements of the quotient space $H(\mathbb{C}^n; p(z))/I$, where I is the ideal generated by the components of the map Φ.

A new approach to the interpolation problem just solved was given in [6]. This paper differs in an essential way from all previous ones, because there one does not only establish the existence of the interpolating functions but one also provides an explicit expression.[5]

Theorem 17 ([6]). *Let $\varphi(z)$ be a nonnegative convex function and let $\Phi=(\Phi_1, \ldots, \Phi_k)\in H_k(\mathbb{C}^n; \varphi)$ be given with $|\Delta_{\Phi,k}(z)|>0$, $\forall z\in \Lambda_\Phi$.*[6] *Moreover, let $f(z)$ be holomorphic on Λ_Φ satisfying with some $B<\infty$ the condition*

$$\int_{\Lambda_\Phi} |f(\zeta)|\, e^{-B\varphi(\zeta)} \frac{dV(\zeta)}{|\Delta_{\Phi,k}(\zeta)|}<\infty,$$

where $dV(\zeta)$ is the Euclidean volume element of the variety Λ_Φ. Define $F(z)$ by the formula

$$F(z)=C(n,k)\int_{\Lambda_\Phi} f(\zeta)\, e^{A\langle\partial\varphi, z-\zeta\rangle}(A\,\partial\bar\partial\varphi)^{n-k}\wedge\mu,$$

where

$$\mu=|\Delta_{\Phi,k}|^{-2}\Big(\sum_j g_i^j\, d\zeta_j\wedge\ldots\wedge\sum_j g_k^j\, d\zeta_j\wedge\overline{\partial\Phi_1}\wedge\ldots\wedge\overline{\partial\Phi_k}\Big)\,\delta_{\Lambda_\Phi}(z)$$

and g_i^j is the coefficient of the Hefer expansion

$$\Phi_i(z)-\Phi_i(\zeta)=\sum g_i^j(\zeta, z)(z_j-\zeta_j),$$

$\delta_{\Lambda\Phi}(z)$ denoting a generalized function (distribution) *given by the formula $(\delta_{\Lambda_\Phi}, \varphi)=\int_{\Lambda_\Phi}\varphi\, dV$, $\forall\varphi\in\mathscr{D}(\mathbb{C}^n)$. Then for A sufficiently large $F(z)$ will be entire with* 1) *$F|\Lambda_\Phi=f$ and* 2) *$F\in H(\mathbb{C}^n; \varphi)$.*

§ 5. Entire Functions of Exponential Type

In various applications of the theory of entire functions one frequently encounters *functions of exponential type* that is, functions of finite type for the order $\rho=1$. To a great extent this is caused by the connection of these functions with Laplace and Fourier transforms and their rôle in questions of completeness and basisness of systems. Let us now in some detail devote us to the corresponding aspects of the multivariate case.

5.1. Entire Functions of Exponential Type and Functions Associated with Them in the Sense of Borel. Let

$$f(z)=\sum_{k\in\mathbb{Z}^n_+} a_k/k!\, z^k,$$

[5] All this refers, of course, to the interpolation problem considered here for classes of entire functions. For bounded domains an explicit construction of interpolation functions was given earlier by G.M. Khenkin.

[6] The notation is the same as before.

where $k!=k_1!\dots k_n!$, be an entire function of exponential type. The *function associated with it in the sense of Borel* is the function

$$F_f(z)=\sum_{k\in\mathbb{Z}_+^n} a_k/z^{k+I},$$

where $I=(1,\dots,1)$, which, as is readily seen, is holomorphic at the point $(\infty,\dots,\infty)$. The functions $F_f(z)$ and $f(z)$ are connected by the following integral relations

$$f(z)=\frac{1}{(2\pi i)^n}\int_{l_1}\dots\int_{l_n} F_f(\zeta)\,e^{\langle z,\zeta\rangle}\,d\zeta_1\dots d\zeta_n,\quad \forall z\in\mathbb{C}^n,$$

$$F_f(z)=\int_{A_\varphi} e^{-\langle z,\zeta\rangle} f(\zeta)\,d\zeta_1\dots d\zeta_n,\quad \forall z\in G_{v,\varphi},\quad v\in T_{-\varphi},$$

Here

$$\varphi\in\mathbb{R}^n,$$

$$A_\varphi=\{\zeta\in\mathbb{C}^n\colon \zeta_j=t_j e^{-i\varphi_j},\ t_j>0, j=1,\dots,n\},$$

$$G_{v,\varphi}=\{z\colon \operatorname{Re} z_j e^{-i\varphi_j}>v_j, j=1,\dots,n\},$$

$$T_\varphi=\{v\in\mathbb{R}^n;\ \exists\, C_{v,\varphi}>0,\ \ln|f(r_1 e^{i\varphi_1},\dots,r_n e^{i\varphi_n})|\leqq C_{v,\varphi}+\langle v,r\rangle,\ \forall r\in\mathbb{R}_+^n\},$$

and $l_1,\dots,l_n$ are simple closed contours in $\mathbb{C}$ bounding domains $D_1,\dots,D_n$ such that $F_f(z)$ has an analytic continuation from a neighborhood of the point $(\infty,\dots,\infty)$ to the domain $\{z\colon z_1\notin\bar{D}_1,\dots,z_n\notin\bar{D}_n\}$.

In order to describe the distribution of the singular points of $F_f(z)$ we denote by $C_f(\varphi)$ the set of points $v\in\mathbb{R}^n$ for which $F_f(z)$ admits an analytic continuation to $G_{v,\varphi}$.

Theorem 18 (V.K. Ivanov [37]). *For each entire function of exponential type $f(z)$ holds the identity*

$$\bar{T}_f(\varphi)=C_f(-\varphi),\quad \forall\varphi\in\mathbb{R}^n.$$

This theorem is the multivariate analogoue of Pólya's well-known theorem on the connection between the growth of an entire function and the distribution of singularities of its associated function. In [37] this theorem is extended to functions of arbitrary order.

5.2. Functions of Exponential Type and Fourier Transforms. In suitable restrictions on the behavior of the function $f(z)$ for $z=x\in\mathbb{R}^n$ its Fourier transform will be a finite function (compact support). For an exact description of this fact one requires the notion of P-indicator (*the Pólya-Plancherel indicator*), which is defined by the formula

$$h_f(\lambda)=\sup_{x\in R^n} h_f^{\cdot}(\lambda,x),$$

where

$$h_f(\lambda,x)=\overline{\lim_{R\to\infty}} R^{-1}\ln|f(x+i\lambda R)|.$$

Theorem 19 (Plancherel, Pólya [37]). *For the function $f(z)$, $z\in\mathbb{C}^n$, to be entire of exponential type satisfying the condition $f(x)\in\mathscr{L}^2(\mathbb{R}^n)$ it is necessary and sufficient that one has the representation*

$$f(z)=\left(\frac{1}{\sqrt{2\pi}}\right)^n \int \Phi(\xi)\, e^{-\langle z,\lambda\rangle}\, d\xi_1 \dots d\xi_n$$

with some $\Phi(\xi)\in\mathscr{L}^2(\mathbb{R}^n)$ such that supp $\Phi\Subset\mathbb{R}^n$. *Then the support function*

$$H_{G_\Phi}(\lambda)=\sup_{x\in G_\Phi}\langle x,\lambda\rangle$$

of the smallest convex set G_Φ outside which $\Phi(\xi)=0$ coincides, for each $\lambda\in R^n$, with the P-indicator $h_f(\lambda)$ of $f(x)$.

Let us remark that the function $f(z)$ entering in this theorem is bounded on $\mathbb{R}^n+i0=\mathbb{R}^n$. Also ([37]) for each fixed λ one has $h_f(\lambda, x)=h_f(x)$ for almost all x, $h_f(\lambda)=\mathscr{L}_f^*(i\lambda)$ and $h_{fg}(\lambda)=h_f(\lambda)+h_g(\lambda)$, where g is any other entire function of exponential type.

Theorem 19 is the multivariate analogue of the corresponding theorem by Paley-Wiener. In the case when instead of the condition $f(x)\in\mathscr{L}^2(\mathbb{R}^n)$ one has $\ln|f(x)|\leqq c_1|x|^p+c_2$, $\forall x\in\mathbb{R}^n$, again $f(z)$ is the Fourier transform of a generalized function (distribution) with compact support (the Paley-Wiener-Schwartz theorem), cf., for instance, [12], [51], [31].

In the general case an entire function $f(z)$ of exponential type is the Laplace transform of an analytic functional carried by a convex set whose support function coincides with the 1-homogeneous minorant of the radial indicator $\mathscr{L}_f^*(z)$ (Martineau, Ehrenpreis [25], [37]).

5.3. Discrete Real Sets of Uniqueness.[7] As is well-known, the zeros of holomorphic functions of several variables cannot be isolated. Nevertheless, the problem of the study of discrete sets of uniqueness for various classes of such functions is a perfectly natural one, because often the study of the completeness of countable systems of functions in several variables leads to such questions. In particular, the study of the completeness of systems of multivariate exponentials $\{e^{\langle\lambda,x\rangle}\}_{\lambda\in E}$ in the space $\mathscr{L}^2(T_r)$, $r\in\mathbb{R}^n$, where $T_r=\{x\in\mathbb{R}^n\colon |x_1|<r_1,\dots,|x_n|<r_n\}$, leads to establishing conditions under which a set $E\subset\mathbb{R}^n$ is a set of uniqueness for entire functions satisfying the condition

$$\ln|f(z)|\leqq\sigma_1|y_1|+\dots+\sigma_n|y_n|+C_f,\quad y=\operatorname{Im} z.$$

A number of results in this direction[8] have been obtained by L.I. Ronkin and by Berndtsson (cf. [38], [5]).

[7] A set E is called a *uniqueness set* for a class of functions M if $f\in M$, $f=0$ on $E\Rightarrow f\equiv 0$.

[8] The first discrete uniqueness sets, not for entire functions but for functions holomorphic in the product of two halfplanes, were considered by Korevaar and Hellerstein [23].

Theorem 20 ([38], [5]). *Let E be a discrete set in $\mathbb{R}^n$ satisfying the condition $\inf\{\|x'-x''\|_\infty: x' \in E, x'' \in E, x' \neq x''\} = h_E > 0$, where $\|x\|_\infty = \max_j |x_j|$, and let $\gamma: E \to \mathbb{N}$ be a function such that*

$$\overline{\lim_{t\to\infty}} \frac{1}{(2t)^n} \sum_{x \in E \cap T_{tI}} \gamma(x) = d_{E,\gamma} > 0.$$

Assume further that the function $f \in H(\mathbb{C}^n)$ satisfies the conditions:

1) $\ln |f(z)| \leqq C_f + \sigma_1 |y_1| + \ldots + \sigma_n |y_n|, \forall z = x + iy \in \mathbb{C}^n$;
2) $\sigma_1 + \ldots + \sigma_n < \pi^n (2^{n-1}(n-1)!)^{-1} h_E^{n-1} d_{E,\gamma}$;
3) $f(z) = 0, \forall z \in E$, *and* $\gamma_f(z) \geqq \gamma(z), \forall z \in E$.

Then $f(z) \equiv 0$.

Corollary. *The system of functions*

$$\{\{x^k e^{i\langle\lambda, x\rangle}\}_{k_1 + \ldots + k_n \leqq \gamma(\lambda) - 1}\}_{\lambda \in E},$$

where E and $\gamma(\lambda)$ are as in Theorem 20, is complete in the space $\mathscr{L}^p(T_\sigma)$, $p > 1$, provided

$$\sigma_1 + \ldots + \sigma_n < \pi^n ((n-1)!\, 2^{n-1})^{-1} h_E^{n-1} d_{E,\gamma}.$$

Without entering into the details concerning other results referring to the question mentioned, let us just remark that necessary conditions for uniqueness are obtained in terms characterizing the size of the given set E in a given direction, that uniqueness sets for entire functions of exponential type without any restriction on their behavior on $\mathbb{R}^n$ are considered and that, finally, in the case $n=2$ the constant $(\pi^n(2^{n-1}(n-1)!)^{-1}|_{n=2} = 2\pi^2$ may be replaced by 2π, which is its best value.

These results are obtained by comparing the estimates from below and from above of the volume of D_f in the band $\{z: x \in \mathbb{R}^n, y \in T_{hI}\}$. This method is based on the utilization of estimates from below of the volume of the divisor mentioned in § 2. By another method a number of results pertaining to discrete uniqueness sets for entire functions of arbitrary order were obtained by V.N. Logvinenko [27].

5.4. Norming Sets and Equivalent Norms. Let $G = \{z: |z_1| + \ldots + |z_n| < 1\}$ and let Γ_n^σ be the set of all entire functions $f(z)$, $z \in \mathbb{C}^n$, of exponential type $\sigma_G \leqq \sigma$. The *set* $E \in \mathbb{R}^n$ is called *norming* for Γ_n^σ if there exists a constant $C = C(n, G, E)$ such that

$$\sup\{|f(x)|: x \in \mathbb{R}^n\} \leqq C \sup\{|f(x)|: x \in E\}, \quad \forall f \in \Gamma_n^\sigma.$$

Note that each norming set is automatically a uniqueness set.

The first theorem on norming sets (for $n=1$) was Cartwright's theorem, stating that the set $\mathbb{Z}$ is a norming set for the class Γ_1^σ for any $\sigma < \pi$. The methods used in the study of norming sets for $n=1$ turned out to be unsuitable for $n > 1$. A new approach (even for $n=1$) was developed by V.N. Logvinenko [27]. His method is based on a certain special kind of approximation of entire functions with polynomials and allows one to obtain a series of propositions on norming sets, among these the following

Theorem 21 [27]. *Let E be a δ-net in $\mathbb{R}^n$.*[9] *Then there exists an integer $p \in \mathbb{N}$ depending only on n such that for each $\sigma < 1/(2p\delta)$ the set E is norming for Γ_n^σ. Also, the corresponding constant C is $\leqq (1-\sigma\delta)^{-1}$.*

The estimate gotten for the number p is, apparently, very far from the optimal one.

The question of norming sets is closely connected with the question of sets defining norms equivalent to the norm in the space $\mathscr{L}^p(\mathbb{R}^n)$. More exactly, we ask for conditions for the set E which guarantee that for $\forall f \in \Gamma_n^\sigma$ holds the inequality

$$\int_{\mathbb{R}^n} |f(x)|^p\,dx \leqq C \int_E |f(x)|^p\,dx, \quad p>0, \tag{8}$$

with a constant C not depending on f.

Theorem 22. 1) *Assuming that a priori $f \in \mathscr{L}^p(\mathbb{R}^n)$, for* (8) *to be fulfilled it is necessary* [33] *that the set E is relatively dense with respect to Lebesgue measure, i.e. that there exist $l>0$ and $\delta>0$ such that*

$$\operatorname{mes}\{E \cap \{x+T_{lI}\}\} \geqq \delta, \quad \forall x \in \mathbb{R}^n.$$

2) *If E is relatively dense with respect to Lebesgue measure, then* (8) *is true for any class Γ_n^σ* [28], [20].

Let us also remark that sets relatively dense with respect to Lebesgue measure are norming sets for each Γ_n^σ (B.Ya. Levin, V.E. Katsnel'son [20]).

§ 6. Other Classes of Entire Functions and Separate Results

6.1. Entire Functions of Completely Regular Growth. An entire functions $f(z)$, $z \in \mathbb{C}$, not worse than of normal type with respect to the order $\rho>0$, is said to be a function of completely regular growth (f.c.r.g.) if $t^{-\rho}\ln|f(te^{i\theta})| \to \mathscr{L}_f(e^{i\theta})$ as $t\to\infty$, $t \in E$, for each set $E \subset \mathbb{R}_+$ satisfying the condition $\lim_{R\to\infty} R^{-1}\operatorname{mes}\{E\cap(0,R)\}=0$ (B.Ya. Levin, Pfluger). A basic fact, which recurs in many investigations of these functions, is the equivalence of the regular growth of a function with a regular distribution for its zero sets, which notion is defined as follows:

1) If ρ is not an integer, the zero set E of the function f is said to have a regular distribution (with respect to ρ) if it has an angular density, that is, if for any θ_1 and θ_2 outside a certain countable set N the limit $\Delta(\theta_1,\theta_2) = \lim_{t\to\infty} t^{-\rho} n_f(t,\theta_1,\theta_2)$ exists, where $n_f(t,\theta_1,\theta_2)$ is the number of zeros of $f(z)$ in the sector $\{z: |z|<t, \theta_1 < \arg z < \theta_2\}$ counted with multiplicity.

[9] Recall that a set E is called a δ-net in $\mathbb{R}^n$ if $\forall x \in \mathbb{R}^n$, $\exists x' \in E$, $|x-x'|<\delta$.

2) If $\rho>0$ is an integer, then E is said to be of regular distribution if it has an angular density and if, in addition, the following limit exists:

$$\lim_{R\to\infty} \sum_{a\in E, |a|<R} a^{-\rho}<\infty.$$

The definition of f.c.r.g. as well as their basic properties have been extended to the case of subharmonic functions in $\mathbb{R}^m$, $m\geqq 2$, by V.S. Azarin.

In the multivariate case f.c.r.g. can be defined in several ways. On the one hand, the definition of f.c.r.g. can be connected with various kinds of indicators. On the other hand, one can in the definition of f.c.r.g. start with either from the global behavior of the function or else from its behavior on a suitable family of complex lines.

Let $f(z)$ be an entire function in $\mathbb{C}^n$ having at most finite type with respect to the order $\rho\in(0,\infty)$. Let us call it of I-completely regular growth if for almost all $z\in\mathbb{C}^n$ the function $f(wz)$, as a function of the variable w, is a f.c.r.g. [15]. We say that $f(z)$ is of II-c.r.g. whenever the limit

$$\overline{\lim_{t\to\infty}} \frac{\ln|f(tz)|}{t^\rho}=\mathscr{L}_f^*(z),$$

exists in the space of distributions $\mathscr{D}'(\mathbb{C}^n)$.

Theorem 23 ([1]). *Every I-c.r.g.f. $f(z)$ is also II-c.r.g.*

P.Z. Agranovich and V.S. Azarin have proved II-c.r.g. for $f(z)$ is equivalent to c.r.g. for $\ln|f(z)|$ as a subharmonic function. It is natural to ask if the notions of I-c.r.g. and II-c.r.g. are equivalent. S.Yu. Favorov [11] show that this is not the case and that there, in fact, exists a function $f(z)$ of II-c.r.g. whose restriction to the complex lines $z=wz^0$, $w\in\mathbb{C}$, is not a f.c.r.g. in w for almost all $z^0\in\mathbb{C}^n$.

As in the one dimensional case, the c.r.g. of a function $f(z)$ entails an especially regular distribution of its zeros. In order to formulate the result, if S is an open subset of the sphere $\{z: |z|=1\}$, set $K_S=\{z: z/|z|\in S\}$, and let μ_f be the measure associated with the function $\mathscr{L}_f^*(z)$ in the sense of Riesz[10], $\sigma_f(t, S)$ being the $(2n-2)$-dimensional volume of the zero set of $f(z)$, counted with multiplicity, in the intersection $K_S^t=K_S\cap B_t$. In other words:

$$\sigma_f(t, S)=\int_{K_S^t\cap D_f} \gamma_f(z)\, dV_{2n-2}.$$

Theorem 24 [1]. *Let $f(z)$ be a function of II-c.r.g.* (in particular, I-c.r.g.). *Then the limit $\lim_{t\to\infty} t^{-\rho-2n+2}\sigma_f(t, S)$ exists for every S such that $\mu_f(\partial K_S)=0$ and one has the formula*

$$\lim_{t\to\infty} t^{-\rho-2n+2}\sigma_f(t, S)=\kappa_n\, \mu_f(K_S^1),$$

[10] I.e. $\mu_f=c_n\Delta\mathscr{L}_f^*(z)$, where Δ is the Laplace operator and c_n a certain constant.

where

$$\kappa_n = 2\pi^{n-1}((n-2)!)^{-1}.$$

An interesting property of f.c.r.g. is the composition of indicators under multiplication. In [9] it is established that $\mathscr{L}^*_{fg} = \mathscr{L}^*_f + \mathscr{L}^*_g$, provided at least one of the functions f and g to be multiplied, of finite type with respect to the order $\rho > 0$, is a function of II-c.r.g. Also the converse is true: if $\mathscr{L}^*_{fg} = \mathscr{L}^*_f + \mathscr{L}^*_g$ for every g with $\sigma(\ln M_{g,G}(R); \rho) < \infty$, then f is of II-c.r.g. [9]. These results are formal analogues of the corresponding results in the one dimensional case.

For $n=1$ a large subclass of f.c.r.g. is constituted by the functions which are Fourier transforms of functions with compact support. In the multivariate case this is not so. There exists a function $f \in H(\mathbb{C}^n)$, $n>1$, which is the Fourier transform of a function $\Phi \in \mathscr{L}^2(\mathbb{R}^n)$ with compact support and which is not of II-c.r.g. [50], and thus also not of I-c.r.g. Such examples are not possible if the convex hull of the support of Φ is a polyhedron [52].

In the study of f.c.r.g. it is likewise natural to ask the question whether there exist f.c.r.g. with a given indicator. The construction by Kiselman and Martineau mentioned in § 1 does not ensure regularity of growth. A positive answer to this question (for f. II-c.r.g.) follows from the fact, as established by Sigurdsson (Sigurdsson, R.: Growth properties of analytic and plurisubharmonic functions of finite order, Math. Scand. *59* (1986), 235–304) that for each function $u(z) \in \mathrm{PSH}(\mathbb{C}^n)$ at most of normal type with respect to the order $\rho > 0$ there exists an entire function $F(z)$ satisfying the condition

$$\lim_{t\to\infty} \frac{1}{t^\rho} \int_{B_1} |u(tz) - \ln|f(tz)|| \, dV_{2n} = 0.$$

Concluding this Section, let us mention that Gruman has considered f.c.r.g. defined in a different way as here. His definition is equivalent to II-c.r.g. In [37] f.c.r.g. are defined with respect to the P-indicator $h_f(\lambda)$ and, finally, in [1] f.c.r.g. are considered with respect to a distinguished variable.

6.2. Quasipolynomials. We denote by $Q_D(\mathbb{C}^n)$, where D is any family of entire (or meromorphic) functions in $\mathbb{C}^n$ at most of minimal (i.e. zero) type with respect to the order 1, the set of all functions of the form

$$f(z) = \sum_{j=1}^{\omega} a_j e^{i\langle \lambda^{(j)}, z\rangle}, \quad a_j \neq 0,$$

where $\omega = \omega(f) < \infty$, $a_j \in D$, $\lambda^{(j)} \in \mathbb{C}^n$. The functions in $Q_D(\mathbb{C}^n)$ will be called D-quasipolynomials or simply quasipolynomials. The case $D = \mathbb{C}$, that is, quasipolynomials with constant coefficients, has a special interest, and likewise the case $D = \mathscr{P}$ where, as before, $\mathscr{P}$ is the set of all polynomials in $z_1, \ldots, z_n$. Quasipolynomials also constitute the class of simplest solutions of partial differential equations with constant coefficients and of convolution equations. In the case $n=1$ the properties of quasipolynomials have been studied in high degree of completeness and it is natural to try to use this in the study of quasipolynomials in several variables. Often one succeeds to do this, exploiting the form of a function which is a quasipolynomials in each variable. A series of results on this issue are found in [34]. Let us state one of them.

Theorem 25. *Let $f(z)$ be a function defined in $\mathbb{C}^n$ such that, for each $j=1, \ldots, n$ and arbitrary fixed $z_1, \ldots, z_{j-1}, z_{j+1}, \ldots, z_n$, $f(z)$ as a function of z_j belongs to the class $Q_{\mathscr{P}}(\mathbb{C})$ (or $Q_{\mathbb{C}}(\mathbb{C})$) then*

$$f(z)=\sum_{k=1}^{\omega} a_k e^{\lambda_k(z)},$$

where the a_k are polynomials (or constants), $\omega=\omega(f)<\infty$ and the $\lambda_k(z)$ polylinear functions in $z_1, \ldots, z_n$.

As a corollary of this theorem and the corresponding known facts in the case $n=1$ we get, for example, results on the divisibility of quasipolynomials.

The simplest theorem of this kind (first obtained without utilizing the preceding result [10]) is the following

Theorem 26 (Avanissian and Gay). *If $f\in Q_{\mathbb{C}}(\mathbb{C}^n)$, $g\in Q_{\mathbb{C}}(\mathbb{C}^n)$ and $f/g\in H(\mathbb{C}^n)$ then $f/g\in Q_{\mathbb{C}}(\mathbb{C}^n)$.*

The strongest results on division of quasipolynomials can be found in [36].

A result close in nature to the theorems of the above type is provided by the following

Theorem 27 ([34]). *If $f=g^q$, where $g\in H(\mathbb{C}^n)$, $q\in N$ and $f\in Q_{\mathbb{C}}(\mathbb{C}^n)$ or $f\in Q_{\mathscr{P}}(\mathbb{C}^n)$ then, correspondingly, $g\in Q_{\mathbb{C}}(\mathbb{C}^n)$ or $g\in Q_{\mathscr{P}}(\mathbb{C}^n)$.*

These results on division and extraction of roots constitute partial solutions to the general question of the form of an entire function which is the solution of algebraic equation with quasipolynomial coefficients. In the case when these coefficients are in $Q_{\mathscr{P}}(\mathbb{C}^n)$ the entire function in question has the form g/f, where $g\in Q_{\mathscr{P}}(\mathbb{C}^n)$ and $f\in\mathscr{P}$ (cf. Levin, B.Ya., Ronkin, A.L.: Asymptotic series and algebroidal functions, Dokl. Akad. Nauk SSSR *280*, 288–291 (1985)).

In the one dimensional case much attention has been devoted to the study of the zeros of quasipolynomials. In the multivariate case the first term of the asymptotic volume of the divisor of a function $f\in Q_{\mathbb{C}}(\mathbb{C}^n)$ in an arbitrary cone is found and estimates for the second one are given in [35]. In [19] the set of common zeros of finite systems of quasipolynomials is investigated.

6.3. Separate Results. We state here some results which are both interesting and easy to formulate.

a) Let $\varphi \in \mathrm{PSH}(\mathbb{C}^n) \cap C^2(\mathbb{C}^n)$. Then the set of entire functions $f(z)$ such that for some $N=N(f)$ holds

$$\int_{\mathbb{C}^n} |f|^2 e^{-\varphi}(1+|z|^2)^{-N} d V_{2n} < \infty,$$

contains a function $f \neq 0$. Moreover, these functions form a dense subspace in $H(\mathbb{C}^n)$ with the topology of uniform convergence on compact sets [17].

b) In papers by Hörmander [18], Kelleher and Taylor [21], Skoda [47] and others, the question on the membership of a function in a given ideal in a ring of holomorphic functions defined by weighted integral estimates is discussed. Let us list one such result, stating it in connection with the case of entire functions.

Theorem 28 ([47]). *Let $\psi \in \mathrm{PSH}(\mathbb{C}^n)$ and let $g \in H^N(\mathbb{C}^n)$. Further, let $\alpha > 1$ and $q = \min(n, N-1)$. Then if the entire function $f(z)$ satisfies the condition*

$$\int_{\mathbb{C}^n} |f|^2 |g|^{-2\alpha q - 2} e^{-\psi} d V_{2n} < \infty,$$

there exists $h \in H^N(\mathbb{C}^n)$ such that $\sum_{j=1}^N h_j g_j = f$ and

$$\int_{\mathbb{C}^n} |h|^2 |g|^{-2\alpha q} e^{-\psi} d V_{2n} \leqq \frac{\alpha}{\alpha - 1} \int_{\mathbb{C}^n} |f|^2 |g|^{-2\alpha q - 2} e^{-\psi} d V_{2n}.$$

c) The function $f \in H(\mathbb{C}^n)$ is said to be irreducible if it cannot be written in the form $f = f_1 \cdot f_2$, with $f_i \in H(\mathbb{C}^n)$ and $\Lambda_{f_i} \neq \emptyset$, $i=1, 2$. In [41] it is shown that for $n \geqq 3$ an entire function $f(z)$ of the form $f_1(z_1) + \ldots + f_n(z_n)$, where $f_i \in H(\mathbb{C})$, $f_i \not\equiv 0$, $i=1, \ldots, p$, $3 \leqq p \leqq n$, must be irreducible.

d) Let $f(z)$ be an entire function of finite type whose zero set is the union of hyperplanes $\sum_{i=1}^n a_{i,k} z_i = b_k$. Then its indicator $\mathscr{L}_f(z)$ has to be continuous [16].

References

For the convenience of the reader, references to reviews in Zentralblatt für Mathematik (Zbl.), compiled using the MATH database, have been included as far as possible.

1. Agranovich, P.Z., Ronkin, L.I.: On functions of completely regular growth of several variables. Ann. Pol. Math. *39*, 239–254 (1981) [Russian]. Zbl. 476.32004
2. Alexander, H.: On a problem of Julia. Duke Math. J. *42*, 327–332 (1975). Zbl. 331.32001
3. Berenstein, C.A., Taylor, B.A.: Interpolation problems in $\mathbb{C}^n$ with applications to harmonic analysis. J. Anal. Math. *38*, 188–254 (1980). Zbl. 464.42003
4. Berenstein, C.A., Taylor, B.A.: On the geometry of interpolating varieties. In: Séminaire Pierre Lelong – Henri Skoda (Analyse), année 1980/81, et Colloque de Wimereux, mai 1981. Lect. Notes Math. 919, 1–25. Berlin etc.: Springer 1982. Zbl. 484.32004
5. Berndtsson, B.: Zeros of analytic functions in several variables. Ark. Mat. *16*, 251–262 (1978). Zbl. 409.32001

6. Berndtsson, B.: A formula for interpolation and division in $\mathbb{C}^n$. Math. Ann. *263*, 399–418 (1983). Zbl. 499.32013 (Zbl. 507.32010)
7. Björk, J.-E.: On extensions of holomorphic functions, satisfying a polynomial growth condition on algebraic varities in $\mathbb{C}^n$. Ann. Inst. Fourier *24*, 157–165 (1974). Zbl. 288.32014 (Zbl. 298.32007)
8. Ehrenpreis, L.: Fourier analysis in several complex variables. New York etc.: Interscience 1970. Zbl. 195, 104
9. Favorov, S.Yu.: On the composition of indicators of entire and subharmonic functions of several variables. Mat. Sb., Nov. Ser. *105* (147), 128–140 (1976) [Russian]. Zbl. 374.32001. Math. USSR, Sb. *34*, 119–130 (1978)
10. Favorov, S.Yu.: On the growth of plurisubharmonic functions. Sib. Mat. Zh. *24*, No. 1 (137) 168–174 (1983) [Russian]. Zbl. 568.32002. Sib. Math. J. *24*, 137–142 (1983)
11. Favorov, S.Yu.: On entire functions of completely regular growth in several variables. Teor. Funkts. Funkts. Anal. Prilozh. *38*, 103–111 (1982) [Russian]. Zbl. 541.32001
12. Gel'fand, I.M., Shilov, G.E.: Generalized functions, Vol. 1–3. Moscow: Fizmatgiz: 1958 [Russian]. Zbl. 91, 111. English transl.: New York and London: Academic Press 1964, 1968, 1967, second edition 1–3 1977. German: Berlin: VEB Verlag 1960, 1962, 1964
13. Geche, F.I.: Study of the growth of entire and holomorphic functions of several complex variables by means of directional characteristics. Dopov. Akad. Nauk Ukr. RSR, Ser. A 1975, 105–110 (1975) [Ukrainian]. Zbl. 313.32003
14. Griffiths, Ph.A.: Entire holomorphic mappings in one and several complex variables. (Ann. Math. Studies No. 85) Princeton: Princeton University Press 1976. Zbl. 317.32023
15. Gruman, L.: Entire functions of several variables and their asymptotic growth. Ark. Mat. *9*, 141–163 (1971). Zbl. 213, 97
16. Gruman, L.: The regularity of growth of entire functions whose zeros are hyperplanes. Ark. Mat. *10*, 23–31 (1972). Zbl. 235.32001
17. Hörmander, L.: An introduction to complex analysis in several variables. Princeton: Van Nostrand 1966. Zbl. 138, 62
18. Hörmander, L.: Generators for some rings of analytic functions. Bull. Am. Math. Soc. *73*, 943–949 (1967). Zbl. 172, 417
19. Kazarnovskiĭ, B.Ya.: On the zeros of exponential sums. Dokl. Akad. Nauk SSSR *257*, 804–808 (1981) [Russian]. Zbl. 491.32002. Sov. Math., Dokl. *23*, 347–351 (1981)
20. Katsnel'son, V.E.: Equivalent norms in spaces of entire functions. Mat. Sb., Nov. Ser *92* (*134*), 34–54 (1973) [Russian]. Zbl. 288.46024. English transl.: Math. USSR, Sb. *21* (1973), 33–55 (1974)
21. Kelleher, J.J., Taylor, B.A.: Finitely generated ideals in rings of analytic functions. Math. Ann. *193*, 225–237 (1971). (Zbl. 214, 383) Zbl. 207, 129
22. Kiselman, C.O.: On entire functions of exponential type and indicators of analytic functionals. Acta Math. *117*, 1–35 (1967). Zbl. 152, 76
23. Korevaar, J., Hellerstein, S.: Discrete sets of uniqueness for bounded holomorphic functions. In: Entire functions and related parts of analysis. Proc. Symp. Pure Math. 11, 273–284. Providence: Am. Math. Soc. 1968. Zbl. 181, 361
24. Lelong, P.: Potentiels canoniques et comparaison de deux méthodes pour la résolution du $\partial\bar{\partial}$ à croissance. In: Séminaire Pierre Lelong – Henri Skoda (Analyse), années 1978–1979. Lect. Notes Math. 822, 144–168. Berlin etc.: Springer 1980. Zbl. 439.32002
25. Lelong, P.: Fonctionnelles analytiques et fonctions entières (*n* variables). Montreal: University Press 1968. Zbl. 194, 388
26. Linnik, Yu.V., Ostrovskiĭ, I.V.: Decomposition of random variables and vectors. Moscow: Nauka 1972 [Russian]. Zbl. 285.60009
27. Logvinenko, V.N.: Theorems of M. Cartwright's type and real uniqueness sets for entire functions of several variables. Teor. Funkts. Funkts. Anal. Prilozh. 22, 85–100 (1975) [Russian]. Zbl. 324.32007
28. Logvinenko, V.N., Sereda, Yu.F.: Equivalent norms in spaces of entire functions of exponential type. Teor. Funkts., Funkts. Anal. Prilozh. no. 20, 102–111 (1974) [Russian]. Zbl. 312.46039

29. Lokshin, B.I.: On sets of reduced order of entire functions in $\mathbb{C}^n$. Teor. Funkts., Funkts. Anal. Prilozh. no. 37, 62–65 (1982) [Russian]. Zbl. 527.32003
30. Maergoiz, L.S.: Functions having the type of an entire function of several variables with respect to its directions of growth. Sib. Mat. Zh. *14*, 1037–1056 (1973) [Russian]. Zbl. 271.32006. English transl.: Sib. Math. J. *14*, (1973), 723–736 (1974).
31. Napalkov, V.V.: Convolution equations in multidimensional spaces. Moscow: Nauka 1982 [Russian]. Zbl. 582.47041
32. Palamodov, V.P.: Linear differential equations with constant coefficients. Moscow: Nauka 1967 [Russian]. English transl.: Grundlehren 168. Berlin etc.: Springer 1970. Zbl. 191, 434
33. Paneyah, B.P.: On some problems in harmonic analyis. Dokl. Akad Nauk SSSR *142*, 1026–1029 (1962) [Russian]. Zbl. 115, 97. English transl.: Sov. Math., Dokl. *3*, 239–242 (1962)
34. Ronkin, A.L.: On quasipolynomials. Funkts. Anal. Prilozh. 12, No. 4, 93–94 (1978) [Russian]. Zbl. 402, 32003. English transl.: Funct. Anal. Appl. *12*, 321–323 (1979)
35. Ronkin, A.L.: Distribution of zeros of quasipolynomials. Funkts. Anal. Prilozh. *14*, No. 3, 91–92 (1980) [Russian]. Zbl. 499.32018. English transl.: Funct. Anal. Appl. *14*, 242–244 (1981)
36. Ronkin, A.L.: Divisibility theorems for quasipolynomials. Teor. Funkts. Funkts. Anal. Prilozh. *34*, 104–111 (1980) [Russian]. Zbl. 439.32004
37. Ronkin, L.I.: Introduction to the theory of entire functions of several variables. Moscow: Nauka 1971 [Russian]; Zbl. 225.32001. English transl.: Providence: Am. Math. Soc. 1974. Zbl. 286.32004
38. Ronkin, L.I.: On discrete uniqueness sets for entire functions of exponential type in several variables. Sib. Mat. Zh. *19*, No. 1, 142–152 (1978) [Russian]. Zbl. 384.32002. English transl.: Sib. Math. J. *19*, 101–108 (1978).
39. Ronkin, L.I.: Some questions on the distribution of zeros of entire functions of several variables. Mat. Sb., Nov. Ser. 87 (129), 351–368 (1972) [Russian]. Zbl. 242.32004. English transl.: Math. USSR, Sb. *16*, 363–380 (1972)
40. Ronkin, L.I.: On continuation with estimates of functions holomorphic on the zero set of a pseudopolynomial. Sib. Mat. Zh. *24*, No. 4, 150–163 (1983) [Russian]. Zbl. 525.32015. English transl.: Sib. Math. J. *24*, 614–625 (1983)
41. Rubel, L.A., Squires, W.A., Taylor, B.A.: Irreducibility of certain entire functions with applications to harmonic analysis. Ann. Math., II, Ser. *108*, 553–567 (1978). Zbl. 402.32002
42. Sadullaev, A., Algebraicity criteria for analytic sets. Funkts. Anal. Prilozh. *6*, No. 1, 85–86 (1972) [Russian]. English transl.: Funct. Anal. Appl. *6*, 78–79 (1972). Zbl. 264.32002
43. Sadullaev, A.: On the canonical expansion of entire functions of n complex variables. Teor. Funkts. Funkts. Anal. Prilozh. 21, 107–121 (1974) [Russian]. Zbl. 324.32004
44. Schopf, G. (≡Shopf, G.): The dependence of the hypersurfaces of conjugate types on the conjugate orders. Izv. Vyss. Uchebn. Zaved. Mat. *4*, 105–121 (1976) [Russian]. Zbl. 335.32003
45. Schopf, G.: Construction of an entire function of several variables with a given asymptotic distribution of its zeros. Ukr. Mat. Zh. *33*, 476–481 (1981) [Russian]. Zbl. 486.32001. English transl.: Ukr. Math. J. *33*, 362–366 (1982)
46. Skoda, H.: Sous-ensembles analytique d'ordre fini ou infini dans $\mathbb{C}^n$. Bull. Soc. Math. Fr. *100*, 353–408 (1972). Zbl. 246.32009
47. Skoda, H.: Applications des téchniques L^2 à la théorie des idéaux d'une algèbre de fonctions holomorphes avec poids. Ann. Sci. Ec. Norm. Sup., IV Sér. *5*, 545–579 (1972). Zbl. 254.32017
48. Skoda, H.: Croissance des fonctions entières s'annulant sur une hypersurface donnée de $\mathbb{C}^n$. In: Sémin. Pierre Lelong – Henri Skoda (Analyse), année 1970/71. Lect. Notes Math. 275, 82–105. Berlin etc.: Springer 1972. Zbl. 258.32001
49. Stoll, W.: Holomorphic functions of finite order in several complex variables. Providence: Am. Math. Soc. 1974. Zbl. 292.32003
50. Vauthier, J.: Comportement asymptotique des fonctions entieres de type exponentiel dans $\mathbb{C}^n$ et bornées dans le domain réel. J. Funct. Anal. *12*, 290–336 (1973). Zbl. 254.32007
51. Vladimirov, V.S., Generalized functions in mathematical physics. Moscow: Nauka 1979 [Russian]. English transl.: Moscow: Mir 1979. Zbl. 515.46033/34
52. Wiegerinck, J.: Growth properties of Paley-Wiener functions on $\mathbb{C}^n$. Nederl. Akad. Wetensch. Proc., Ser. A, *87*, No. 1, 95–112 (1984) (=Indagationes Math. 46). Zbl. 571.32001

II. Multidimensional Value Distribution Theory

I.M. Dektyarev

Translated from the Russian
by J. Peetre

Contents

§ 1. Introduction. Motivation

The study of value distribution theory may be considered to have its origin in the famous Sokhotskiĭ-Weierstrass theorem [1]: the set of values of a nonconstant holomorphic function $w=f(z)$, defined in the entire z-plane, is everywhere dense in the w-plane. In fact, all values are assumed with the exception of at most one (Picard). These results and several subsequent ones, which at first sight looked so isolated, turned out to be part of a rather deep and elegant theory, known as value distribution theory or, after its founder, Nevanlinna theory. In particular it follows from the First Main Theorem of Nevanlinna that a meromorphic map not only takes almost all values but indeed takes them, in some sense, equally often. And if some values are

[1] *Translator's note.* =Casorati-Weierstrass!

taken too sparsely then this is compensated by the fact that, as z tends to infinity, $f(z)$ can be approximated with these values on large subsets.

In order to have a quantitative characterization of what we above referred to as "too sparsely", Nevanlinna introduced the notion of defect. The defect of "typical" values equals zero and if a value is not assumed at all then the defect equals one. The defect of all other values lies between 0 and 1. From his Second Main Theorem Nevanlinna derived the remarkable defect relation: the sum of the defects of an arbitrary family of values does not exceed 2.

The two Main Theorems of Nevanlinna and the consequences derived from them constitute the core of the classical value distribution theory but quite many theorems on the behavior and properties of meromorphic functions are on the same level or otherwise connected with the theory.

In its broad outlines the picture is as follows. The behavior of a meromorphic function $w=f(z)$ can be studied from several points of view. First, for each value a of f one considers quantities describing the distribution and the number (with multiplicities taken into account) in a disc of radius r. Other quantities describe the behavior of f on a circumference of radius r. Thereby one takes mainly into account the degree of closeness of the values of the function to the point a. In particular, if $a=\infty$ these quantities characterize the growth of the function. Besides characteristics connected with individual values one also puts into play their averages. The theory addresses all kind of connections between these quantities and the obstacles towards the existence of various functions arising from these connections.

Such a look at the subject under study leads naturally to a clarification of the question from which properties of the analytic structure of the complex plane these connections arise. In particular, in which conditions can we obtain analogous results in higher dimensions or, in the general case, for maps of manifolds?

§ 2. The Examples of Fatou-Bieberbach and Cornalba-Schiffman

The Sokhotskiĭ-Weierstrass theorem has no immediate extension to the case of several variables. This was made clear already in 1922 by Fatou (in the same year as the first papers by Nevanlinna were published). He found an example of two entire functions in two complex variables with the Jacobian identically equal to one and a not everywhere dense range of values. Today many analogous examples are known. Let us give one of the methods for constructing such examples.

Consider an automorphism $S\colon \mathbb{C}^2 \to \mathbb{C}^2$ given by the functions

$$f_1 = \lambda_1 z_1 + (\text{terms of higher order})$$

$$f_2 = \lambda_2 z_2 + (\text{terms of higher order})$$

where $|\lambda_1|=|\lambda_2|=\rho<1$. Pick a real number $\alpha>1$ such that $\alpha^2\rho<1$ and a number r such that $|Sz|<\alpha\rho|z|$ for $|z|<r$. Let L be the linear part of S. Then, for the same values of z, we have the inequality $|z-L^{-1}Sz|<M|z|^2$. Hence

$$\begin{aligned}&|L^{-m}S^m z-L^{-(m+1)}S^{m+1}z|=\\&\quad=|L^{-m}(S^m z-L^{-1}SS^m z)|<\rho^{-m}|S^m z-L^{-1}SS^m z|<\rho^{-m}M|S^m z|^2<\\&\quad<Mr^2(\alpha^2\rho)^m.\end{aligned}$$

Thus, the sequence of maps $\{L^{-m}S^m\}$ converges uniformly in the ball $|z|<r$. The linear part T of the limit map at the origin 0 is the identity and therefore the limit map is invertible in some neighborhood of the origin.

The map T satisfies the identity $L^{-1}TS=T$, whence $T^{-1}=S^{-k}T^{-1}L^k$ for all k. Using this relation T^{-1} extends in an obvious way to a biholomorphic equivalence between the entire space $\mathbb{C}^2$ and the domain of attraction at the origin of the map S, i.e. the union of all sets $S^{-m}U$, where U is a small neighborhood of 0.

With obvious modifications this argument carries over to any attractive fixed point (at least, if the linear part of the map at such a point can be put in diagonal form, with numbers of the same modulus on the diagonal).

If, for example, S is given by the formula

$$(z_1, z_2)\to(z_1, \lambda^2 z_2+(1-\lambda^2)(z_2-\varphi(z_2))),$$

then for each a such that $\varphi(a)=0$ and $\varphi'(a)=1$ the point with coordinates (a, a) will satisfy this desideratum and therefore for each such point there will be a corresponding domain of attraction. These domains, of course, do not intersect and each of them is biholomorphically equivalent to $\mathbb{C}^2$. Despite the simplicity of these examples, only little is known about the geometry of the domains which are biholomorphic to $\mathbb{C}^2$. Thus, it is not hard to see that for any two disjoint lines in $\mathbb{C}^2$ at least one must intersect the domain. But does there exist one single line which does not intersect the domain? It is plausible that there are no such lines and it would be nice to have a proof of this.

Let us consider one more result of the classical theory and the corresponding counterexamples in the multidimensional case.

From Jensen's formula, which is one of the oldest results of value distribution theory and a key device in the proof of the First Main Theorem, one easily gets the following remarkable inequality, connecting the quantities $n_f(r)$ – the number of zeros of the entire function f in a disc of radius r – and $M_f(r)$ – the maximum modulus of f in the same disk:

$$n_f(r)\leqq\ln M_f(er).$$

Contrary to this Cornalba and Schiffman showed that there exist pairs $F=(f, g)$ of analytic functions in two variables such that, although both functions have zero order, the growth of the quantity $n_F(r)$, describing the distribution of their common zeros, may be arbitrarily fast.

The idea of the proof is extremely simple. Let us consider a function $\varphi(z)$ in one variable of slow growth and with a sparse distribution of first order zeros $a_1, a_2, \ldots$ Put $\varphi_k(z)=\varphi(z)/(z-a_k)$ and let $P_1, P_2, \ldots$ be a sequence of polynomials such that the number of their zeros grows rather fast with k while at the same time all roots are sufficiently close to the origin. Let us set

$$\Psi(z_1, z_2)=\sum \alpha_k \varphi_k(z_1) P_k(z_2),$$

where the coefficients $\alpha_1, \alpha_2, \ldots$ converge so rapidly to 0 as to guarantee the required small speed of the growth of the function $\Psi(z_1, z_2)$. The common zeros of the pair of functions $(\varphi(z_1), \Psi(z_1, z_2))$ are precisely those points for which the first component is one of the numbers a_k and the second one is some root of the corresponding polynomial P_k. The choice of the sequence $\{a_k\}$, the polynomials $\{P_k\}$ and the coefficients $\{\alpha_k\}$ secures all our desiderata.

§ 3. Exhaustions

In order to describe the behavior of the quantities which characterize a holomorphic map it is necessary to introduce a parameter with respect to which one considers its behavior. In the classical case a natural parameter is $|z|$. As we will witness in the sequel, it is from the point of view of the general theory often better to consider $\ln|z|$, because this is a harmonic function. These examples indicate that the most expedient way of introducing the appropriate parameter is using exhaustion functions.

Definition. A smooth real function τ on a noncompact manifold X is called an *exhaustion* if the domains of lower values of this function are all compact and if all critical points are situated within a compact subset.

In what follows, whenever X is equipped with an exhaustion τ, then we denote by X_r the set $\{x \in X: \tau(x)<r\}$. By definition X_r is relatively compact in X. That a manifold admits an exhaustion with some particular properties turns out to be a very important property of the manifold. For example, one can show (I.M. Dektyarev) that if $f: X \to M$ is a holomorphic map of n-dimensional manifolds, and if the Laplacean of the exhaustion τ, computed in the induced metric, is nonpositive, then the set of values of f which are taken sufficiently sparsely has Hausdorff dimension not exceeding $2n-1$ (analogue of the Sokhotskiĭ-Weierstrass theorem). This assumption is fulfilled, in particular, if the Levi form $d d^c \tau$ of the exhaustion τ is non-positive definite and $f: X \to M$ is an arbitrary nondegenerate holomorphic map into an n-dimensional Kähler manifold. This motivates the following

Definition. An *exhaustion* τ is called *concave* if its Levi form $d d^c \tau$ is non-positive definite off some compact set. If it is nonnegative definite then we say that we have a *convex* exhaustion.

In order to get theorems of the Sokhotskiĭ-Weierstrass type for maps of manifolds which admit a convex exhaustion one has to impose supplementary conditions on the maps. As we shall see in what follows, these conditions amount explicitly or implicitly to the following. If there is given, in some way or other, an Hermitean metric on the image manifold, then the holomorphic map induces a pseudo-metric (i.e. the positivity condition may fail) on the preimage. The condition says now that the $(2n-1)$-dimensional volume, taken in this pseudometric, on the manifold ∂X_r should be small, as compared to the $2n$-dimensional volume of X_r.

If the manifold admits a concave exhaustion then such a condition is automatically fulfilled.

Griffiths and King introduced, in [9], exhaustion functions which they called special. Their definition differs only in inessential details from the following.

Definition. An *exhaustion* τ on an n-dimensional manifold X is termed *special* if

1) its Levi form $d\,d^c\,\tau$ is nonnegative definite;
2) $(d\,d^c\,\tau)^n \equiv 0$;
3) $(d\,d^c\,\tau)^{n-1} \not\equiv 0$ on the maximal complex tangent space to ∂X_r.

Let us recall once more that all these conditions are required to hold off a compact set. The special exhaustions of Griffiths-King are specially well adapted to the situation under study when instead of preimages of points one considers preimages of divisors. In order to describe quantities of such preimages one uses integrals extended over them with respect to the form $(d\,d^c\,\tau)^{n-1}$ and the condition $(d\,d^c\,\tau)^n \equiv 0$ makes the exhaustion similar to a concave one.

In the same paper it is shown that special exhaustions exist on arbitrary smooth affine algebraic varieties. More exactly, the following is proved. For any smooth n-dimensional affine algebraic variety X there exits a projection $\pi\colon X \to \mathbb{C}^n$, constituting a finitely sheeted covering, such that all critical points of the function $\tau(x) = \log|\pi(x)|$ are confounded to a compact part of X. All the other requirements are automatically fulfilled. Let us remark that in their definition Griffiths and King allow τ to be $-\infty$. We are obliged to smooth it near such points but this will have no influence in what follows.

Rather close to the special exhaustions of Griffiths-King are the parabolic ones studied by Stoll.

Definition. An *exhaustion* τ in an n-dimensional manifold X is called *parabolic* if $(d\,d^c\,\tau)^n \not\equiv 0$ but $(d\,d^c \log\tau)^n \equiv 0$. The exhaustion (as well as the manifold) are called *strictly parabolic* if the form $d\,d^c\,\tau$ is positive definite.

As in the case of special exhaustions, one shows that each smooth affine variety admits a parabolic exhaustion.

Both types of exhaustions considered are symmetrized variants of a more general type, which still has many useful properties and which can be applied not only to divisors but also in the case of higher codimensions.

Definition. Let χ be a nonnegative definite (p, p)-form such that on p-dimensional complex tangent spaces to ∂X_r it takes positive values and assume further that the exhaustion τ has the property that the form $dd^c\tau \wedge \chi$ is nonpositive definite. Then τ is called a *χ-concave exhaustion.*

§ 4. Multiplicity of Holomorphic Maps

In constructing quantities describing preimages of points these preimages must be counted with multiplicity. For holomorphic maps one has several different definitions of this notion. Here is a definition based on some notions and results from local algebra.

Let $f=(f_1, \ldots, f_n)$ be a germ of a holomorphic map $\mathbb{C}^m \to \mathbb{C}^n$ with $f(0)=0$. We denote by Q_f the *local ring* of this *map*, i.e. Q_f is the quotient ring of holomorphic functions at the origin in $\mathbb{C}^m$, by the ideal generated by the functions $f_1, f_2, \ldots, f_n$. Let m_x be the maximal ideal in the ring of germs of holomorphic functions. It is well-known that the dimension of the quotient ring $Q_f/m_x^k Q_f$ considered as a vector space over $\mathbb{C}$ for large k is a polynomial in k (the so-called *Hilbert polynomial*). If the degree of this polynomial equals $m-n$ (this happens in the case when $\dim f^{-1}(0)=m-n$), then the coefficient of the top term, multiplied with $(m-n)!$, is called the *multiplicity of the ring* Q_f and this multiplicity may be regarded as the *multiplicity of the germ of f* at the origin.

There exists also a geometric definition of multiplicity. This definition can be formulated as follows.

Let again $f\colon U \to V$ be a holomorphic map of neighborhoods of the origins in $\mathbb{C}^m$ and $\mathbb{C}^n$ $(m \geqq n)$ with $f(0)=0$. Assume that the preimage of the origin has dimension $m-n$. This is the case iff f is open at the origin (Remmert). Pick an n-dimensional subspace H in $\mathbb{C}^m$ and consider the restriction of f to the set $H \cap U$. If H is in "general position" then the origin is an isolated point in the set $f^{-1}(0)$ in $H \cap U$. If we choose U such that the origin is the only point in this set then for all points sufficiently close to the origin $f^{-1}(0)$ is a discrete set in $H \cap U$ and for almost all a it consists of the same number of points. This number is then taken to be the *multiplicity* of f at the origin. As observed by V.P. Palamodov, this definition is equivalent to the previous one.

What is unsatisfactory with these definitions from the point of view of value distribution theory (especially the first one of the two) is that they are rather difficult to use in concrete computations "taking into account the multiplicity". Now we give a definition which in our opinion is far more suitable for such purposes.

Let $f\colon X\to M$ be a continuous map of n-dimensional real manifolds and let X_0 be an isolated point of the preimage of the point $a=f(x_0)$. In this situation one has a very convenient *topological definition of the multiplicity* of f at the point x_0. We take small punctured (at these points) neighborhoods of x_0 and a, which have the homotopy type of the $(n-1)$-dimensional sphere. If s and σ are the generators of the corresponding $(n-1)$-dimensional homologies, then $f_* s=k\sigma$. The coefficient k is taken to be the sought multiplicity.

The dual definition is even more useful in the applications.

Let λ_a be a closed $(n-1)$-form on $U\setminus a$ such that $\int_\sigma \lambda_a=1$ (i.e. the generating form in the $(n-1)$-dimensional de Rham cohomology). Then (in the case of a differentiable map) the multiplicity k equals the integral $\int_s f^* \lambda_a$.

An important observation is the following. Let us extend λ_a in an arbitrary fashion to the full manifold $M\setminus a$. Suppose that this form and the map f are such that $f^* \lambda_a$ is locally in L^1 and thus defines a 1-dimensional current on X, written $[f^* \lambda_a]$. On the other hand, in view of the smoothness of λ_a, the form $d f^* \lambda_a$ is smooth off the point $f^{-1}(a)$ and vanishes in a neighborhood of it. Extend it to be zero at $f^{-1}(a)$. The 0-dimensional current thus obtained is denoted $[d f^* \lambda_a]$. If now φ is a test function on X whose support (compact) contains only isolated points of the set $f^{-1}(a)$ then it is easy to see that the current $d[f^* \lambda_a]-[d f^* \lambda_a]$ on this function equals $\sum \varphi(x_i)$, where the summation extends over all x_i such that $f(x_i)=a$ and each x_i is taken as many times as the multiplicity of that point.

These considerations justify the following definition, which makes sense also for noncoinciding dimensions.

Definition. Let there be given on the manifold M an $(n-1)$-form λ_a, which is smooth off a, closed in a punctured neighborhood of a and such that its integral over a small sphere about the point a equals one. Assume that $f\colon X\to M$ is such that the form $f^* \lambda_a$ is locally in L^1. Then the *current* $d[f^* \lambda_a]-[d f^* \lambda_a]$ will be called the *multiplicity* of a for the map f.

What is somewhat strange in this definition is that now one does not assign a multiplicity to each individual point of the preimage. However, one can show that in "good" cases the above multiplicity is a *holomorphic chain*, i.e. a current which turns out to be the operation of integration over the regular part of the analytic set $f^{-1}(a)$. The components of that set are taken with integer coefficients, these coefficients being equal to the multiplicities in the previous sense.

The proof of this fact is based on a series of estimates which allow one to obtain the continuous dependence of the multiplicity (as well as of the current) on the point $a\in M$.

Moreover, with the aid of the same estimates one can prove the continuous dependence in a for other important quantities in value distribution theory. Some of the estimates have also an independent interest. To give precise estimates we require some notation.

Let U and V be bounded domains in $\mathbb{R}^n$ and let K be a compact subset of U. Denote by d_V and v_V the diameter respectively the volume of V, analogously letting d_K and v_K be the diameter and the volume of K. If $F=(f_1, \ldots, f_n)$ is a continuously differentiable map from U into V, we denote by $n_f(a, K)$ the number of points (possibly infinite) in the intersection $F^{-1}(a) \cap K$. This counting is done without taking account of multiplicity or orientation. It follows from Sard's theorem that this function is everywhere finite except for a set of measure zero (the set of critical values). It is also clear that it is lower semicontinuous.

The quantity $1/v_V \int_V n_f(a, K)\, d a$ is called the *average number of preimages* of the set V in K. In view of the above discussion the integral in the definition exists. It is not hard to see that it equals the integral of the modulus of the Jacobian of F, extended over $K \cap F^{-1}(V)$.

Theorem 1 (on the average number of preimages). *Let s be an integer $\geqq 0$ and A a real constant. Consider the class of maps such that all partial derivatives up to order $s+1$ of the functions making up the map are bounded in modulus by A. Let F be in this class. Denote by ρ the distance from K to the boundary of U and assume that $v_V < \rho^{n+s}$. Then there exists a constant L, depending only on n and s, such that, in the above hypotheses, the average number of preimages of the set V in K does not exceed the number*

$$L(n, s) \cdot A^n \cdot v_K \cdot v_V^{-n/(n+s)}.$$

Thus, although the average number of preimages when V decreases may grow to infinity, this growth is for smooth maps rather moderate.

This theorem is crucial. The remaining estimates either follow rather easily from it or else by "general nonsense".

Next, let $U \subset \mathbb{R}^n$, $V \subset \mathbb{R}^p$ and $W \subset R^m$ be neighborhoods of the origin in the corresponding spaces, assuming $p>0$, $n+p \geqq m$. Denote by B_r the ball of radius r in $\mathbb{R}^m$ and assume that $F\colon U \times V \to W$ is a smooth map such that for some constants c and α the volume of the projection onto V of set $F^{-1}(B_r)$, with r sufficiently small, is $\leqq c r^\alpha$. Pick ρ such that W is contained in a ball of radius ρ, set $s=[2n/\alpha]+1$ and denote by $\{f_{i_1}, \ldots, f_{i_n}\}$ a subset of the family of functions constituting the map F and by $y_1, \ldots, y_p$ the coordinates in $\mathbb{R}^p$.

Theorem 2. *The integral*

$$\int_{U \times V} \|F\|^{-n}\, d f_{i_1} \wedge \ldots \wedge d f_{i_n} \wedge d y_1 \wedge \ldots \wedge d y_p$$

is absolutely convergent and $\leqq L \cdot \rho^{\alpha/2}$ where L is a constant depending only on α, c and the maximum modulus of the derivatives up to order s of the functions delivering the map F.

It is now plain that for the existence of the current defining the multiplicity and for the proof that the latter depends continuously on a it is necessary to choose the corresponding form λ_a in such a way that its behavior near

a satisfies corresponding estimates. The proof of the continuity of the other quantities is based on the same device.

§ 5. Unintegrated and Integrated First Main Theorem

Formally speaking, multidimensional value distribution theory has three aspects.

First, one studies pre-images of divisors. The theory of holomorphic curves belongs here. Until these days the deepest results on the defect relation pertain precisely to this case. We will speak of these matters later.

The second aspect concerns a circle of problems connected with preimages of points under holomorphic maps. Despite a difference in the formulation of the problems and, in part, also the methods of approach, we will in what follows see that, essentially, the second approach to the subject allows one to study both aspects from a unified point of view. In any case, all notions and results can be reformulated in a corresponding way.

Many things have a purely geometric character and hold true for smooth maps of real manifolds as well. In this approach (this is a constituent of the third aspect) one obtains not only more general results but to some extent widens the understanding of the complex case too. We will begin our investigation with the real manifolds.

The First Main Theorem of value distribution theory is, by and large, just Stokes's theorem applied to certain special forms with singularities. Then what regards the singularities these cause standard difficulties which one also resolves in the standard way. The whole affair is to construct forms with the desired properties.

Let ω be the normalized volume form on a compact n-dimensional oriented manifold. Assume that for each $a \in Y$ there is a given smooth $(n-1)$-form λ_a on $Y \backslash a$, enjoying the following property. The form $|x-a|^{n-1} \lambda_a$ has, in terms of local coordinates near a, bounded coefficients and, moreover, off a it is required that $d\lambda_a = \omega$. Let there now be given a smooth orientation preserving map $f\colon X \to Y$ from an oriented $(n+p)$-dimensional manifold X $(p \geqq 0)$. Take in X a relatively compact subset G with piecewise smooth boundary and assume that the point a is a regular value for the restriction of f to G, the submanifold $f^{-1}(a)$ being transversal to each boundary component of G.

Theorem 1. *For each closed p-form κ defined in a neighborhood of G we have*

$$\int_G f^* \omega \wedge \kappa = \int_{\partial G} f^* \lambda_a \wedge \kappa + \int_{f^{-1}(a) \cap G} \kappa .$$

This is the weakest form of the *unintegrated First Main Theorem.* The requirement of regularity of the value a and the transversality of $f^{-1}(a)$

to the boundary are imposed in order to avoid difficulties connected with the application of Stokes's theorem.

In the complex case one chooses the forms in such a way that they have additional properties which allow us, on the one hand, to eliminate superfluous restrictions and, on the other hand, to get more subtle estimates.

Let X and Y be complex manifolds of complex dimensions $n+p$ and n. Let us first assume that there is given a Hermitean metric on Y, denoting the corresponding distance by ρ. One can choose the metric in such a way that the (1, 1)-form λ associated with it enjoys the following property: its n-th exterior power λ^n gives the normalized volume element ω whereas its $(n-1)$-th power $\mu=\lambda^{n-1}$ satisfies the relation $d d^c \mu \equiv 0$.

Theorem 2. *One can choose a smooth function $\gamma(a, y)=\gamma_a(y)$, defined in the complement of the diagonal of $Y\times Y$, such that for each $a\in Y$ one has $d d^c(\gamma_a \mu) =\omega$ and in addition holds the estimate*

$$\gamma_a(y) \leqq \frac{1}{(n-1)\cdot 2^{n-1}\cdot \pi^n \cdot \rho(a, y)^{2n-2}} + O(\rho(a, y)^{3-2n}).$$

On the other hand, let us assume that on X there is given an exhaustion τ and a closed non-negative definite (p, p)-form χ.

Let us now apply the previous unintegrated First Main Theorem to the following situation. For G we take, in the direct product $X\times \mathbb{R}$, a set bounded "from above" by the level set $X\times\{r\}$ and "from below" by the graph of τ. The form κ will in our case be $\chi\wedge d\tau$ and the family of forms λ_a of the form $d^c(\gamma_a \mu)$. Instead of the map f we take the composition of the projection of the product $X\times\mathbb{R}$ onto the first component with a holomorphic map $F: X\to Y$. If a is a regular value of F, then it is easy to check that all requirements are fulfilled and we obtain the identity

$$\int_G f^*\omega\wedge\chi\wedge dt = \int_{\partial G} d^c f^*(\gamma_a\mu)\wedge\chi\wedge dt + \int_{f^{-1}(a)\cap G} \chi\wedge dt.$$

The expression

$$\int_{\partial G} d^c f^*(\gamma_a\mu)\wedge\chi\wedge dt,$$

entering in the right hand side of this expression, can by not too complicated manipulations with differential forms be put in the form

$$\int_{\partial X_r} d^c\tau\wedge F^*(\gamma_a\mu)\wedge\chi - \int_{X_r} d d^c\tau\wedge F^*(\gamma_a\mu)\wedge\chi.$$

The first integral here will be written $m_F(r, a)$ and the second $\Delta_F(r, a)$. For the remaining terms of this identity one has also standard notations. The integral $\int_G f^*\omega\wedge\chi\wedge dt$, equal to $\int_{-\infty}^r v(t)\,dt$, where $v(t)$ is the integral $\int_{X_t} F^*\omega\wedge\chi$, is usually written $T(r)$ and is called the Nevanlinna *characteristic function.* As regards the second integral to the right, it is not hard to see that it equals the value of the current $\nu_F(a)$, giving the multiplicity of the value a for the map F, on the form $(r-\tau(x))\cdot\chi$. This quantity again is written $N_F(r, a)$. The function $m_F(r, a)$ is usually called the *approximation function* and $N_F(r, a)$ the *counting function*. We get the First Main Theorem in unintegrated form.

First Main Theorem. *If the value a is such that the map F is open at points of the set $F^{-1}(a)\cap X_r$ then*

$$T(r)=N_F(r, a)+m_F(r, a)-\Delta_F(r, a).$$

The previous reasoning shows how to prove this identity for regular values. The estimates at the end of the last Section allow us to conclude that on the set of those a for which our map is open at points in $F^{-1}(a)\cap X_r$ all expressions in the right hand member depend continuously on a. Therefore the identity of the First Main Theorem holds for all such a.

Remarkably many papers on value distribution theory (if not the majority) address themselves to the situation when one considers not preimages of points in Y but preimages of divisors or sometimes, more generally, preimages of submanifolds. There are also papers where instead of maps f: $X\to Y$ one considers the family of sections of a vector bundle over X and instead of preimages (of points or submanifolds) one takes zeros of the sections. In all these situations we can likewise prove a First Main Theorem; then one either gives appropriate forms on Y and pulls them back to X or else, in the second case, the forms are constructed directly on X using the given holomorphic section. It is, however, not hard to see that the same results can be obtained by the scheme outlined above (in any case, in all what concerns the First Main Theorem and its consequences).

Consider three compact manifolds A, M and Y, assuming that M lies in the direct product $Y\times A$, that the projections p: $M\to A$ and π: $M\to Y$ are surjective holomorphic maps and that the rank of π at each point coincides with the dimension of Y. It is clear that for each $a\in A$ the set $Y_a=\pi\circ p^{-1}(a)$ is a submanifold of Y (possibly with singularities) so we have a holomorphic family of submanifolds parametrized by A. In papers by Griffiths, Griffiths-King, Schiffman and many others one considers an ample positive line bundle L over Y and for A one takes the complete linear system of divisors $|L|$. In this case one can take as M the set of pairs (y, σ) where y is a point in Y and σ a section of L vanishing at y (we do not discriminate between sections proportional to each other and identify them with their zero divisor). It is easy to check all the desired requirements.

If now f: $X\to Y$ is a holomorphic map then we can consider the set $\tilde{X}$ in the direct product $X\times A$ consisting of pairs $\tilde{x}=(x, a)$ such that $(f(x), a)$ belongs to M. This condition is equivalent to $f(x)\in Y_a$. Consider the projection $\tilde{f}$: $\tilde{X}\to A$. It is evident that $\tilde{f}(\tilde{x})$ equals a iff $\tilde{x}=(x, a)$, i.e. $f(x)\in Y_a$.

Thus, the question of the distribution of the preimages of the manifold Y_a under the map f blows down to studying the preimage of a under $\tilde{f}$.

Denoting by $p\colon \tilde{X}\to X$ the projection of the set $\tilde{X}\subset X\times A$ onto the first factor, then one must take as the exhaustion $\tilde{\tau}$ on $\tilde{X}$ the function $p^*\tau$ and as $\tilde{\chi}$ the form $p^*\chi$. In this case $\tilde{\tau}$ is of exactly the same type as τ (concave, convex, χ-concave etc.).

In the case at hand certain relations must hold between the types of the forms and the dimensions of the manifolds. Namely, if X has dimension m and χ has type (k, k) then the codimension of each manifold Y_a has to be $m-k$. The extremal case is when k equals the difference between the dimensions of X and Y. In this case the submanifolds are just points. Let us also mention that using "fiberwise integration" one can reduce the First Main Theorem, written down for the map $\tilde{f}$, to the case of the map f. The differential forms obtained in this way may differ somewhat from the ones usually employed but this difference is unessential. Let us remark that it is not always worth while to perform such a reduction.

§ 6. Simplest Consequences of the First Main Theorem. Theorems on Equidistribution

Two properties of the quantities appearing in the First Main Theorem play a key rôle in the proof of equidistribution theorems.

The first is the fact that the number

$$m_F(r, a)=\int_{\partial X_r} d^c\tau\wedge F^*(\gamma_a\mu)\wedge\chi.$$

is nonnegative. Indeed, from the fact that the forms $F^*\mu$ and χ are nonnegative definite follows that the form $d\tau\wedge d^c\tau\wedge F^*(\gamma_a\mu)\wedge\chi$ gives, everywhere where it is not degenerate, rise to a positive volume element. We use here the usual convention regarding orientation.

The second is the geometrically evident statement that $T_F(r)$ is the mean value of $N_F(r, a)$ with respect to a.

Let us show at the hand of examples how these facts are used.

Let τ be a χ-concave exhaustion, i.e. the form $-d\,d^c\tau\wedge\chi$ is positive definite off a compact set, which we may assume to be contained in the set X_{r_0}. Then we have the estimate $\Delta_F(r, a)\leqq\Delta_F(r_0, a)$. The right hand side here is a continuous function of a on a compact manifold. Therefore $\Delta_F(r, a)$ is in this case bounded, so from the nonnegativity of $m_F(r, a)$ we obtain a uniform in a estimate (Nevanlinna's inequality)

$$T_F(r)\geqq N_F(r, a)+O(1),$$

i.e. no individual (in a) value of $N_F(r, a)$ can be very much larger than $T_F(r)$. On the other hand, the average (in a) value of $N_F(r, a)$ equals $T(r)$ and this finally shows that only on a set of measure zero one can have

$\overline{\lim}\, N(r, a)/T(r) < 1$. This is a crude result. A considerable strengthening of it can be obtained with the aid of potential theory.

Let us integrate the First Main Theorem with respect to a, using the volume form ω on Y. The integral of $N_F(r, a)$ equals $T(r)$ so we get

$$\int_{\partial X_r} \lambda(x)\, d^c\tau \wedge F^*\mu \wedge \chi = \int_{X_r} \lambda(x)\, d\, d^c\tau \wedge F^*\mu \wedge \chi,$$

where $\lambda(x)$ stands for the value of the function $\varphi(y) = \int_Y \gamma(a, y)\,\omega_a$ at the point $y = F(x)$. Now $\varphi(y)$ is continuous and positive on a compact manifold so

$$\int_{\partial X_r} \lambda(x)\, d^c\tau \wedge F^*\mu \wedge \chi < c \cdot \int_{X_r} d\, d^c\tau \wedge F^*\mu \wedge \chi,$$

where the expression to the right, again, is bounded.

If a set $D \subset Y$ has a positive 2-capacity, then the integral of the function $\gamma(a, y)$ with respect to the equilibrium measure over D is a continuous and therefore bounded function. If we have $T(r) - N(r, a) \to \infty$ on such a D, then integrating the First Main Theorem with respect to this measure and taking into account all what has been said above, we would arrive at a contradiction. Therefore holds:

Theorem 1 (I.M. Dektyarev). *If the exhaustion is χ-concave, then $T(r) - N(r, a) \to \infty$ can hold only on a set of Newtonian capacity* 0, *i.e. on sets of Hausdorff dimension not greater than $2n-2$.*

If we take as X a smooth affine algebraic variety and use the special exhaustion of Griffiths-King on it then the latter is χ-concave for $\chi = (d\, d_c \tau)^{n-1}$. From the relations between the dimensions and the types of the forms at the end of the last Section one sees that on Y one has to consider divisors and we find that a set of divisors which sufficiently sparsely intersects $f(X)$ has Hausdorff dimension $2n-2$, which answers a question posed by Griffiths and King. Further results of this type in various versions were obtained by L.I. Ronkin and A. Sadullaev.

Just as χ-concavity of an exhaustion on X guarantees *equidistributionality* (i.e. the "smallness" of the set of "sparsely" assumed values) for arbitrary nondegenerate maps of the manifold X into an arbitrary compact complex manifold Y, in order to get analogous results in other cases one must require supplementary conditions to be fulfilled.

Let us consider the case when there is given on X a χ-convex exhaustion. Here one can repeat the same reasoning for the measure given by the form ω also with the equilibrium measure. However we cannot prove that the quantities $\int_{\partial X_r} d^c\tau \wedge F^*\mu \wedge \chi$ and $\int_{X_r} d\, d^c\tau \wedge F^*\mu \wedge \chi$ are bounded. Therefore we have to impose the auxiliary assumption that the growth of one of them as $r \to \infty$ is sufficiently small. No matter which of the two, since also in this case the ratio of this two numbers will for $r > r_0$ lie between two constants. Let us give a more precise formulation of the result obtained (I.M. Dektyarev).

Let $f\colon X \to Y$ be a holomorphic map of an $(n+p)$-dimensional manifold X on which there is given a nonnegative definite (p, p)-form and a χ-convex exhaustion on the n-dimensional manifold Y. The forms ω and μ on Y have the same meaning as above. If for $r \to \infty$ we have the condition

$$\varliminf \frac{\int_{\partial X_r} d\,d^c \tau \wedge f^* \mu \wedge \chi}{T(r)} = 0$$

then the set of those $a \in Y$ for which $\varlimsup N(r, a)/T(r) < 1$ has zero 2-capacity.

Weaker variants of this theorem (measure zero instead of 2-capacity zero and the additional hypothesis that Y is Kähler) were originally obtained by Wu and Stoll. Even earlier Chern had such a result (still with measure 0) in the case when Y is a projective space.

If the exhaustion τ is not of definite sign, additional complications arise and the results obtained are more weak. Taking the average with respect to a in the First Main Theorem now does not give anything for the comparison of

$$\int_{\partial X_r} d^c \tau \wedge f^* \mu \wedge \chi$$

and

$$\int_{X_r} d\,d^c \tau \wedge f^* \mu \wedge \chi,$$

as the integrand in the second integral may not have constant sign. This has also the effect that the reasoning with integration with respect to equilibrium measure is not applicable. Here it turns out to be expedient to consider the unintegrated form of the First Main Theorem and then one gets results in the case when X and Y have equal dimension.

Let us use the following notation. Let $f\colon X \to Y$ be a continuously differentiable map of n-dimensional real manifolds, assuming that both manifolds are orientable and that f preserves orientation where it is nondegenerate.

We assume that Y is compact and that there is given on it a Riemannian metric $d s^2$ which is normalized so that the total volume of Y is 1. We assume further that there is given an exhaustion τ on X. Let $v(r)$ denote the volume of X_r and $s(r)$ the $(n-1)$-dimensional volume of the boundary of this domain. Both volumes are taken with respect to the pseudo-metric $f^* d s^2$. We denote by $n(r, a)$ the number of points of the set $f^{-1}(a) \cap X_r$. For discrete points in the pre-image we take account of the multiplicities. If the pre-image is not discrete we set $n(r, a) = \infty$.

One can prove (I.M. Dektyarev) that if $\varliminf s(r)/v(r) = 0$ then the set of values $a \in Y$ for which $\varlimsup n(r, a)/v(r) < 1$ has Hausdorff dimension not exceeding $n-1$.

In the case of holomorphic maps of complex manifolds the condition $\varliminf s(r)/v(r) = 0$ is guaranteed if we assume that

$$\int_{\partial X_r} d^c \tau \wedge f^* \mu < c\, v(r)^{1-\varepsilon}.$$

for all sufficiently large r and suitable constants c and ε. Let us remark that this assumption is stronger than the condition

$$\underline{\lim} \frac{\int_{\partial X_r} d^c \tau \wedge f^* \mu}{T(r)} = 0$$

but we impose here no conditions whatsoever on τ. The question is: is it possible, in the complex case in the hypothesis of one of the above requirements, to prove that the set of "sparsely attained values" has Hausdorff dimension $n-2$? It is plausible that the answer is negative and that the resulting estimate of the dimension is best possible if no additional conditions are imposed on the exhaustion.

Concluding this Section we say a few words concerning the problem which has given rise to the whole complex of ideas treated here, namely what can be said about holomorphic maps of a vector space into itself?

Here as before little is known. Formally speaking, a holomorphic map $f: \mathbb{C}^n \to \mathbb{C}^n$ may be viewed as a map in $\mathbb{C}\mathbb{P}^n$ so one can apply the apparatus developed for this case, using a convex exhaustion on the source manifold (unfortunately, there are no concave ones). However, it turns out that the conditions which imply equidistribution are rather hard to verify and few classes of maps are known for which it is fulfilled. The most explicit result is due to Sibony and P.M. Wong.

Theorem 2. *Let there be given a family of holomorphic functions* $F = (f_1, \ldots, f_n)$ *defining a nondegenerate map from* $\mathbb{C}^n$ *into* $\mathbb{C}^n$. *Set*

$$M_q(r) = \max_{\|z\| \leq r} \log |f_q(z)|.$$

Assume that to $q = 1, 2, \ldots, n-1$ *there correspond numbers* $\varepsilon_1, \ldots, \varepsilon_{n-1}$ *such that* $\sum \varepsilon_q \leq 1$ *and*

$$\limsup \frac{M_q(r)}{(\log r)^{1+\varepsilon_q}} < \infty.$$

Then the complement of the image of this map has measure zero.

Notice that we do not put any condition on the last function f_n.

§ 7. The Second Main Theorem. Defect Relation

The deepest and most subtle facts of the multidimensional value distribution theory (as, besides, also in the one dimensional case) are connected with the Second Main Theorem. For this profoundness one has, however, to pay for by the general character of the results obtained. The theory developed until now allows one only to study the distribution of preimages of divisors. For sets of higher codimension (and, in particular, for points) one does not have any workable approach so far.

The problem to be solved can be formulated as follows. We saw in the last Section that, in suitable hypotheses on the exhaustion, the average value of the difference $T(r)-N(r, a)$ for all a in a sufficiently large set (for instance, of positive 2-capacity) can be estimated from below by a constant depending on the set of values a and a measure on this set, but not on r. But is it possible to find nontrivial estimates of this average of values for a not in that large set? For instance, for a finite set? In which assumptions can this be done?

Until now satisfactory results have been obtained in three essentially different situations, and for each of these situations special methods and ideas are required.

The first group of results pertains to the case when one considers maps into the complex projective space $\mathbb{C}\mathbb{P}^n$.

This comprises the theory of holomorphic curves and its "natural" generalizations. A distinctive feature of this theory (as compared to the two other situations) is an essential use of the linear structure of the space of homogeneous coordinates defining the projective space $\mathbb{C}\mathbb{P}^n$.

The Second Main Theorem for holomorphic curves was first proved by H. Cartan. Instead of a holomorphic map $f\colon \mathbb{C}\to\mathbb{C}\mathbb{P}^n$ he considered, expressing the latter in homogeneous coordinates, the map $\tilde{f}\colon \mathbb{C}\to\mathbb{C}^{n+1}\setminus\{0\}$ where the family $\{f_0, f_1, \ldots, f_n\}$ of coordinate maps is assumed to be linearly independent. Every hyperplane in $\mathbb{C}\mathbb{P}^n$ is given by a linear form in $\mathbb{C}^{n+1}$ and if one considers hyperplanes in general position, then any $n+1$ among these forms are linearly independent. Let $A_1, \ldots, A_q$ be an arbitrary collection of such linear forms. Consider the family of functions $\{g_i = A_i \circ \tilde{f}\}$. Simple estimates based on the elementary theory of determinants give that

$$(q-(n+1))\log\|\tilde{f}\| \leqq \log\left|\frac{g_1\cdot g_2\cdot\ldots\cdot g_q}{W}\right| + \log|D| + c,$$

where W is the Wronskian of the family $\{f_0, \ldots, f_n\}$ and D is the determinant

$$\begin{vmatrix} 1 & & 1 \\ g_k'/g_k & \cdots & g_l'/g_l \\ \vdots & & \vdots \\ g_k^{(n)}/g_k & \cdots & g_l^{(n)}/g_l \end{vmatrix}.$$

We let here $\{g_k, \ldots, g_l\}$ be a subsystem of $\{g_1, \ldots, g_q\}$ consisting of $n+1$ functions. We take a different subsystem for each z chosen in such a way that the following condition is fulfilled: if g_i is in the subsystem but not g_j then $|g_i(z)| \leqq |g_j(z)|$. One can show that the average over the circumference $|z| = r$ of $\log\|\tilde{f}\|$ equals up to $O(1)$ the characteristic function $T_f(r)$. By Jensen's formula the average of $\log|g_i|$ equals $N_{g_i}(0, r) = N(A_i, r)$. As a result we get the estimate

$$\sum(T_{\tilde f}(r)-N(A_j,r))\leqq(n+1)\,T_{\tilde f}(r)-N(R_{\tilde f},r)+S(r).$$

Here $N(R_{\tilde f},r)$ is the counting function for the zeros of Wronskian. As for the remainder term $S(z)$, which is the average of $\log|D|$ over $|z|=r$, one can show, using a remarkable lemma on the logarithmic derivative (established by Nevanlinna), that if r tends to infinity then, omitting a set of finite logarithmic length, one has $S(r)=o(T(r))$.

If one defines the *defect* $\delta(A_j)$ of the hyperplanes A_j as $1-\lim N(A_j,r)/T_f(r)$ then it follows from the Second Main Theorem proved above that one has the *defect relation*

$$\sum_{j=1}^{q}\delta(A_j)\leqq n+1.$$

Subsequently Cartan's theorem was generalized by Ahlfors to adjoint curves, by H. and J. Weyl to curves where the domain of definition is a Riemann surface, by Kneser and Stoll to maps from a manifold of higher dimension with given exhaustion into projective space. But in all these papers a fundamental rôle was occupied by the linear structure of $\mathbb{C}^{n+1}$, as mentioned above. Roughly speaking, this structure allows one to reduce the situation to the case of functions. Therefore in this context one cannot replace hyperplanes by hypersurfaces of higher degree. Of course, using the Veronese map, one can imbed $\mathbb{C}\mathbb{P}^n$ into a projective space of sufficiently high dimension such that the hypersurfaces of given degree d correspond to hyperplane sections. However, then in the right hand side of the defect relation there stands a much too large quantity.

In view of what has been said it seems that it must be hard to prove the following

Griffith's Conjecture. *Let $f\colon \mathbb{C}\to\mathbb{C}\mathbb{P}^n$ be a holomorphic curve and assume that $f(\mathbb{C})$ is not contained in any algebraic hypersurface. Let $D_1,\ldots,D_q$ be smooth hypersurfaces in $\mathbb{C}\mathbb{P}^n$ of degree d and with normal crossings. Then*

$$\sum\delta(D_i)\leqq(n+1)/d.$$

In Griffith's original formulation he had the weaker condition that $f(\mathbb{C})$ is not contained in any hypersurface of degree d. However, in this case Biancafiore has recently given a counterexample. He also displayed a class of curves for which Griffith's conjecture is true but its description is rather cumbersome.

The second situation, where one can prove a Second Main Theorem, was studied in papers by Carlson-Griffiths-King, Shiffman, and others. These authors succeeded to reshape in a rather deep way Ahlfors's original idea concerning maps into manifolds with negative curvature. All these considerations are based on the existence of a *singular volume form* on projective varieties having singularities on prescribed divisors and such that its Ricci curvature is positive. In the case of Riemann surfaces this condition is equivalent to the negativity of the Gauss curvature.

Let V be a smooth n-dimensional projective variety and K_V its canonical bundle, i.e. a line bundle whose holomorphic sections are the globally defined holomorphic n-forms on V. As usual, let $c(L)$ be the Chern class of the line bundle L. The following result holds true

Theorem 1 (Carlson-Griffiths-King). *Assume that the line bundle L on V satisfies the condition $c(L)+c(K_V)>0$. Let D be a divisor in the complete linear system $|L|$ such that $D=D_1+\ldots+D_k$ where the divisors D_i are without singularities and assume also that all selfintersections of D are normal. Then there exists a singular volume form Ψ, singular on D, such that its Ricci form* Ric Ψ *satisfies the conditions*

$$\operatorname{Ric}\Psi>0; \quad (\operatorname{Ric}\Psi)^n \geqq \Psi; \quad \int_{V\setminus D} (\operatorname{Ric}\Psi)^n < \infty.$$

In this theorem also the character of the singularities of Ψ near the divisors D_i is specified.

It is not hard to construct such a form Ψ. Let Ω be an arbitrary volume form on V. By assumption there exists a metric ω on L with Chern form Ω such that $\omega+\operatorname{Ric}\Omega>0$. For each D_i let L_i be the line bundle defined by it. Thus $L=L_1\otimes\ldots\otimes L_k$. Pick a section σ_i of L_i such that D_i is its zero divisor. Then $\sigma=\sigma_1\otimes\ldots\otimes\sigma_k$ defines D. Define metrics on $L_1, \ldots, L_k$ such that if $\xi_1, \ldots, \xi_k$ is any family of sections of $L_1, \ldots, L_k$ then

$$|\xi_1\otimes\ldots\otimes\xi_k| = |\xi_1|\cdot\ldots\cdot|\xi_k|.$$

Now we take

$$\Psi=\left(\prod_{j=1}^{k}(\log|\sigma_j|^2)^2\,|\sigma_j|^2\right)^{-1}\cdot\Omega.$$

All the desiderata are verified by straightforward computations.

Let us now consider a holomorphic map $f\colon \mathbb{C}^n \to V$ which is nondegenerate at at least one point; V is as before a smooth projective n-dimensional variety equipped with a positive line bundle. Let ω be a positive (1, 1)-form representing the Chern class $c(L)$. It may be obtained as the Ricci curvature of an Hermitean metric on L.

We observed in Sec. 5 that the First Main Theorem can be written not only for preimages of points but also for preimages of submanifolds belonging to a given family and, in particular, thus for divisors in the complete linear system $|L|$. If we do not carry out the reduction described there, then the expressions for $T(r)$, $N(r, a)$ and $m(r, a)$ can be written directly in terms of the situation at hand. More precisely, we consider on $X=\mathbb{C}^n$ the exhaustion $\tau=\ln|z|$ (by necessity, it has to be redefined at the origin) and set $\chi=(d\,d^c\,\tau)^{n-1}$. The function $T(r)$ is defined as

$$\int_0^r\Big(\int_{X_t} f^*\omega\wedge\chi\Big)\,dt,$$

while $N(r, D)$ is given by the integral

$$\int_0^r \Big(\int_{X_t \cap f^{-1}(D)} \chi \Big) dt$$

and $m(r, D)$ by the integral

$$\int_{\partial X_t} (d^c \tau \wedge \chi) \cdot f^* \log 1/|\sigma|^2.$$

Here $|\sigma|$ denotes the Hermitean norm (in the bundle L) of a section σ defining the divisor D. We do not write down the last term of the First Main Theorem, as in view of the χ-concavity the latter must be $O(1)$.

The First Main Theorem

$$T(r) = N(r, D) + m(r, D) + O(1)$$

is proved directly by integrating twice the following equation for currents (Poincaré's formula):

$$d\,d^c f^* \log |\sigma|^2 = f^{-1} D - f^* \omega.$$

Let us now write the preimage of the above singular volume form Ψ in the form $f^* \Psi = \xi \Phi$ where Φ the Euclidean volume form in $\mathbb{C}^n$. Since f is not identically degenerate, ξ can not be identically zero. It is everywhere nonnegative and vanishes along the zero divisor R_f of the Jacobian of f (the Jacobian depends on the local coordinates but R_f is invariantly defined). Along $D_f = f^{-1}(D)$, ξ becomes infinite.

Integrating the current equation

$$d\,d^c \log \xi = f^* \operatorname{Ric} \Psi + R_f - D_f$$

twice we get, as in the First Main Theorem,

$$\int_0^r \Big(\int_{X_t} \operatorname{Ric} \Psi \wedge \chi \Big) dt + N(r, R_f) = N(r, D_f) + \int_{\partial X_t} (d^c \tau \wedge \chi) \log \xi + O(1).$$

Let the divisor D equal the sum $D_1 + \ldots + D_k$ where all the D_i are divisors of one and the same linear system $|L|$, L being a positive line bundle, and assume that all previous requirements are fulfilled. Then

$$\operatorname{Ric} \Psi = k\omega + \operatorname{Ric} \Omega - \sum_{j=1}^{k} d\,d^c \log(\log |\sigma_j|^2)^2$$

and we obtain the identity

Second Main Theorem (provisional form).

$$kT(r)+\int_0^r\Big(\int_{X_t} f^*\operatorname{Ric}\Omega\wedge\chi\Big)\,dt+N(r,R_f)=$$

$$=\sum_{i=1}^k N(r,D_i)+\int_{\partial X_r}(d^c\tau\wedge\chi)\log\xi+$$

$$+\sum_{i=1}^k\int_0^r\Big(\int_{X_t}\chi\wedge dd^c f^*\log(\log|\sigma_i|^2)^2\Big)\,dt+O(1).$$

This theorem can be reformulated if we estimate some of the integrals therein.

The integrals

$$S_i(r)=\int_0^r\Big(\int_{X_t}\chi\wedge dd^c f^*\log(\log|\sigma_i|^2)^2\Big)\,dt$$

are easy to estimate. Integrating by parts using Stokes's formula and the logarithmic convexity leads to the following estimate

$$S_i(r)\leqq\log m(r,D_i)+O(1)\leqq\log T(r)+O(1).$$

It is considerably harder to estimate

$$\mu(r)=\int_{\partial X_t}(dd^c\tau\wedge\chi)\log\xi.$$

It is precisely for this estimate that one requires the burdensome assumption that the image $f(\mathbb{C}^n)$ contains an open set. In any case, one has to show that if r tends to infinity and in addition does not take values in a suitable subset of finite length then $\mu(r)=o(T(r))$. The final result can be stated as follows.

Second Main Theorem (final formulation).

$$\sum_{j=1}^k(T(r)-N(r,D_j))+N(r,R_f)\leqq\int_0^r\Big(-\int_{X_t}f^*\operatorname{Ric}\Omega\wedge\chi\Big)\,dt+\varepsilon(r),$$

where the remainder term $\varepsilon(r)$ behaves as in the classical theory.

Recall that $\operatorname{Ric}\omega$ is the Chern form of the canonical bundle K_V so that $-\operatorname{Ric}\Omega$ is the Chern form of the dual bundle K_M^*. If we choose α such that $\alpha\cdot c(L)\geqq c(K_V^*)$ then it follows from the Second Main Theorem that

Theorem 2 (defect relation).

$$\sum\delta(D_i)\leqq\alpha.$$

If the divisors D_i are hyperplanes in $\mathbb{C}\mathbb{P}^n$ then we can take $\alpha=n+1$, which gives the same defect relation as for holomorphic curves. However, we get a weaker statement as there is a supplementary restriction on the dimension. However, if D_i are hypersurfaces of degree p, we get $\alpha=(n+1)/p$, which so far has not been obtained by any other method.

This Second Main Theorem and the defect relation have been generalized in several directions: Griffiths and King replaced $\mathbb{C}^n$ by an arbitrary manifold with a special exhaustion (possibly of high dimension), Shiffman succeeded in eliminating partially the hypothesis that the divisors were without singularities and the requirement about normal selfintersections etc. However, in all these papers one keeps the assumption that the image contains an open set.

A cardinal unsolved problem in multidimensional value distribution theory is to prove the defect relation (and the Second Main Theorem) without this assumption.

There is yet another situation when one can prove an analogue of the Second Main Theorem and the defect relation can be established, but on a somewhat different level. In the previous discussion we have assumed that the divisors under study were in "general position".

There are papers where this requirement is somewhat relaxed. However, as simple examples show, if one completely drops this condition, then for finite or countably infinite families of divisors it is not possible to have any nontrivial estimates. For instance, Ya.I. Savchuk showed that if a set A in the space of hyperplanes, i.e. in the dual projective space, consists of the union of a countable number of linear subspaces, then there exists a holomorphic curve such that precisely all the hyperplanes in A are defect for it (no others).

Thus, if we completely drop the assumption of general position then we have to try to get an estimate of the average of $T_f(r)-N_f(r,a)$ with respect to a measure defined for a comparatively large set of points a. At the same time let us remember that if this set has positive 2-capacity and the average is taken with respect to the equilibrium measure then in the case of a χ-concave exhaustion the average can be estimated by a constant. Therefore the supports of the averaging measure must be sets of 2-capacity 0. Indeed, the desired results can be claimed if the measure (denote it by m) satisfies a certain special condition which guarantees the convergence of the integral with respect to the exponential of the potential of the measure, taking as kernel the function $\gamma(a,y)$ (cf. § 5).

The condition amounts to requiring that if for some value a the function $\gamma_a(y)$ takes on a set $U\subset Y$ values less than t^{-1} then the m-measure of U cannot exceed $t/(n-1)+o(t)$. An equivalent condition: if U is contained in a geodesic ball of radius r then its m-measure is $\leqq 2^{n+1}\pi^n r^{2n-2}+o(r^{2n-2})$.

Assume now that the conditions, under which the First Main Theorem holds true, are fulfilled and that the exhaustion τ is χ-concave. This means, in particular, that the identity in the First Main Theorem can be written as

$$T(r)-N(r,a)=\int_{\partial X_r} d^c\tau\wedge f^*(\gamma_a\mu)\wedge\chi+O(1).$$

Integrating this with respect to the measure m, subject to the above conditions, then after a series of transformations the right hand side can be rewritten in the form of a sum of two terms one of which does not depend on m and the other, depending on m, has an estimate of the form $O(\ln T(r))$. As usual, this estimate holds off intervals of finite total length. The Second Main Theorem in these hypotheses takes the form:

Theorem 3 (I.M. Dektyarev).

$$\int_Y (T(r)-N(r,a))\,m(d\,a) = -\int_{\partial X_r} \log\kappa\cdot d^c\tau\wedge f^*\mu\wedge\chi + O(\ln T(r)),$$

where κ is a function given by the relation

$$\kappa\, d\tau\wedge d^c\tau\wedge f^*\mu\wedge\chi = f^*\omega\wedge\chi.$$

If all these considerations are applied in the case when one considers preimages of divisors, as described in Sec. 5, then the expression to the right can be estimated by $T(r)$. More precisely, there exists a constant d, not depending on f, but which depends on the collection {manifold + a family of divisors on it}, such that the integral

$$-\int_{\partial X_r} \log\kappa\cdot d^c\tau\wedge f^*\mu\wedge\chi$$

does not exceed $d\cdot T(r)+O(r)$. In some cases one can compute d. This estimate leads to a defect relation for maps such that $r/T(r)\to 0$:

$$\int_Y \delta(a)\,m(d\,a)\leqq d.$$

§ 8. More on the Functions $T(r)$ and $N(r, a)$

Besides the results which immediately follow from the Main Theorems there are also many other interesting properties shared by $T(r)$ and $N(r, a)$. Let us consider some of these.

First, if f is a holomorphic map $\mathbb{C}^n\to\mathbb{C}^n$ then, as in the one dimensional case, $T(r)$ is closely connected with quantities describing the speed of the growth of f. More exactly, it is here not question of one single function $T(r)$ but rather a family of such functions, so that one can study preimages of subspaces of different codimensions. Let us recall that these functions, for $k=1, \ldots, m$, are defined by the following identity

$$T_k(r)=\int_0^r\Big(\int_{X_t} (f^*\omega)^k\wedge(d\,d^c\tau)^{n-k}\Big)\,d\,t.$$

In $\mathbb{C}^n$ we use as exhaustion the function $\tau=\ln\sum|z_i|^2$. As χ we take the form $(d\,d^c\tau)^{n-k}$. As for the form ω in $\mathbb{C}^n$ we obtain it as the restriction of the Study-Fubini metric $dd^c \ln(1+\sum |w_i|^2)$, $\mathbb{C}^n$ being considered as imbedded in $\mathbb{C}\mathbb{P}^n$. The function $T_k(r)$ corresponds to the distribution of the pre-images of subspaces of codimension k. Let us observe that here we have deviated from the previous line of development in order to be able to take into consideration the manifold of subspaces and to redefine in a corresponding manner the characteristics of maps (cf. Sec. 5).

Simple transformations, using Stokes's formula and simple facts from function theory, allow us to obtain for each $\alpha > 1$ the following inequality due to Carlson,

$$T_k(r) \leqq \frac{1}{\ln \alpha} T_{k-1}(\alpha r) \log \max_{|z|=r} \sqrt{1+\sum |f_i|^2}.$$

Iterating this estimate, we arrive at an inequality connecting $T_k(r)$ with $T_1(r)$. For $T_1(r)$ we get by similar methods

$$T_1(r) \leqq \log \max_{|z|=r} \sqrt{1+\sum |f_i|^2}.$$

Using Poisson's formula, it is likewise easy to get a lower estimate for $T_1(r)$:

$$\log \max_{|z|=r} \sqrt{1+\sum |f_i|^2} \leqq \frac{R^{2n-2} \cdot (R+r)}{(R-r)^{2n-1}} \cdot T_1(r),$$

where $R > r$ is an arbitrary number.

As for the functions $T_k(r)$ with $k > 1$, one has for them in general no lower estimates in terms of the growth of the functions. In fact, consider an automorphism $f\colon \mathbb{C}^2 \to \mathbb{C}^2$. By definition it is clear that $T_2(r)$ grows as r but one can choose the automorphism such that one of the coordinate functions has an arbitrary fast growth.

For some classes of functions it is nevertheless possible to have lower estimates. Thus P.V. Dektyar' introduced a class of maps, which he called maps of q-regular growth, and proved that if the order of such a map equals ρ then for $k \leqq m+1-q/2$

$$T_k(r) \geqq H_f^{(k)} \cdot r^{k\rho} + o(r^{k\rho}).$$

Here $H_f^{(k)}$ is a constant. Notice that from the estimates obtained for this class of maps it is easy to get theorems on the everywhere density of the image.

Now a few words on $N(r, a)$. One can consider it even without reference to maps. Let A be an analytic subset of $\mathbb{C}^n$ (it is not important if it is the preimage of anything). Denote by $n(t, A)$ the integral of the form $(dd^c \ln \sum |z_j|^2)^k$ extended over the intersection of A with the ball of radius t, k being the complex dimension of A. The function $N(r, A) = \int_0^r n(t, A)\, d\log t$ differs only in an inessential way from the one introduced earlier. It turns out that the behavior of $N(r, A)$ is a most useful characteristic of the set A. For example, if $N(r, A)/\log r$ is bounded then, as Stoll has shown, A is algebraic (the converse is trivial). Subsequently various variants of this theorem were established, by Griffiths and King, L.I. Ronkin, A. Sadullaev and others.

An interesting property of the function $N(r, a)$ was discovered by L.I. Ronkin. He proved that its behavior is completely specified by the intersections of A with a rather small family of lines. Here is the exact formulation.

Let $\tau=(\tau_1, \ldots, \tau_n)$ be a family of positive real numbers and $\kappa=(\kappa_1, \ldots, \kappa_n)$ a family of numbers of unit modulus, i.e. κ is a point of the skeleton of the unit polydisk. Let us define the line $\mathbb{C}_{\tau,\kappa}$ as the set

$$\{w\in\mathbb{C}^n\colon w_1=z\tau_1\kappa_1, \ldots, w_n=z\tau_n\kappa_n;\ z\in\mathbb{C}\}$$

and for an analytic subset $A\subset\mathbb{C}^n$ of codimension 1 and not passing through the origin define $N_{\tau,\kappa}(r, A)$ by the same rule as the function $N(r, A)$, but counting only points of the intersection of A with $\mathbb{C}_{\tau,\kappa}$. Let $N_\tau(r, A)$ be the average of $N_{\tau,\kappa}(r, A)$ when κ runs through the entire skeleton of the unit polydisk.

Theorem 1 (L.I. Ronkin). *The function $N_\tau(r, A)$ has for each τ the same growth behavior (order, type etc.) as $N(r, A)$.*

Gruman has carried over this theorem to the case when A has higher codimension than one.

An interesting analogue of L.I. Ronkin's theorem has been found by Sibony and P.M. Wong. There A still has codimension one but for the directors of the conical surface one takes not the skeleton of the polydisk but a set E of positive Γ-capacity.

Theorem 2. *For each such E there is a $\theta<1$ such that for all r holds*

$$\sup_{z\in\mathbb{C}^{n-1}} N(\theta r, A\cap L_z)\leqq\sup_{z\in E} N(r, A\cap L_z).$$

Here L_z is the line passing through the point z and the origin.

Corollary. *If $A\cap L_z$ is finite for each $z\in E$ then A is an algebraic set.*

A further strengthening of these results was obtained by Alexander. He replaced the requirement of positive Γ-capacity of E by the requirement of positive projective capacity. Projective capacity is somewhat more adapted to the purposes of complex analysis. For example, a set of zero Γ-capacity is not necessarily polar but a set of zero projective capacity surely is.

An interesting class of analytic sets arises if we, instead of the boundedness of the ratio $N(r, A)/\log r$, implying that A is algebraic, impose that

$$\liminf N(r, A)/(\log r)^2<\infty.$$

Sibony and P.M. Wong proved that if an irreducible analytic set satisfies this condition then there exist no nonconstant bounded analytic functions on that set.

Stoll's *algebraicity criterion* for analytic sets, just discussed, leads to the following question. Let us consider a purely p-dimensional analytic set $A \subset \mathbb{C}^n$ such that $N(r, A)$ has finite order ρ. Is it possible to represent this set as the common zero set of a family $\{f_1, \ldots, f_k\}$ of functions none of which is of an order exceeding λ? The canonical Weierstrass product provides a positive answer to this question in the classical case $n=1$. For an arbitrary n but A of codimension 1, a positive answer was given by Stoll. The case of higher codimension is, as usual, much harder. Pan solved the problem for a zero dimensional set A and likewise for intermediate values of the codimension but under supplementary conditions on the behavior of A at infinity. The best result obtained so far is due to Skoda. He showed that in the hypothesis considered there is a family of entire functions $\{f_1, \ldots, f_{n+1}\}$ such that A coincides with the set of its common zeros and that on a sphere of radius r holds

$$\ln \sup |f_i(z)| \leqq c(\ln r)^2 \cdot n(r+\varepsilon r, A),$$

this for any i, ε being an arbitrary small positive number.

What can one say about the converse statement? Does the common set of holomorphic functions of order not excelling λ have a counting function $N(r, a)$ of order not greater than λ? The Coralba-Shiffman example reproduced in § 2 shows that this is not the case.

This problem is connected with the so-called "transcendental" Bezout problem, the study of which was initiated by Griffiths and which is closely interwoven with value distribution theory. There is now a considerable literature about this circle of ideas and it is not possible for us to treat it even very briefly.

§ 9. Applications

We begin with the so-called *uniqueness problem*. A polynomial in one variable, as is well-known, is almost completely determined by the family of points where it vanishes. If the roots are given with their multiplicities, then in order to determine the coefficient of the highest term one requires just one more value. For entire functions two values are too little. And three are also unsufficient. But four values suffice. For meromorphic functions one requires five points. In fact, one has the following remarkable

Theorem 1 (Nevanlinna). *Let f and g be two meromorphic functions such that for five points a_1, a_2, a_3, a_4, a_5 holds $f^{-1}(a_i)=g^{-1}(a_i)$. Then $f(z)\equiv g(z)$.*

Fujimoto has devoted a whole series of papers to the problem of extending this and similar results to the multidimensional case. Let us state two of the most explicit of the theorems established by him.

Theorem 2 (Fujimoto). *Let f_1 and f_2 be meromorphic maps from $\mathbb{C}^n$ into $\mathbb{C}\mathbb{P}^N$ and let $H_1, H_2, \ldots, H_{3N+1}$ be hyperplanes in general position such that $\nu(f_1, H_i) = \nu(f_2, H_i)$, where $\nu(f_j, H_i)$ is the preimage of the divisor H_i under the map f_j. We require further that at most one of the maps f_1 and f_2 is algebraically nondegenerate, i.e. that the image is not contained in any algebraic submanifold of $\mathbb{C}\mathbb{P}^N$. Then f_1 and f_2 coincide identically.*

For $N > 1$ the number $3N+2$ is not the best possible. Thus, for $N=2$ Fujimoto proved that 7 hyperplanes suffice. However, in other cases one has so far not succeeded in improving the number $3N+2$. It would be interesting to find the least possible value.

Theorem 3 (Fujimoto). *Let F be a family of nondegenerate meromorphic maps $\mathbb{C}^n \to \mathbb{C}\mathbb{P}^N$ such that for any $N+2$ hyperplanes $H_1, \ldots, H_{N+2}$ in general position their preimages are fixed divisors in $\mathbb{C}^n$. Then F is finite and cannot contain more than $N+1$ algebraically independent maps.*

It is interesting to note that for the proof of these theorems Fujimoto uses, besides combinatorial computations, only É. Borel's theorem (the first theorem in the theory of holomorphic curves). This is an analogue of Picard's theorem and says that if the entire functions $f_0, f_1, \ldots, f_n$ are linearly independent and do not vanish, then any linear combination with nonvanishing coefficients by necessity must take the value zero.

In another direction, using more recent developments in the multidimensional theory, Drouilhet has generalized Nevanlinna's theorem. The formulation of his theorem in maximal generality is rather cumbersome. So we give only one of the most important special cases.

Theorem 4 (Drouilhet). *Let A be a hypersurface of degree $\geqq k+4$ in $\mathbb{C}\mathbb{P}^k$ with normal crossings and let f and g be two nondegenerate meromorphic mappings such that the set $f^{-1}(A)$ coincides with $g^{-1}(A)$ and f and g coincide on that set. Then they coincide everywhere.*

Let us point out that Fujimoto does not make such an assumption (coincidence of the maps on $f^{-1}(A) = g^{-1}(A)$). Instead he imposes rather subtle conditions on A.

Now an application to algebraic geometry.

Theorem 5 (Kodaira). *If the first Betti number of a projective surface W vanishes or if W contains $\mathbb{C}^2$ as an open subset then W is a rational surface.*

The proof is based on two facts. First: a regular surface is rational iff all its plurigenera vanish, that is, no power of its canonical line bundle has nontrivial holomorphic sections. This is entirely a result in algebraic geometry.

Instead the second fact belongs to value distribution theory. Let $f\colon \mathbb{C}^n \to W$ be a nondegenerate holomorphic map into an n-dimensional projective manifold W, assuming that some power of the canonical bundle of that manifold has a nontrivial section. Let ω be the volume form on W and B_r the ball of radius r in $\mathbb{C}^n$. It is then the statement that $\int_{B_r} f^* \omega$ tends to infinity as $r \to \infty$. This implies that f cannot be an imbedding, because otherwise the integral, apparently, would be finite.

The idea of the proof of the last statement is as follows. The existence of a nontrivial section of a power of the canonical bundle shows that there exists on W a volume form with nonnegative definite Ricci form $\operatorname{Ric}\omega$. Let the function ξ on $\mathbb{C}^n$ be defined by the formula $f^* \omega = \xi \Omega$, where Ω is the volume form in C^n. Denote by dS_r the area element of the sphere of radius r. Using methods of value distribution theory Kodaira now proves the fundamental relation

$$\int_{\partial B_r} \log \xi \, dS_r \geqq 0.$$

The rest is easy. In view of the convexity of the logarithm we get

$$\int_{\partial B_r} \log \xi \, dS_r \leqq \log(S(r)^{-1} \int_{\partial B_r} \xi \, dS_r),$$

where $S(r)$ is the area of the sphere of radius r in $\mathbb{C}^n$. As

$$\int_{\partial B_r} \log \xi \, dS_r$$

is nonnegative then

$$\int_{\partial B_r} \xi_r \, dS > S(r).$$

Integrating this with respect to r we get

$$\int_{B_r} f^* \omega \geqq \int_{B_r} \Omega,$$

as requested.

Another application to algebraic geometry is due to Griffiths and King. They prove the following

Theorem 6. *Let A be a smooth affine variety and V a variety of general type. Then each holomorphic map f: $A \to V$, whose image contains an open subset of V, is by necessity rational.*

Recall that a smooth projective n-dimensional variety V is called a *variety of general type* if

$$\limsup_{k \to \infty} \frac{\dim H^0(V, K_V^k)}{k^n} > 0,$$

where, as usual, K_V is the canonical bundle of V and $H^0(V, K_V^k)$ the space of holomorphic sections of its k-th power K_V^k. For instance, varieties with positive canonical bundles are of general type.

Here is another application. This time to number theory. To tell the truth, the result which we have in mind was first obtained without use of value distribution theory. It is due to Bombieri, who in the general line of thoughts of M. Schneider and Serge Lang carried over some theorems of the formers to the multidimensional case. A new proof of Bombieri's theorem, with some strengthenings, was obtained by Demailly, who moreover used a technique developed by him in the spirit of value distribution theory.

We display now some of Demailly's results not so much because of their applications to Bombieri's theorem but because of their independent beauty. In the theorems below new connections between the growth of an entire function of several variables and the distribution of its zero set are given.

Consider a function f holomorphic in the entire space $\mathbb{C}^n$. Let T be its zero divisor, with multiplicities of zeros taken into account; when integrating over T or a portion of T we likewise pay attention to this multiplicity. Let $P_1, \ldots, P_N$ be polynomials of degree δ in n variables ($N>n$) such that their homogeneous parts do not have any common roots, except zero of course. It follows from this, in particular, that the set of common zeros of $P_1, \ldots, P_N$ is discrete. Let us set $\varphi(z)=\sum|P_j(z)|^2$, $X=\{z\in\mathbb{C}^n\colon\ \varphi(z)<t\}$ and $\beta=i/2\cdot d\,d^c\,\varphi(z)$. Moreover, let $M_r=\sup|f(z)|$ where the supremum extends over all z subject to $|z|\leqq r$. Now we can state Demailly's first theorem.

Theorem 7. *There exists a constant* $c(0<c\leqq 1)$ *independent of the special choice of* $P_1, \ldots, P_N$ *such that for any* R *and* r $(R\geqq r\geqq 1)$ *we have*

$$\log\frac{M_R}{M_r}\geqq\frac{1}{(2\delta)^n\pi^{n-1}}\int_{r^{2\delta}}^{cR^{2\delta}} dt/t\Big(\int_{T\cap X_t}\beta^{n-1}\Big).$$

Demailly's second theorem allows one to transform the right hand side using estimates of the expression under the integral sign.

Theorem 8. *In the previous hypotheses let* z_0 *be a common root of the polynomials* $P_1, \ldots, P_N$ *and* ω *a sufficiently small neighborhood of* z_0. *Pick a number* r *less than the infimum of* $\varphi(z)$ *over the boundary of* ω. *Then one of the components* X_r *must be entirely contained in* ω. *Denote this component by* ω_r. *Among the numbers* $s_1, \ldots, s_N$ *giving the multiplicity with which the polynomials* $P_1, \ldots, P_N$ *vanish at the point* z_0 *select* $n-1$ *numbers such that the numbers so selected are the least possible. Moreover, let* s *be the multiplicity of the zero of* f *at the point* z_0. *Then the following inequality is fulfilled:*

$$\frac{1}{(2\pi r)^{n-1}}\int_{\omega_r\cap T}\beta^{n-1}\geqq s\cdot s_1\cdot\ldots\cdot s_{n-1}.$$

Combining Theorem 7 and Theorem 8 and selecting with the aid of elementary algebraic considerations a suitable system of polynomials, we obtain

Theorem 9. *Let* S *be an arbitrary subset of* $\mathbb{C}^n$ *and denote by* $\omega_1(S)$ *the least possible degree of an algebraic hypersurface containing* S (it may happen that $\omega_1(S)$ is infinite). *Let* δ *be a natural number not surpassing* $\omega_1(S)$. *Then there exists a constant* c *such that for each entire function* F *having at the points of the set* S *zeros of multiplicity not less than* t *and for any* R *and*

r $(R \geqq r \geqq 1)$ *holds*

$$\log \frac{M_R}{M_r} \geqq t \frac{\delta \cdot (\delta+1) \dots (\delta+n-1)}{n! \, \delta^{n-1}} \log \frac{R}{c r}.$$

And, finally, the application promised.

Theorem 10 (Bombieri-Demailly). *Let $f_1, \dots, f_p$ be a family of meromorphic functions in $\mathbb{C}^n$. Let d of these functions $(d > n)$, say $f_1, \dots, f_d$, be algebraically independent over the field of rational numbers and of orders $\rho_1, \dots, \rho_d$ respectively. Assume that each partial derivative of any of the functions $f_1, \dots, f_p$ can be expressed as a polynomial in these functions with coefficients in a given algebraic extension K of the field $\mathbb{Q}$ of rational numbers. Let S be the set of points in $\mathbb{C}^n$ such that all the $f_1, \dots, f_p$ take values in K on this set. Then S is contained in a hypersurface of degree δ where δ is an arbitrary number subject to the inequality*

$$\frac{\delta \cdot (\delta+1) \cdot \dots \cdot (\delta+n-1)}{n! \, \delta^{n-1}} \geqq \frac{\rho_1 + \dots + \rho_d}{d-n} [[K : \mathbb{Q}]].$$

Here $[[K : \mathbb{Q}]]$ equals $[K : \mathbb{Q}]$ if K is contained in the field of real numbers and equals $1/2 \cdot [K : \mathbb{Q}]$ in the opposite case.

The idea of the proof is straightforward. Assume that for some δ there is no hypersurface of degree δ containing S. Then one can find a family of $m(\delta) = (\delta+1) \dots (\delta+n-1)/n!$ points contained in S and having the same property. Now for a suitable sequence $t \to \infty$ one can find a sequence of functions F_t which are polynomials in $f_1, \dots, f_N$, requiring that the coefficients of these polynomials are chosen such that at the points selected F_t has a zero of order t and, moreover, that for a certain fixed number r and a specially chosen subsequence of numbers R_t the sequence of numbers $m_{r,t} = \sup_{|z|=r} |F_t(z)|$ does not drop off sufficiently fast while, on the other hand, the sequence $M_{R_t,t} = \sup_{|z|=R_t} |F_t(z)|$ does not grow too fast. The choice of the sequence R_t and the exact meaning of the words "not too fast" are such that if one applies Theorem 9 to the function F_t then one finds that for sufficiently large values of t the number δ does not satisfy the inequality imposed.

References

As a general source of bibliographical material on multidimensional value distribution theory, we mention the monograph [16]. There one can find more detailed discussion of many of the results described in this Part. Other aspects of multidimensional value distribution theory are highlightened in the collection [13].

The counterexamples in §2 can be found in [16]. Our construction of the Fatou-Bieberbach example is similar to the one in [12].

Analogues of the Sokhotskiĭ-Weierstrass theorem for holomorphic maps of manifolds with concave exhaustions are proved in [2] and, in a more general situation, in [5]. Moreover, in [5] one studies maps of manifolds with nonpositive Laplacean of the exhaustion. In [11] the notion of special exhaustion function is introduced and its existence on any smooth affine variety is secured.

The definition of the multiplicity of a holomorphic map as a current is given in the note [7]. The other definitions of multiplicity mentioned in Sect. 4, apparently, are very well-known.

Various variants of the First Main Theorem are given in [5], [2], [19]–[22], [23], [24]. The function used in our presentation of this result, defined on a compact manifold and having singularities of the prescribed type, was constructed in [6].

The theorems in § 6 on the structure of defect values for maps of complex manifolds were proved in [6]. The results of that Section for maps of real manifolds in a somewhat different form can be found in [5] while the theorem on maps of $\mathbb{C}^n$ into itself is from [18].

A modern account of the theory of holomorphic curves can be found in the book [25] and in [4]. The partial solution of Griffith's problem in § 7 is given in [1]. A treatment of the Second Main Theorem and the defect relation for divisors in general position can be found in [3], [11], [17] and variants of these results without the general position assumption in [7].

The estimates for $T_k(r)$ in § 8 are set forth in the monograph [16]. There one finds also a criterion for the algebraicity of analytic sets. Estimates of $N(r, a)$ in terms of characteristics of the intersection of the set A with lines are given in [14].

The cycle of papers [9]–[10] is devoted to the problem whether a holomorphic map is uniquely defined by pre-images of divisors. The applications of value distribution theory to algebraic geometry are taken from [11] and [12] and the number theory applications are from [8].

For the convenience of the reader, references to reviews in Zentralblatt für Mathematik (Zbl.), compiled using the MATH database, have been included as far as possible.

1. Biancofiore, A.: A hypersurface defect relation for a class of meromorphic maps. Trans. Am. Math. Soc. *270*, 47–60 (1982). Zbl. 515.32009
2. Bott, R., Chern, S.-S.: Hermitian vector bundles and the equidistribution of the zeroes of their holomorphic sections. Acta Math. *114*, 71–112 (1965). Zbl. 148, 319
3. Carlson, J.A., Griffiths, Ph.A.: A defect relation for equidimensional holomorphic mappings between algebraic varieties. Ann. Math., II. Ser. *95*, 557–584 (1972). Zbl. 248.32018
4. Cowen, M., Griffiths, Ph.A.: Holomorphic curves and metrics of negative curvature. J. Anal. Math. *29*, 93–153 (1976). Zbl. 352.32014
5. Dektyarev, I.M.: Problems of value distributions in dimensions greater than unity. Usp. Mat. Nauk *25*, No. 6 (156), 53–84 (1970) [Russian]. Zbl. 207, 378. English transl.: Russ. Math. Surv. *25*, No. 6, 51–82 (1970)
6. Dektyarev, I.M.: The structure of defect sets in multidimensional value distribution theory. Funkts. Anal. Prilozh. *6*, No. 2, 32–40 (1972) [Russian]. Zbl. 246.32024. English transl.: Funct. Anal. Appl. *6*, 112–118 (1972)
7. Dektyarev, I.M.: The multidimensional defect relation without the assumption of general position. Dokl. Akad. Nauk SSSR *273*, 276–279 (1983) [Russian]. Zbl. 577.32025. English transl.: Sov. Math., Dokl. *28*, 611–614 (1983)
8. Demailly, J.-P.: Formules de Jensen en plusieurs variables et applications arithmétiques. Bull. Soc. Math. Fr. *110*, 75–102 (1982). Zbl. 493.32003
9. Fujimoto, H.: The uniqueness problem of meromorphic maps into complex projective space. Nagoya Math. J. *58*, 1–23 (1975). Zbl. 313.32005
10. Fujimoto, H.: Remarks to the uniqueness problem of meromorphic maps into $\mathbb{P}^N(\mathbb{C})$, I–IV. Nagoya Math. J. *71*, 13–24, 25–41 (1978); *75*, 71–85 (1979); *83*, 153–181 (1981). Zbl. 358.32021. Zbl. 358.32022. Zbl. 431.32021. Zbl. 431.32022
11. Griffiths, Ph., King, J.: Nevanlinna theory and holomorphic mappings between algebraic varieties. Acta Math. *130*, 145–220 (1973). Zbl. 258.32009
12. Kodaira, K.: Holomorphic mappings of polydiscs into compact complex manifolds. J. Differ. Geom. *6*, 33–46 (1971). Zbl. 227.32008

13. Kujala, R.O., Vitter III, A.L. (eds.): Value distribution theory. Pure Appl. Math. *25*. New York: Dekker 1974. Zbl. 276.00006
14. Ronkin, L.I.: Introduction to the theory of entire functions of several variables. Moscow: Nauka 1971 [Russian]. Zbl. 225.32001. Engl. transl.: Transl. Math. Monogr. *44*, Providence, R.I.: Am. Math. Soc. 1974
15. Savchuk, Ya.I.: On the set of defect vectors of entire curves. Ukr. Mat. Zh. *35*, 385–389 (1983) [Russian]. Zbl. 518.30030. English transl.: Ukr. Math. J. *35*, 334–338 (1983)
16. Shabat, B.V.: Distribution of values of holomorphic maps. Moscow: Nauka 1982 [Russian]. English transl.: Transl. Math. Monogr. 61. Providence: Am. Math. Soc. 1985. Zbl. 537.32008
17. Shiffman, B.: Nevanlinna defect relations for singular divisors. Invent. Math. *31*, 155–182 (1975). Zbl. 436.32022
18. Sibony, N., Wong, P.-M.: Remarks on the Casorati-Weierstrass theorem. In: Proc. Sym. Pure Math. 35, 91–95. Providence: Am. Math. Soc. 1979. Zbl. 429.32033
19. Stoll, W.: A general first main theorem of value distribution. I, II. Acta Math. *118*, 111–146, 147–191 (1967). Zbl. 148, 319
20. Stoll, W.: About value distribution of holomorphic maps into projective space. Acta Math. *123*, 83–114 (1969). Zbl. 177, 113
21. Stoll, W.: Value distribution of holomorphic maps into compact complex manifolds. Lect. Notes Math. 135. Springer: Berlin etc. 1970. Zbl. 195, 367
22. Stoll, W.: Value distribution of parabolic spaces. Lect. Notes Math. 600. Springer: Berlin etc. 1977. Zbl. 367.32001
23. Tung, C.-C.: The first main theorem of value distribution on complex spaces. Atti Accad. Naz. Lincei, Mem., Cl. Sci. Fis. Mat. Nat., VIII. Ser., Sez. I, *15*, 93–261 (1979). Zbl. 496.32018
24. Wu, H.-H.: Remarks on the first main theorem in equidistribution theory. I–IV. J. Differ. Geom. *2*, 197–202 (1968); *2*, 369–384 (1968); *3*, 83–94 (1969); *3*, 433–446 (1969). Zbl. 164, 381. Zbl. 177, 113. Zbl. 182, 415. Zbl. 192, 439
25. Wu, H.-H.: The equidistribution theory of holomorphic curves. Ann. Math. Stud. 64. Princeton: Princeton University Press 1970. Zbl. 199, 409

III. Invariant Metrics

E.A. Poletskiĭ, B.V. Shabat

Translated from the Russian
by J. Peetre

Contents

Introduction

The idea to construct metrics invariant for suitable maps goes back to Bernhard Riemann. Among many others ideas which came to determine the subsequent development of the geometric approach to mathematics and its application it was expressed in his famous lecture "Über die Hypothesen, welche der Geometrie zu Grunde liegen" (1854). In a more precise form it is formulated in the so-called Erlanger program of Felix Klein in the latters inaugural lecture "Vergleichende Betrachtungen über neuere geometrische Forschungen" (1872), where invariants of various transformation groups are exhibited.

The first concrete results, pertaining to the problem of interest to us, appear toward the end of the last century and are due to Henri Poincaré. In his study of automorphic functions he constructed a model of Lobachevskiĭ (non-Euclidean) geometry on the unit disk $\Delta=\{\zeta\in\mathbb{C}: |\zeta|<1\}$ in the complex plane, where an important rôle is played by the metric

$$\rho(\zeta',\zeta'')=\frac{1}{2}\ln\frac{1+\left|\dfrac{\zeta'-\zeta''}{1-\bar{\zeta}'\zeta''}\right|}{1-\left|\dfrac{\zeta'-\zeta''}{1-\bar{\zeta}'\zeta''}\right|},$$

which is invariant for Lobachevskiĭ motions, i.e. conformal automorphisms of the disk. This is the *Poincaré metric.*

This Part is devoted to metrics which are invariant with respect to biholomorphic maps of complex manifolds.[1]

§ 1. The Carathéodory Metric

Historically the first of metrics considered here is the one introduced by Constantin Carathéodory [11] in 1927.

1.1. Definition and General Properties.

Definition 1.1. Let there be given a complex manifold M and two points p, q in it. The *Carathéodory distance* between p and q is the number

$$c_M(p,q)=\sup_f \rho(f(p),f(q)), \tag{1.1}$$

[1] We are obliged to M.G. Zaĭdenberg for reading the manuscript carefully and for several useful suggestions.

where the least upper bound is taken over all holomorphic maps f from M into the unit disk $\Delta=\{\zeta\in\mathbb{C}: |\zeta|<1\}$ and $\rho(\zeta', \zeta'')$ denotes the Poincaré distance between the points $\zeta', \zeta''\in\Delta$.

It is not hard to see that c_M possesses all the usual properties of a semimetric:

$$\begin{aligned} &\text{a)} && c_M(p, q)\geqq 0, \\ &\text{b)} && c_M(p, q)\leqq c_M(p, r)+c_M(r, q), \\ &\text{c)} && c_M(q, p)=c_M(p, q) \end{aligned} \tag{1.2}$$

(here p, q, r are arbitrary points in M). However, generally speaking, this is not a metric, i.e. it is not necessarily true that $c_M(p, q)>0$ if $p\neq q$.

For example if $M=\mathbb{C}^n$ then $c_M\equiv 0$ by Liouville's theorem. Similarly, $c_M\equiv 0$ on each compact complex manifold M, as by the maximum principle each holomorphic function on M is constant. On the other hand, c_M apparently is a metric if M is a bounded domain in $\mathbb{C}^n$.

Likewise it is easy to establish the *contraction property* of this semimetric with respect to holomorphic maps.

Theorem 1.1. *If f: $M\to N$ is a holomorphic map from a complex manifold M into a complex manifold N, then for any two points p, $q\in M$*

$$c_N(f(p), f(q))\leqq c_M(p, q). \tag{1.3}$$

In particular, this shows that c_M is invariant under biholomorphic maps and further that $c_M(p, q)\leqq c_N(p, q)$ if $M\subset N$ (for the proof apply (1.3) to the imbedding map i: $N\to M$).

For the ball $B=\{z\in\mathbb{C}^n: |z|<1\}$ and the polydisk $U=\{z\in\mathbb{C}^n: |z_k|<1, k=1, \dots, n\}$ the Carathéodory metric is computed with the aid of Schwarz's lemma: in particular we have

$$c_B(0, z)=\frac{1}{2}\ln\frac{1+|z|}{1-|z|}, \qquad c_U(z)=\frac{1}{2}\max\ln\frac{1+|z_k|}{1-|z_k|}. \tag{1.4}$$

However, for some simple domains its computation is difficult. For example, as Simha [66] has shown, for a circular annulus $R=\{z\in\mathbb{C}: 1/r<|z|<r\}$ one has $c_R(z, w)=1/2\cdot\ln(1+q)/(1-q)$, where

$$q=q(z, w)=\frac{1}{r|w|}\left|g(z, w)\, g\left(\frac{1}{|z|}, -|w|\right)\right|,$$

where

$$g(z, w)=\left(1-\frac{z}{w}\right)\prod_{k=1}^{\infty}\left(1-\frac{1}{r^{4k}}\frac{w}{z}\right)\left(1-\frac{1}{r^{4k}}\frac{z}{w}\right)\Big/$$

$$\prod_{k=1}^{\infty}\left(1-\frac{zw}{r^{4k-2}}\right)\left(1-\frac{1}{r^{4k-2}zw}\right). \tag{1.5}$$

Besides Definition 1.1 it is expedient to consider an infinitesimal definition of the Carathéodory metric, having it origin in the latters paper [11].

Definition 1.2. Let M be a complex manifold, p a point in M, $v \in T_p(M)$ a tangent vector. The *Carathéodory seminorm* is defined as

$$C_M(p, v) = \sup_f |f_*(v)|, \tag{1.6}$$

where the greatest lower bound extends over all holomorphic maps $f\colon M \to \Delta$ and $f_* = df$ is the differential of f at p.

This differential can be expressed in local coordinates; by the theorem on the invariance of the differential, $df(v)$ does not depend on the choice of these coordinates. If in local coordinates

$$v = \sum_{k=1}^{n} v_k \frac{\partial}{\partial z_k}$$

then

$$df(v) = \sum_{k=1}^{n} \frac{\partial f}{\partial z_k} v_k;$$

in other words, $df(v) = v(f)$ is derivation in the direction of v. Let us further remark that in (1.6) we may restrict ourselves to maps such that $f(p)=0$: if $f(p)=\alpha \neq 0$, then upon replacing f by the composition $g \circ f$, where $g(\zeta)=(\zeta-\alpha)/(1-\bar{\alpha}\zeta)$, $|f_*(v)|$ gets multiplied by $(1-|\alpha|^2)^{-1}>1$, that is, it increases.

Example 1.1. For the ball B

$$C_B^2(z, v) = \frac{|v|^2}{1-|z|^2} + \frac{|(z, v)|^2}{(1-|z|^2)^2}, \tag{1.7}$$

where (z, w) stands for the Hermitean inner product, and for the polydisc U

$$C_U(z, v) = \max_k \left\{\frac{|v_k|}{(1-|z_k|^2)}\right\}. \tag{1.8}$$

We further mention [12] that for domains $D=\{z \in \mathbb{C}^n\colon \varphi(z)<0\}$, where $\varphi(z) = -2\operatorname{Re} z_n + H(z, z)$ and H is a positive definite Hermitean form, holds

$$C_D^2(z, v) = \frac{H(v, v)}{-\varphi(v)} + \left|\frac{H(v, z)-v_n}{\varphi(z)}\right|^2. \tag{1.9}$$

In the same paper [11] Carathéodory proved that C_M satisfies the triangle inequality: for arbitrary $p \in M$ and $u, v \in T_p(M)$

$$C_M(p, u+v) \leqq C_M(p, u) + C_M(p, v), \tag{1.10}$$

and that it also has the contraction property: for each holomorphic map $f: M \to N$ and arbitrary $p \in M$, $v \in T_p(M)$

$$C_N(f(p), f_*(v)) \leqq C_M(p, v). \tag{1.11}$$

From this property it follows that C_M is invariant under biholomorphic maps. This may be viewed as a multivariate generalization of *Schwarz's lemma in invariant formulation*, stating that for each holomorphic map $f: \Delta \to \Delta$

$$\frac{|f'(z)|\,|dz|}{1-|f(z)|^2} \leqq \frac{|dz|}{1-|z|^2}. \tag{1.12}$$

Let us also remark that c_M is continuous on $M \times M$ and that C_M is continuous on the tangent bundle $T(M)$. It is likewise easy to see that extremal functions exist for any of the two variational problems (1.1) and (1.6) and that these, if they are nontrivial, map M onto a dense subset of Δ.

1.2. Further Properties

Definition 1.3. Let there be given a manifold M equipped with a semimetric d. The *length* of an arc $\gamma: [0, 1] \to M$ in this semimetric is defined to be

$$|\gamma|_d = \sup \sum_{k=1}^{m} d(\gamma(t_k), \gamma(t_{k+1})), \tag{1.13}$$

where the least upper bound is taken over all families $\{t_k\}$, $t_k \in (0, 1)$, $t_k \leqq t_{k+1}$ for all k; $1 \leqq k \leqq m$. The *inner distance*, in a semimetric d, between two points p, q in M is the least upper bound $d^i(p, q)$ of the lengths $|\gamma|_d$ in this semimetric of all paths connecting p and q. If $d \equiv d^i$, we say that d is an *inner semimetric.*

The Carathéodory semimetric is in general not inner. For example, for a spherical shell $D = \{z \in \mathbb{C}^n: 1 < |z| < 2\}$ with $n > 1$ by the theorem on removal of compact singularities (cf. [62]) c_D coincides with the metric c_B of the ball $B = \{|z| < 2\}$, so that the inner distance $c_D^i(p, q)$ is larger than $c_D(p, q)$ for suitable p, q. Vigué [71] has found a more subtle example: if D is the domain of holomorphy $\{z \in \mathbb{C}^2: |z_1| + |z_2|, |z_1 z_2| < 1/16\}$ and one takes the points $0 = (0, 0)$, $z = (z_1, z_2)$, with $1/8 < |z_1| < 1/16$, then $c_D(0, z) < c_D^i(0, z)$.

In this sense the Carathéodory seminorm has a preference compared to the semimetric; for it is clear that the seminorm is connected with the corresponding inner semimetric. More exactly, for any complex manifold M holds

$$c_M^i(p, q) = \inf_\gamma \int_0^1 C_M(\gamma(t), \gamma'(t))\, dt, \tag{1.14}$$

where the greatest lower bound is taken over all smooth paths $\gamma: [0, 1] \to M$ connecting p and q. In cases when it is nondegenerate, the inner metric always induces the standard topology. Vigué [72] gave an example of a complex space with a nondegenerate Carathéodory metric such that the induced topology differs from the standard one. It is not known whether similar examples for complex manifolds exist.

In the one dimensional case the Carathéodory seminorm is connected with the analytical capacity introduced by Ahlfors in 1947. For our purposes it is sufficient to assume that $E \subset \bar{\mathbb{C}}$ is a closed set with connected complement $D = \bar{\mathbb{C}} \setminus E$; the *analytic capacity* of this set is defined to be $\gamma(E) = \sup |\operatorname{res}_\infty f|$, where the least upper bound is taken over all holomorphic functions in D such that $|f(z)| < 1$ everywhere in D. Thus $C_D(z, e) = \gamma(E_z)$ where $E_z = f_z(E)$, $f_z(w) = (w-z)^{-1}$, e a unit vector.

Ahlfors proved that if $\gamma(E_z) = 0$ for some z then $\gamma(E_z) \equiv 0$ for all z. This means that if $n=1$ then the Carathéodory norm is either nondegenerate in the domain or else nowhere nondegenerate; the Carathéodory metric enjoys an analogous property. For $n > 1$ this is not the case.

Example 1.2. In the domain $D = \{z \in \mathbb{C}^2 \colon |z_1 z_2| < 1\}$ one has $C_D(0, v) = 0$ for all $v \in \mathbb{C}^2$ but at other points $z \in D$ the seminorm does not vanish identically. One has $c_D(z, w) = 0$ if $z = 0$ or $w = (z_1, 0)$ or $(0, z_2)$ but $c_D(z, w) \not\equiv 0$.

The following theorem gives a characterization of the degeneracy set of the metric.

Theorem 1.2 (Vesentini [70]). a) *For any holomorphic vector field v in a domain $D \subset \mathbb{C}^n$, $n > 1$, the function $u(z) = \ln C_D(z, v(z))$ is plurisubharmonic in D.* b) *The function $\ln c_D(z, w)$ is plurisubharmonic in $D \times D$.*

A *set* such that there exists a plurisubharmonic function, not identically $-\infty$, which takes the value $-\infty$ on that set, is termed *pluripolar*; in some sense such sets are thin (cf. A. Sadullaev's Part in Vol. 8). It follows from Theorem 1.2 that the degeneracy sets of Carathéodory metrics and Carathéodory norms either coincide with their entire domain of definition or else are pluripolar.

Not much is known about intrinsic properties of the Carathéodory metric. Vesentini [70] introduced the notion of *complex geodesic* in the metric c_M on a complex manifold M, as a holomorphic curve $f \colon \Delta \to M$ such that $c_M(f(\zeta', \zeta'')) = \rho(\zeta', \zeta'')$ (cf. Introduction) for all $\zeta', \zeta'' \in \Delta$. In the case of convex domains in $\mathbb{C}^n$ he proved the existence of such geodesics connecting any two given points. Let us mention two more of his results [70].

Theorem 1.3. *If $f \colon \Delta \to M$ is a holomorphic map such that either*

a) $C_M(f(0), f'(0)) = 1$,

or

b) $c_M((0), f(\zeta)) = \rho(0, \zeta)$ *for some point* $\zeta \in \Delta \setminus \{0\}$,

then f is a complex geodesic in the metric c_M.

Theorem 1.4. *Let f be a holomorphic map f of the bidisk Δ^2 into itself, other than the identity, with at least one fixed point. Then either there is only this fixed point or else the set of fixed points is the image of a complex geodesic.*

In many questions of complex geometry it is of importance to know the boundary behavior of the Carathéodory metric. Let $D \subset \mathbb{C}^n$ be a bounded domain and $\delta(z, \partial D)$ the Euclidean distance from the point z to the boundary of the domain. Applying Schwarz's lemma to the restriction of the map f: $D \to \Delta$ to the complex line passing through the point $z \in D$ in the direction of the vector $v \in \mathbb{C}^n$, it is not difficult to get the estimate

$$C_D(z, v) \leqq \frac{|v|}{\delta(z, \partial D)}. \tag{1.15}$$

For a strictly pseudoconvex domain in $\mathbb{C}^n$ (cf. S.I. Pinchuk's Part in this Volume) S.I. Pinchuk and G.M. Khenkin independently have obtained a much more refined estimate, which takes into account the position of the vector v with respect to the complex structure of the boundary of the domain.

Theorem 1.5. *For each strictly pseudoconvex domain $D \subset \mathbb{C}^n$ there exists positive constants k_1 and k_2 such that for all points $z \in D$, sufficiently close to the boundary ∂D, and all $v \in \mathbb{C}^n$*

$$k_1 \leqq C_D(z, v) \left(\frac{|\pi_T(v)|}{\sqrt{\delta(z, \partial D)}} + \frac{|\pi_N(v)|}{\delta(z, \partial D)} \right)^{-1} \leqq k_2, \tag{1.16}$$

where $\pi_T(v)$ and $\pi_N(v)$ are the projections of v onto the complex tangent plane and the complex normal to ∂D at the point $a = a(z)$ closest to z.

In the case of the ball (cf. Example 1.1) such a different behavior of $C_D(z, w)$ near the boundary in different directions is evident. There is a refinement of Theorem 1.5, due to Graham [24], which we will discuss in Sec. 2.2.

1.3. Domains of Bounded Holomorphy and Completeness. It is clear from the definition of the Carathéodory metric and the Carathéodory norm that these are notions which are connected with bounded holomorphy. The collection of functions which are bounded and holomorphic in a subset E of a complex manifold will be denoted by $H^\infty(E)$; if E is not open, then by holomorphic function we intend a function which has a holomorphic continuation to an open subset containing E.

Definition 1.4. A domain $D \subset \mathbb{C}^n$ is called a *domain of bounded holomorphy* if for each point $a \in \partial D$ there exists a function $g \in H^\infty(D)$ whose restriction to the intersection of D with a suitable neighborhood of a cannot be continued holomorphically across that point.

It is natural to compare this notion with H^∞-convexity, i.e. convexity with respect to the class $H^\infty(D)$. If $n=1$, a domain $D \subset \mathbb{C}$ is a domain of bounded holomorphy iff it is H^∞-convex (Ahern and R. Schneider [1]). If $n>1$, every bounded domain of bounded holomorphy is H^∞-convex (cf. [62], p. 232) but the converse is in general not true.

Example 1.3 (Sibony [64]). Let $\{\alpha_k\}$ be a sequence of points in Δ such that for each point $\zeta \in \partial\Delta$ one can find a subsequence which converges to ζ nontangentially. Pick constants $\gamma_k > 0$ such that the series

$$\varphi_1(\zeta) = \sum_{k=1}^{\infty} \gamma_k \ln \left| \frac{\zeta - \alpha_k}{2} \right|$$

converges uniformly on compact subsets of $\Delta \setminus \{\alpha_k\}$. The function φ_1 is negative and subharmonic in Δ. Therefore $\varphi(\zeta) = \exp \varphi_1(\zeta)$ too is subharmonic in Δ and $0 \leqq \varphi(\zeta) \leqq 1$. The domain

$$D = \{(z_1, z_2) \in \mathbb{C}^2 : |z_2| < e^{-\varphi(z_1)}\}$$

is H^∞-convex but it is not a domain of bounded holomorphy: every function $g \in H^\infty(D)$ can be continued to a function holomorphic in the bidisk Δ^2.

A sufficient condition for a domain to be of bounded holomorphy is provided by the property of completeness of the Carathéodory metric, which notion is of importance also in other contexts.

Definition 1.5. A complex manifold M is called *complete in the sense of Carathéodory* if for each point $p \in M$ the ball $\{q \in M: c_M(p, q) < r\}$ for any $r > 0$ is relatively compact in M.

The following (cf. [64]) holds true.

Theorem 1.6. *If the domain $D \subset \mathbb{C}^n$ is complete in the sense Carathéodory then it is H^∞-convex and is a domain of bounded holomorphy.*

Let us also remark that H^∞-convexity does not guarantee completeness in the sense of Carathéodory.

Example 1.4 (Ahern – R. Schneider [1]). Let a_k, $c_k \in \mathbb{R}_+$ be such that a_k 0, $a_k - a_{k-1} > c_k + c_{k-1}$ and $\sum_{k=1}^{\infty} c_k/(a_k - c_k) < 1/2$. Remove from $\bar{\mathbb{C}}$ the point 0 and the disks $\{|\zeta - a_k| \leqq c_k\}$, which are disjoint in view of our choice of a_k and c_k; the remaining domain is denoted D. This domain is H^∞-convex, because the functions $(\zeta - \alpha)^{-1}$, where α belongs to the omitted disks, separate each compact set from the boundary of the domain (i.e. the values of the function on a compact set are in modulus less than on the boundary). On the other hand, it is easy to show that, for all ζ with $\operatorname{Re}\zeta < 0$, each function $f \in H^\infty(D)$, $f(\infty) = 0$, satisfies the inequality $|f(\zeta)| \leqq 1/2 \cdot \|f\|_D$, where $\|f\|_D = \sup_{\zeta \in D} |f(\zeta)|$. It follows from this that the Carathéodory ball with center at the point $\zeta = \infty$ and radius 1/2 is not compact in D.

By Theorem 1.5 it is clear that each strictly pseudoconvex domain in $\mathbb{C}^n$ is complete in the sense of Carathéodory. Kobayashi [43] introduced the notion of *generalized analytic polyhedron*, as a bounded domain $D \subset \mathbb{C}^n$ such that there exists a domain $G \supset \bar{D}$ and a family of functions $f_{jk} \in \mathcal{O}(G)$ such that

$$D=\left\{z\in G\colon \sum_{j=1}^{m_k} |f_{jk}(z)|<1,\ k=1,\ldots,l\right\}.$$

In the same paper he shows:

Theorem 1.7. *Each generalized analytic polyhedron is complete in the sense of Carathéodory.*

Another sufficient condition for Carathéodory completeness is due to Kim [41].

Theorem 1.8. *Let $D\subset\mathbb{C}^n$ be a bounded domain and assume that there exists a set $K\Subset D$ such that for each point $z\in D$ one can find a biholomorphic automorphism $f\in\operatorname{Aut}D$ mapping z onto a point in K. Then D is complete in the sense of Carathéodory.*

1.4. Negligibility and Removable Singularities

Definition 1.6. A *subset A* of a complex manifold M is termed *negligible in the sense of Carathéodory* if $c_{M\setminus A}\equiv c_M$. A *subset A* is termed *H^∞-removable* if each function $f\in H^\infty(M\setminus A)$ extends to a function holomorphic in M.

It is easy to see that if a subset A of a domain D in $\mathbb{C}^n$ is removable then under continuation of functions $f\in H^\infty(D\setminus A)$ the quantity $\sup_{D\setminus A}|f|$ does not increase. In fact, if there were a point $a\in A$ such that $|f(a)|=m>1$ then for the function $g=2f/(l+m)$ we would have $\sup_{D\setminus A}|g|=2l/(l+m)<1$, whereas $|g(a)|=2m/(l+m)>1$. Then $h(z)=\sum_{k=1}^{\infty}[e^{-i\alpha}g(z)]^k$ where $\alpha=\arg g(a)$, would belong to $H^\infty(D\setminus A)$ but does not extend to the point a, i.e. A cannot be H^∞-removable. This shows that every H^∞-removable set must be negligible in the sense of Carathéodory.

For $n=1$ it follows from the properties of analytic capacity of sets, as established by Ahlfors (cf. Sec. 1.1) that in any domain $D\subset\bar{\mathbb{C}}$ a relatively compact set A is H^∞-removable iff it is negligible in the sense of Carathéodory, which again is equivalent to $\gamma(A)=0$, i.e. that the Carathéodory metric degenerates on $\bar{\mathbb{C}}\setminus A$.

For $n>1$ the situation is more complicated and in the general case it is not known if the two notions of Carathéodory negligibility and H^∞-removability coincide. In $\mathbb{C}^n$ all compact subsets are H^∞-removable and further all subsets of vanishing $(2n-1)$-dimensional Hausdorff measure. The set $A=\Delta^2\setminus D$ in Example 1.3 is H^∞-removable in Δ^2, because every function in $H^\infty(\Delta^2\setminus A)$ extends to one holomorphic in Δ^2. It is clear that the notions of H^∞-removability and negligibility in the sense of Carathéodory coincide for sets of measure 0.

1.5. Applications. The geometrical applications of the Carathéodory metric rely on the fact that it is invariant under biholomorphic maps and does not increase under arbitrary holomorphic maps. The first results in this direction are due to Carathéodory himself. Here is one of them.

Theorem 1.9. *Let $D \subset \mathbb{C}^n$ be convex and circular* (i.e. together with each point of the domain also each circumference $\{z e^{i\theta}: 0 \leqq \theta \leqq 2\pi\}$ belongs to it). *Then* a) *the unit ball in the Carathéodory norm $\{v \in \mathbb{C}^n: C_D(0, v) < 1\}$ coincides with the domain itself*; b) *every biholomorphic automorphism of the domain which preserves the origin is linear.*

The invariance of the Carathéodory metric is used in an essential manner in several results on the biholomorphic nonequivalence of domains of different types. Among these let us mention S.I. Pinchuk's theorem [55] on biholomorphic nonequivalence of two bounded pseudoconvex domains in $\mathbb{C}^n$, $n > 1$, one of which has C^2-smooth boundary and the other piecewise C^2-smooth but non-smooth boundary, and G.H. Khenkin's theorem [37] on the biholomorphic nonequivalence of analytic polyhedra of a sufficiently general type with domains whose boundaries contain a nonempty open subset of points of strict pseudoconvexity.

Using the contraction property for the Carathéodory metric S.I. Pinchuk and G.M. Khenkin independently of each other and in various assumptions on the boundaries of the domains proved the extendability to the closures of the domains of proper holomorphic maps between them, the continuation satisfying a Hölder condition of exponent 1/2.

Let us further mention some results pertaining to *manifolds hyperbolic in the sense of Carathéodory*, i.e. complex manifolds whose universal covering has a nondegenerate inner Carathéodory metric.

Theorem 1.10 (Urata [69]). *For any compact complex manifold M, the number of different holomorphic maps from M onto a compact manifold N, hyperbolic in Carathéodory's sense, is finite.*

It follows from this that if M is an arbitrary compact manifold and N is compact and hyperbolic in Carathéodory's sense, $A \subset M$ and $B \subset N$ being closed analytic subsets, then the number of holomorphic maps f: $M \to N$ such that $f(A) = B$ is finite.

Another result is due to A. Borel and Narasimhan [7].

Theorem 1.11. *Let M be a complex manifold such that each bounded from above real plurisubharmonic function on it is constant and let N be hyperbolic in the sense of Carathéodory. If two holomorphic maps f, g: $M \to N$ coincide at some point $p \in M$ and induce the same map of fundamental groups: $\pi_1(M, p) \to \pi_1(N, p)$, then $f \equiv g$.*

Let us point out that here Carathéodory hyperbolicity can not be replaced by Kobayashi hyperbolicity (cf. Sec. 2.3 below). This follows from an example by V.V. Rabotin [57].

We conclude by E.M. Chirka's multidimensional generalization of Lindelöf's theorem to the effect that each function $f \in H^\infty(\Delta)$, which has a radial limit as $\zeta \to a \in \partial \Delta$, also has the same limit as $\zeta \to a$ along each path nontangential to $\partial \Delta$. For $n>1$ it turns out that one can allow a tangential approach in certain directions but not others. Consider a bounded domain $D=\{z\in\mathbb{C}^n: \varphi(z)<0\}$, where $\varphi\in C^2(\bar{D})$ with gradient $\nabla\varphi=(\partial\varphi/\partial z_1, \ldots, \partial\varphi/\partial z_n)\neq 0$ on $\partial D=S$; then at a point $a\in S$ the unit vector $v=\bar{\nabla}\bar{\varphi}/|\nabla\varphi|$ is in the direction of the normal, while $\tau=iv$ belongs to the tangent plane $T_a(S)$ and is orthogonal to the complex tangent plane $T_a^c(S)$.

Let $\tau(z)=|\operatorname{Re}(z-a, v)|$ be the distance of z from $T_a(S)$ and put

$$K_a=\{z\in D: |(z-a, v)|<(1+\alpha)\,\delta(z), |z-a|<k\,\delta^{1/2+\varepsilon}(z)\}, \tag{1.17}$$

where α, $k>0$, $0<\varepsilon<1/2$, denoting by (z, a) the Hermitean inner product in $\mathbb{C}^n$. The domain K_a is tangent to $T_a(S)$ in directions of the complex tangent plane $T_a^c(S)$ and gives an acute angle in the orthogonal direction τ.

Theorem 1.12 (Chirka [14]). *If the function $f\in H^\infty(D)$ has a limit as $z\to a\in\partial D$ in the direction of the normal v then it has the same limit as $z\to a$ through the points of K_a.*

The proof is based on the remark that for functions $f\in H^\infty(D)$, $|f|\leq M$ for any two points z, $w\in D$ one has the estimate

$$|f(z)-f(w)|\leq 2M\,\frac{e^{c_D(z,w)}-1}{e^{c_D(z,w)}+1}\,.$$

It follows from this estimate that if a function $f\in H^\infty(D)$ has a limit as $z\to a$ through some set A then it has the same limit as $z\to a$ through an arbitrary set $K\supset A$ such that $c_D(z, a)\to 0$ as $z\to a$, $z\in K$. In the case of Chirka's theorem the rôle of A is played by the normal $\{z=a+tv: t\in\mathbb{R}\}$ and it remains to show that the set K_a in (1.16) satisfies the desired condition. To this end one uses (1.15). Notice that in (1.16) it is not possible to omit the term ε: the function $f(z)=z_2^2/(1-z_1^2)$ is holomorphic and bounded in the ball $B=\{z\in\mathbb{C}^2: |z|<1\}$ and has the limit 0 as $z\to a=(1, 0)$ along the normal. It is easy to see that it tends to 0 as $z\to a$ iff $|z-a|=o(\sqrt{\delta(z)})$.

§ 2. The Kobayashi Metric

This metric, introduced by Kobayashi [42] in 1967, is in a some sense dual to the Carathéodory metric – it is defined not by maps from the manifold into the disk but by maps from the disk into the manifold.

2.1. Definition and General Properties. Let M be a complex manifold and Δ be the unit disk equipped with the Poincaré metric ρ. By a *holomorphic chain* on M connecting the points p and q one intends a finite family of holomorphic disks $f^j: \Delta\to M$ and pairs of points $\{\zeta_j', \zeta_j''\}$ in Δ $(j=1, \ldots, m)$ such that

1) $f^1(\zeta_1')=p, f^m(\zeta_m'')=q$
and
2) $f^j(\zeta_j'')=f^{j+1}(\zeta_j')$ for all $j=1, \ldots, m-1$.

Definition 2.1. The *Kobayashi semimetric* on M is the function k_M: $M \times M \to \mathbb{R}^+$ defined for points $p, q \in M$ as

$$k_M(p, q)=\inf \sum_{j=1}^{m} \rho(\zeta_j', \zeta_j''), \tag{2.1}$$

where the greatest lower bound is taken over all holomorphic chains in M connecting p and q.

It is easy to see that k_M enjoys the same properties of a semimetric as described in (1.2) but, in general, it is not a metric (for instance, $k_M \equiv 0$ for $M=\mathbb{C}^n$). If one in Definition 2.1 instead of chains takes single disks, that is, puts

$$\tilde{k}_M(p, q)=\inf \rho(\zeta', \zeta'') \tag{2.2}$$

with the greatest lower bound over all holomorphic maps f: $\Delta \to M$, $f(\zeta')=p$, $f(\zeta'')=q$, then in general the triangle inequality is not fulfilled and $\tilde{k} \neq k$ (cf. Example 2.3 below). However, as has been demonstrated by Lempert [50], if D is a strongly linear convex domain in $\mathbb{C}^n$ with smooth boundary then the triangle inequality holds for $\tilde{k}_D$ and $\tilde{k}_D \equiv k_D$. A *domain* $D \subset \mathbb{C}^n$ is called *strongly linearly convex* if at each point $a \in \partial D$ it is touched by a complex hyperplane Π lying outside D such that $\delta(z, \Pi) \geqq c\,\delta^2(z, a)$ for all $z \in \delta D$ and some $c>0$ (δ = Euclidean distance).

The Kobayashi semimetric, like c_M, does not increase under holomorphic maps and is invariant under biholomorphic transformations. The difference of this new metric as compared to the old one is seen, in particular, at the hand of the following propositions: a) the Kobayashi semimetric on a complex manifold M is the largest among all metrics which do not increase under holomorphic maps f: $\Delta \to M$, exactly as the Carathéodory semimetric is the smallest one which does not increase under holomorphic maps f: $M \to \Delta$; b) on any complex manifold M holds $c_M(p, q) \leqq k_M(p, q)$ for all $p, q \in M$.

Let us also remark that the Kobayashi metric is preserved under passage to coverings; if π: $\tilde{M} \to M$ is a holomorphic cover then

$$k_M(p, q)=\inf k_{\tilde{M}}(\tilde{p}, \tilde{q}), \tag{2.3}$$

where for a fixed point $\tilde{p} \in \pi^{-1}(p)$ the greatest lower bound is taken over all $\tilde{q} \in \pi^{-1}(q)$. For the Carathéodory seminorm this is not always true: the sphere $\bar{\mathbb{C}}$ minus three points, for which this semimetric is trivial, is covered by the unit disk Δ with nondegenerate c_Δ. For the Cartesian product of two complex manifolds one has:

$$k_{MM'}((p, p'), (q, q'))=\max\{k_M(p, q), k_{M'}(p, q)\}. \tag{2.4}$$

All these properties are easy to prove and can be found in Kobayashi's paper [43] (cf. also [44]). In 1971 Royden gave the following infinitesimal definition of the Kobayashi metric.

Definition 2.2. Let M be a complex manifold, p a point in M and $v\in T_p(M)$ a tangent vector. The *Kobayashi seminorm* is defined as

$$K_M(p, v)=\inf_f\left\{\frac{1}{r}: f(0)=p, f'(0)=v\right\}, \tag{2.5}$$

where the greatest lower bound is taken over all holomorphic maps f of disks $\Delta_r=\{\zeta\in\mathbb{C}: |\zeta|<r\}$ into M, with the normalization indicated.

This definition can be given in another two different forms if we instead consider holomorphic maps $f\colon \Delta\to M$ of the unit disk into M:

a) $K_M(p, v)=\inf r^{-1}$ over all maps with the normalization $f(0)=p$, $f'(0)=r v$, where $r>0$.

b) $K_M(p, v)=\inf \|u\|_\Delta$ over all maps with the normalization $f(0)=p$, $f'(0)=v$, where u is the preimage of the vector $v\in T_p(M)$ under the map $f_*=df$, $\|u\|_\Delta$ denoting its length in the Poincaré metric.

Royden's definition has its classical source in the *Schottky-Landau theorem*, according to which for any holomorphic map f of the disk $\Delta_r=\{|\zeta|<r\}$ into to the Riemann sphere $\bar{\mathbb{C}}$ with three points a, b, c removed, with $f'(0)\neq 0$, the radius r is bounded from below by a constant depending only on a, b, c and the values of $f(0)$, $f'(0)$.

In [60] Royden proved that K_M, considered on $T_p(M)$, has the *contraction property* under holomorphic maps: if $f\colon M\to N$ is a holomorphic map then for each $v\in T_p(M)$

$$K_N(f(p), f_*(v))\leqq K_M(p, v), \tag{2.6}$$

He also established properties analogous to (2.3) and (2.4).

Example 2.1. For the upper halfplane $H=\{z\in\mathbb{C}: \operatorname{Im} z>0\}$ the differential form of the Poincaré metric is given by $K_H(z, v)=|v|/\operatorname{Im} z$. The upper half-plane covers the punctured disk $\Delta^*=\{\zeta\in\mathbb{C}: 0<|\zeta|<1\}$ with the projection $\zeta=e^{2\pi i z}$ and so, by the differential analogue of property (2.3), we find

$$K_{\Delta^*}(\zeta, v)=\frac{|v|}{|\zeta|\ln|\zeta|}. \tag{2.7}$$

For the disk $\Delta=\{|\zeta|<1\}$ we have $K_\Delta(\zeta, v)=|v|/(1-|\zeta|^2)$ so that the exclusion of the point $\zeta=0$ has an influence on the Kobayashi metric, while at the same time the Carathéodory metric does not change. Let us further remark that in the metric k_{Δ^*} the length of the circumference $\{|\zeta|=r\}$ tends to 0 as $r\to 0$, whereas the area of $\{0<|\zeta|<r\}$ is finite. This remark is often used in the applications.

In the same paper [60] it is proved:

Theorem 2.1 (Royden). *For each complex manifold M, k_M is upper semicontinuous on $T(M)$ and*

$$\inf_{\gamma} \int_0^1 K_M(\gamma(t), \gamma'(t))\, dt = k_M(p, q), \tag{2.8}$$

where the greatest lower bound is taken over all smooth paths γ: $[0, 1] \to M$ connecting the points p and q.

Thus, the Kobayashi metric k_M, in contradistinction to the Carathéodory metric, is always inner (cf. Sec. 1.1) and K_M is indeed the infinitesimal form of k_M.

Example 2.2. In the general case the Kobayashi seminorm $K_M(p, v)$ with p fixed may vanish in some directions but may be different from zero in others, i.e. the balls in this seminorm are not convex by necessity. Consider once more the domain $D = \{z \in \mathbb{C}^2 : |z_1 z_2| < 1\}$ in Example 1.2. For the point $z = 0$ and the vectors $v_1 = (1, 0)$ and $v_2 = (0, 1)$ the quantity $K_D(0, v_j)$, $j = 1, 2$, clearly is zero. However, as we now show, $K_D(0, v) = 1$ for $v = (1, 1)$. Indeed, let $f = (f_1, f_2)$: $\Delta \to D$ be a holomorphic map with the normalization $f(0) = 0$, $f'(0) = r v$, where $r > 0$. We have $f_j(0) = 0$, $f_j'(0) = r$ for $j = 1, 2$ and $|f_1(\zeta) f_2(\zeta)| < 1$ for all $\zeta \in \Delta$. Applying Schwarz's lemma in one dimensional function theory to the function $g(\zeta) = f_1(\zeta) f_2(\zeta)$, we find that $\lim_{\zeta \to 0} |g(\zeta)/\zeta^2| = r^2 \leqq 1$. Thus $r \leqq 1$ and by the form a) of Definition 2.1 we find $K_D(0, v) \geqq 1$. This lower bound is reached for the map $f_0(\zeta) = (\zeta, \zeta)$, with $f_0(0) = 0$, $f_0'(0) = v$, so that $K_D(0, v) = 1$. Generalizing the above, we see that, generally speaking, the level lines $\{z_1 z_2 = c,\ |c| < 1\}$ are biholomorphically equivalent to $\mathbb{C}^* = \mathbb{C} \setminus \{0\}$ and that at a point $z \in D$ one has $K_D(z, v) = 0$ for vectors v parallel to these level lines and $K_D(z, v) \neq 0$ for all other vectors.

Example 2.3. By a small modification of the preceding example one can obtain a domain D such that k_D differs from the quantity $\tilde{k}_D$ as defined in (2.2). Take $D = \{z \in \mathbb{C}^2 : |z_1 z_2| < 1,\ |z_2| < M\}$ and consider the points $z = (1, 0)$, $w = (0, 1)$. Clearly $k_D(0, z) = 0$, while $k_D(0, w) = \rho(0, 1/M) = 1/2 \cdot \ln(M + 1)/(M - 1)$, because for these points the extremal map is given by $f_0(\zeta) = (0, M\zeta)$, with $f_0(0) = 0$ and $f_0(1/M) = w$. By the triangle inequality for k_D we therefore have $k_D(z, w) \leqq 1/2 \ln(M + 1)/(M - 1)$.

On the other hand, for an arbitrary map $f = (f_1, f_2)$: $\Delta \to D$, with $f(0) = z$ and $f(\alpha) = w$ (we may assume that $0 < \alpha < 1$), we have $f_1(\zeta) = (\alpha - \zeta)\, \varphi_1(\zeta)$, $\varphi_1(0) = \alpha^{-1}$ and $f_2(\zeta) = \zeta\, \varphi_2(\zeta)$, $\varphi_2(\alpha) = \alpha^{-1}$. By Schwarz's lemma we find from $|f_1(\zeta) f_2(\zeta)| < 1$ for all $\zeta \in \Delta$ that

$$|\varphi_1(\zeta)\, \varphi_2(\zeta)| < \frac{1}{1 - \alpha}, \tag{2.9}$$

and from $|f_2(\zeta)| < M$ that $|\varphi_2(\zeta) - \varphi_2(0)| < 2M |\zeta|$. Putting $\zeta = \alpha^{-1}$ in the last inequality we get $|\varphi_2(0)| \geqq \alpha^{-1} - 2M\alpha$ so that by (2.9) then $1 - 2\alpha^2 M \leqq \alpha^2 (1$

$-\alpha$). For $M \geqq 10$ we conclude from this that for $\alpha \in (0, 1)$ by necessity $\alpha \geqq 2/M$ and, consequently, $\tilde{k}_D(z, w) \geqq 1/2 \ln(M+2)/(M-2) > k_D(z, w)$.

2.2. Other Properties. Royden [60] and Barth [3] showed that the semi-metric k_D is always continuous on $M \times M$ and if it is nowhere degenerate, then it induces the standard topology on M. By contrast, the seminorm, which by Theorem 1.1 always is upper semicontinuous, need not be continuous. For unbounded domains this is seen at the hand of Example 2.2: the function $K_D(z, v)$, where $v=(1, 1)$, equals 1 at the point $z=(0, 0)$ but is 0 for $z=(a, -a)$, $a \neq 0$. Here is a more interesting example.

Example 2.4. Consider in Δ the function

$$\varphi_1(\zeta) = \sum_{k=2}^{\infty} \gamma_k \ln\left|\frac{\zeta - k^{-1}}{2}\right|, \tag{2.10}$$

where all γ_k are <0 and such that the series converges for all $\zeta \in \Delta \setminus \{k^{-1}\}$, assuming also that $\varphi_1(0)=1$. This function is superharmonic in Δ, while the function $\varphi(\zeta)=\min\{\varphi_1(\zeta), 4\}$ is continuous in $\Delta \setminus \{0\}$. Pick the vector $v=(0, 1)$ and the points $z^k=(k^{-1}, 0)$, $k=2, 3, \ldots$, in the domain $D=\{z \in \mathbb{C}^2: |z_1|<1, |z_2|<\varphi(z_1)\}$. The for all these points $K_D(z^k, v)=1/4$, because we have $\varphi(k^{-1})=4$ and D contains the disks $\{z_1=k^{-1}, |z_2|<4\}$. The points z^k tend to 0 as $k \to \infty$ but, as we will see in an instant, $K_D(0, v) \geqq 1/3$, so that $K_D(z, w)$ is discontinuous at the point $z=0$.

Let $f=(f_1, f_2)$ be a holomorphic map $\Delta \to D$ such that $f(0)=0$ and $f'(0) = rv = (0, r)$; without loss of generality we may assume that f is continuous in $\bar{\Delta}$ and then $|f_2(\zeta)| \leqq \varphi \circ f_1(\zeta)$ everywhere in $\bar{\Delta}$. By the superharmonicity of $\varphi \circ f_1$

$$\varphi \circ f_1(0) = 1 \geqq \frac{1}{2\pi} \int_0^{2\pi} \varphi \circ f_1(e^{i\theta})\, d\theta.$$

As $\varphi \circ f_1(\zeta) \geqq 0$ everywhere on $\partial\Delta$, it follows from this that the measure of the set $E=\{\zeta \in \partial\Delta: \varphi \circ f_1(\zeta) > 2\}$ does not exceed π. Taking account of this, we get from Cauchy's formula

$$r = |f_2'(0)| \leqq \frac{1}{2\pi} \int_0^{2\pi} |f_2(e^{i\theta})|\, d\theta \leqq \frac{1}{2\pi} \int_0^{2\pi} \varphi \circ f_1(e^{i\theta})\, d\theta \leqq$$

$$\leqq \frac{1}{2\pi} \Big\{ \int_{\partial\Delta \setminus E} 2\, d\theta + \int_E 4\, d\theta \Big\} \leqq 3.$$

Thus, $K_D(0, v) = \inf r^{-1} \geqq 1/3$.

Notice that D is a domain of holomorphy, because it is the range of the negative values of the plurisubharmonic function $u(z)=|z_2|-\varphi(z_1)$. However, this function is not continuous in $\bar{D}$: it is discontinuous at the boundary point $(0, 1)$.

Let us mention two sufficient conditions for the continuity of the Kobayashi norm. The first of them is established in the above cited paper by Lempert: the function $K_D(z, v)$ is continuous on $D \times \mathbb{C}^n$ for each strongly linear convex domain $D \subset \mathbb{C}^n$ with smooth boundary. In the second condition the assumptions have been only slightly strengthened, as compared to Example 2.4, where K_D was not continuous.

Theorem 2.2. *Let D be a bounded domain of the form* $D=\{z \in \mathbb{C}^n: \varphi(z)<r\}$, *with* φ *plurisubharmonic in D and continuous in* $\bar{D}$, *or a domain obtained from such a domain by exclusion of an analytic subset A of codimension* 1. *Then* $K_D(z, v)$ *is continuous in* $D \times \mathbb{C}^n$.

The idea of the proof is as follows. Taking account of Theorem 2.1 it suffices to prove that for each sequence $\{z^\nu\} \subset D$, $z^\nu \to z \in D$, and any sequence $v^\nu \to v$

$$K_D(z, w) \leqq \varliminf_{\nu \to \infty} K_D(z^\nu, v^\nu). \tag{2.11}$$

Let us fix $\varepsilon > 0$ and consider a sequence of holomorphic maps $f_\nu: \Delta \to D$, $f_\nu(0) = z^\nu$, $f_\nu'(0) = r_\nu v^\nu$ such that $r_\nu^{-1} \leqq K_D(z^\nu, v^\nu) + \varepsilon$. By Montel's theorem we may assume, without loss of generality, that $f_\nu \to f$ uniformly on compact subsets of Δ and that $r_\nu \to r$. Then $f(0) = z$, $f'(0) = r v$ and $f(\Delta) \subset \bar{D}$. The function $\varphi \circ f$ is subharmonic and nonpositive in Δ and $\varphi \circ f < 0$. Consequently, $\varphi \circ f(\zeta) < 0$ everywhere on Δ and $f(\Delta) \subset D$, provided there is no set A. But if there is such a set then, as D is a domain of holomorphy, it is defined by an equation $g(z) = 0$, where $g \in \mathcal{O}(D)$. For every ν we have $g \circ f_\nu(\zeta) \neq 0$ and, therefore, by Hurwitz's theorem also $g \circ f \neq 0$ in Δ, i.e. as before $f(\Delta) \subset D$. The existence of such a map f shows that $K_D(z, v) \leqq r^{-1} \leqq \lim_{\nu \to \infty} K_D(z^\nu, v^\nu) + \varepsilon$. As ε is arbitrary, this establishes (2.1).

Let us further remark that, as proved by Royden [60], $K_M(z, v)$ is continuous on complete hyperbolic manifolds (cf. Sec. 3.3 below).

For complete circular domains (cf. [62]) Barth [55] has found the following analogue of Carathéodory's Theorem 1.9. Let $p(z) = \inf\{\alpha \in \mathbb{R}: z/\alpha \in D\}$ be the Minkowski functional of such a domain D; then D is convex iff $p(z)$ is a seminorm. Barth proved that a) D is pseudoconvex iff p is plurisubharmonic; b) if $B^c(D)$ and $B^k(D)$ are the unit balls in the Carathéodory and the Kobayashi semimetrics then $B^c(D) \supset B^k(D) \supset D$; c) $B^k(D) = D$ if D is pseudoconvex, while the equality $B^c(D) = B^k(D)$ holds iff D is convex. From this he concluded that every circular domain, which is biholomorphically equivalent to a convex domain, is itself convex.

Definition 2.3. A *subset* A of a complex manifold M is called *Kobayashi negligible* if $k_{M \setminus A} \equiv k_M$.

Campbell, Howard and Ochiai [9] showed that all analytic subsets $A \subset M$ of codimension greater than 1 are negligible (in particular, this shows that for A in Theorem 2.2 we may take an analytic subset of D of arbitrary dimension). By a slight variation of their reasoning, we can prove the following result.

Theorem 2.3. *Every subset A of an n-dimensional Stein manifold, such that its $(2n-2)$-dimensional Hausdorff measure $H_{2n-2}(A)$ vanishes, is negligible in the sense of Kobayashi.*

For analytic sets of codimension 1 or subsets with $H_{2n-2}(A)>0$, this theorem need not be true (cf. Example 2.1). Nor need it be true if M is not a manifold but just a complex space with singularities [10]. The question of removability will be treated below (cf. Sec. 2.4).

The boundary behavior of the Kobayashi metric is described by the following two theorems. The first of them refers to bounded domains of rather general type and is based on direct estimates.

Theorem 2.4. *Let $D \subset \mathbb{C}^n$ be a bounded domain and $a \in \partial \Delta$ a point such that D can be reached from the exterior by a circular cone V with non-zero angle at the vertex. If v is a unit vector in the direction of V's axis in the exterior of D, then for each point $z=a-tv$, $t>0$, holds*

$$K_D(z, v) \geqq \frac{c}{\sqrt{|z-a|}}, \tag{2.12}$$

where c depends only on D.

One can show that this estimate, as far as the order goes, is exact: for the domain $D=\{z \in \mathbb{C}^2 \colon |z_1|<5,\ |z_2|^2+(\operatorname{Im} z_1)^2>(\operatorname{Re} z_1)^2 \text{ for } \operatorname{Re} z_1>0\}$ and the vector $v=(0,1)$ holds $K_D(-tv, v) \leqq 1/\sqrt{t}$ for $0<t<1$. The second theorem gives much more exact information for strictly pseudoconvex domains.

Theorem 2.5 (Graham [24]). *Let $D \subset \mathbb{C}^n$ be a strictly pseudoconvex domain and φ a defining function for it such that at some point $a \in \partial D$ the gradient $\nabla \varphi$ equals 1 in modulus. Assume that the normal component $v_N(a)$ of the vector v* (with respect to the normal to ∂D at the point a) *is $\neq 0$. Then*

$$\lim_{z \to a} K_D(z, v)\, \delta(z, \partial D) = \tfrac{1}{2}\, |v_N(a)|\,. \tag{2.13}$$

If however $v_N(a)=0$, then

$$\lim_{\substack{z \to a \\ z \in V}} [K_D(z, v)]^2\, \varphi(z, \partial D) = \tfrac{1}{2} H_a(\varphi, v), \tag{2.14}$$

where V is a cone with vertex a and axis, directed into the interior of D, nontangential to ∂D,

$$H_a(\varphi, v) = \sum_{j,k=1}^{n} \frac{\partial^2 \varphi}{\partial z_j\, \partial \bar{z}_k}\, v_j \bar{v}_k$$

being the Levi form of φ. (This result holds also with C_D.)

Almost nothing is known about the sets where the Kobayashi metric or the Kobayashi norm degenerates. Our description of the degeneracy sets of the Carathéodory metric was based on the fact that the function $\ln C_D(z, v(z))$ was plurisubharmonic for holomorphic vector fields v in domains of $\mathbb{C}^n$, but this does not extend to the case of the Kobayashi metric. For example, if $D = \{z \in \mathbb{C}^2: 1 < |z| < 2\}$ then for the holomorphic vector field $v(z) \equiv z$ one has, in view of Theorem 2.4, $\varphi(z) = \ln K_D(z, v(z)) \to \infty$ as $z \to \partial D$. This shows that φ cannot be plurisubharmonic, for D is not a domain of holomorphy (cf. [62], p. 260). Let us further remark that Diederich and Sibony [18] have found an example of a holomorphy domain $D \subset \mathbb{C}^n$ such that $k_D(z, w) \equiv 0$ but $K_D(z, v) \not\equiv 0$, all bounded holomorphic functions in D being constant.

2.3. Hyperbolic Manifolds

Definition 2.4. A complex *manifold M* is called *hyperbolic* if k_M is a metric, i.e. if $k_M(p, q) > 0$ for all $p \neq q$. A hyperbolic *manifold M* is termed *complete* if every ball in the Kobayashi metric is relatively compact.

Apparently, all bounded domains in $\mathbb{C}^n$ are hyperbolic, as they are contained in a ball. Further all bounded domains of holomorphy with sufficiently smooth boundaries are hyperbolic complete (Kerzman and Rosay [36]). Also $\mathbb{C}$ with three points omitted is a complete hyperbolic manifold and, more generally, the projective space $\mathbb{C}\mathbb{P}^n$ with $2n+1$ hyperplanes in general position removed. The last result goes back to Bloch (1926); in the last years several proofs of this have been given (Kiernan, Fujimoto, Green).

Let us give some sufficient conditions for hyperbolicity of manifolds.

Theorem 2.6 (Kiernan [38]). *If the basis and the fiber of a holomorphic bundle are hyperbolic then the bundle space too is hyperbolic.*

This theorem was generalized by M.A. Illarionov [33].

Theorem 2.7 (Eastwood [20]). *Let N be a complex manifold covered by a family $\{U_\alpha\}$ of open sets and let f: $M \to N$ be a holomorphic map. If all preimages $f^{-1}(U_\alpha)$ are (complete) hyperbolic then the (complete) hyperbolicity of N entails the (complete) hyperbolicity of M.*

By the classical Liouville's theorem, each holomorphic map from $\mathbb{C}$ into a bounded domain is constant. By an *entire curve* on a complex manifold M we intend a nonconstant holomorphic map $\mathbb{C} \to M$. We also say that M does not contain entire curves if each holomorphic map f: $\mathbb{C} \to M$ is constant. Manifolds, possessing this property, are thus in an obvious way relatives of bounded domains.

This property enters into the following theorem, which is a variant of the previous one.

Theorem 2.8 (M.G. Zaĭdenberg [75]). *If f: $M \to N$ is a proper holomorphic map of complex manifolds and if for each point $q \in N$ the preimage $f^{-1}(q)$*

does not contain entire curves then the (complete) hyperbolicity of N entails the (complete) hyperbolicity of M.

The notion of hyperbolicity is connected with the classical theorems of Picard. The little *Picard theorem* says that each holomorphic map from $\mathbb{C}$ into $\mathbb{C}\mathbb{P}^1$ minus three points is constant. An immediate generalization of it is the proposition that each holomorphic map from $\mathbb{C}$ into an arbitrary hyperbolic manifold is constant (indeed, the Kobayashi metric on $\mathbb{C}$ is trivial, and by the contraction property we get $0 \leqq k_M(f(\zeta'), f(\zeta'')) \leqq k_{\mathbb{C}}(\zeta', \zeta'') = 0$ for any two point $\zeta', \zeta'' \in \mathbb{C}$ and so by the hyperbolicity $f(\zeta') = f(\zeta'')$). A different formulation of the same result: hyperbolic manifolds do not contain entire curves. The converse is not true.

Example 2.5. The domain $D = \{z \in \mathbb{C}^2 \colon |z_1 z_2| < 1,\ |z_1| < 1$ and $|z_2| < 1$ for $z_1 = 0\}$ is not hyperbolic, as $k_D(0, z) = 0$ if $z = (0, z_2)$. The map $g\colon (z_1, z_2) \to (z_1, z_1 z_2)$ maps D into the bidisk Δ^2 and is one-to-one except for $z_1 = 0$. If $f\colon \mathbb{C} \to D$ is a holomorphic map, then by Liouville's theorem $g \circ f = \text{const}$. Consequently, f is either constant or maps $\mathbb{C}$ into $D \cap \{z_1 = 0\}$, i.e. the disk $\{z_1 = 0, |z_2| < 1\}$, so again $f = \text{const}$.

The converse remains in force if additional assumptions of compactness of the manifold are imposed.

Theorem 2.9 (Brody [8]). *Every compact complex manifold, which does not contain any entire curves, is hyperbolic.*

The existence of nontrivial a holomorphic map $f\colon \mathbb{C} \to M$ for nonhyperbolic M in this theorem is proved on the basis of the so-called Brody representation lemma, which is useful also in other applications. To formulate it, let us recall that a *differential metric* on a complex manifold M is a nonnegative function H on the tangent bundle TM of M such that $H(p, \lambda v) = |\lambda|\, H(p, v)$ for all $p \in M$, $v \in T_p(M)$ and $\lambda \in \mathbb{C}$.

Lemma (Brody [8]). *Let M be a complex manifold, possibly with boundary, and let H be a differential metric on M. Then for each holomorphic map f: $\Delta_r \to M$ such that $H(f(0), f'(0)) \geqq c > 0$ one can find another holomorphic map $\tilde{f}$: $\Delta_r \to M$ such that $\tilde{f}(\Delta_r) \subset f(\Delta_r)$ and* a) $H(\tilde{f}(0), \tilde{f}'(0)) = c$, b) $H(\tilde{f}(\zeta), \tilde{f}'(\zeta)) \leqq c r^2/(r^2 - |\zeta|^2)$ *for each $\zeta \in \Delta_r$.*

With the aid of this lemma Green [26] established the hyperbolicity of $\mathbb{C}\mathbb{P}^n$ with removed hyperplanes (cf. below) and also that a closed complex subspace of a complex torus ($\mathbb{C}^n$ factored by a lattice of maximal rank) is hyperbolic iff it does not contain any complex subtori. Let us mention yet another of his results.

Theorem 2.10 (Green [25]). *The image $f(\mathbb{C}^m)$ of $\mathbb{C}^m$ under a holomorphic map $f\colon \mathbb{C}^m \to \mathbb{C}\mathbb{P}^n \setminus \{n+k$ hyperplanes in general position$\}$ lies in a linear subspace of $\mathbb{C}\mathbb{P}^n$ of dimension equal to the integer part of n/k. This estimate of the dimension of the subspace is sharp.*

Another classical result, which is connected with the notion of hyperbolicity, is *Montel's theorem* for normal families of holomorphic functions. Let us denote by $\mathcal{O}(M, N)$ the space of holomorphic maps of the complex manifold M into the complex manifold N with the topology of uniform convergence on compact subsets. Generalizing the definition of normality, we are lead to the following

Definition 2.5. A complex manifold M is said to be a *Montel manifold* if each sequence $\{f^\nu\} \subset \mathcal{O}(\Delta, M)$ contains a subsequence $\{f^{\nu_k}\}$ which either a) converges uniformly on compact subsets of Δ or else b) compactly diverges (i.e. for arbitrary compact sets $K \subset \Delta$ and $K' \subset M$ there exist a k_0 such that $f^{\nu_k}(K) \cap K' = 0$ for all $k \geqq k_0$). In the literature such manifolds are also termed *taut*.

One easily proves:

Theorem 2.11 (Kiernan [39]). *Every Montel manifold is hyperbolic and every complete hyperbolic manifold is Montel.*

The converse of these statements is not true: a) the domain D in Example 2.4 is hyperbolic, because it is bounded, but from the sequence $f^\nu\colon \Delta \to D$, $\nu = 2, 3, \ldots$, where $f^\nu(\zeta) = (\nu^{-1}, 4\zeta)$ it is not possible to select a subsequence which converges uniformly on compact sets, nor one which is compactly divergent; b) Rosay [59] gave an example of a Montel domain in $\mathbb{C}^n$ which is not complete hyperbolic.

A simple variant of Montel's theorem states that a family of holomorphic functions of one variable is relatively compact iff it is uniformly bounded. A multidimensional generalization of this result holds for maps into hyperbolic manifolds.

Theorem 2.12 ([44]). *If M is an arbitrary complex manifold and N a hyperbolic manifold then a family $F \subset \mathcal{O}(M, N)$ is relatively compact* (in the topology of that space) *iff for each point $p \in M$ the set $\{q: q = f(p), f \in F\}$ is relatively compact in N.*

2.4. Hyperbolic Imbedding. Here we consider some generalizations and strengthenings of the notions introduced in Section 2.3.

Definition 2.6. A complex *manifold M* is called *hyperbolic modulo* a subset $A \subset M$ if the relation $K_M(p, q) = 0$ entails that $p = q$ or else $p, q \in A$.

It follows from the contraction property of the Kobayashi metric that a holomorphic map from an arbitrary manifold with degenerate metric either is constant or else its image is contained in A.

Of far greater importance is the following notion due to Kobayashi and Ochiai (1971).

Definition 2.7. A *submanifold M* of a complex manifold $\tilde{M}$ is called *hyperbolically imbedded* in $\tilde{M}$ if 1) M is hyperbolic, 2) for any two sequences $\{p_\nu\}$,

$\{q_\nu\} \in M$ such that $p_\nu \to p$, $q_\nu \to q$ ($p, q \in \tilde{M}$) and $k_M(p_\nu, q_\nu) \to 0$ it follows that $p = q$ and 3) $M \Subset \tilde{M}$.

The point of this definition is that k_M not only separates points of M but also points of the closure (the metric k_M does not degenerate if we approach the boundary). An equivalent form of the definition is the following:

For any point $p \in M$ and an arbitrary neighborhood U of p in $\tilde{M}$ there is a smaller neighborhood $V \subset U$ such that $k_M(V \cap M, M \setminus U) > 0$.

A simple example is the manifold $M = \bar{\mathbb{C}} \setminus \{0, 1, \infty\}$, which is hyperbolically imbedded in $\bar{\mathbb{C}}$. More generally, the manifold $M = \mathbb{C}\mathbb{P}^n \setminus \{2n+1$ complex hyperplanes in general position$\}$ is hyperbolically imbedded in $\mathbb{C}\mathbb{P}^n$.

The notion of hyperbolic imbedding is connected with the big *Picard theorem*, stating that each holomorphic map of the punctured disk $\Delta^* = \{\zeta \in \mathbb{C}: 0 < |\zeta| < 1\}$ into $\bar{\mathbb{C}}$ minus three disjoint points extends holomorphically to the point $\zeta = 0$. The point is that hyperbolicity of M alone does not guarantee that a map $f: \Delta^* \to M$ extends holomorphically to Δ.

Example 2.6. The domain $D = \{z \in \mathbb{C}^2: 0 < |z_1| < 1, |z_2| < |\exp(1/z_1)|\}$ is hyperbolic, because $(z_1, z_2) \to (z_1, z_2 \exp(-1/z_1))$ maps it biholomorphically onto the bounded domain $\Delta^* \times \Delta$, but the map $f(\zeta) = (\zeta, 1/2 \cdot e^{1/\zeta})$ does not extend holomorphically to the point $\zeta = 0$.

However, hyperbolic imbedding guarantees that such a continuation is possible and one has the following analogue of the big Picard theorem. For its formulation, recall that an analytic *subset with normal selfintersections* ("crossings") is an analytic subset A, $\operatorname{codim} A = 1$, of a complex manifold M which is locally given by the equation $z_1 \dots z_m = 0$, $m \leqq n$, where $(z_1, \dots, z_n)$ are coordinates on M.

Theorem 2.13 (Kiernan [39]). *Let A be an analytic subset with normal selfintersections of a complex manifold M and let N be a manifold hyperbolically imbedded in $\tilde{N}$. Then each holomorphic map $f: M \setminus A \to N$ extends to a holomorphic map $f: M \to \tilde{N}$.*

Notice that the assumption of normality of the selfintersections is essential here: if $M = \mathbb{C}^2$ and $A = \{z_1 z_2 (z_1 - z_2) = 0\}$ an analytic subset, which at the origin $z = 0$ does not fulfill this condition, then $f(z) = z_2/z_1$ maps $\mathbb{C}^2 \setminus A$ holomorphically into $\bar{\mathbb{C}} \setminus \{0, 1, \infty\}$, which manifold is hyperbolically imbedded in $\bar{\mathbb{C}}$, but the map cannot be continued holomorphically to $\mathbb{C}^2$. A special case of this theorem, when A is a complex submanifold (without singularities), was earlier established by Kobayashi.

Even before that Kwack [49] proved that if A is a closed subset of a complex manifold M and N is a hyperbolic manifold, then a holomorphic map $f: M \setminus A \to N$ extends holomorphically to M if either: a) N is compact or b) complete hyperbolic and $\operatorname{codim} A \geqq 2$. However, this result does not contain the big Picard theorem in the classical case, as there $N = \bar{\mathbb{C}} \setminus \{0, 1, \infty\}$ is noncompact and $\{\zeta = 0\}$ is a subset of Δ of codimension 1.

Kiernan [39] connected the notion of hyperbolic imbedding with topological properties of the space of holomorphic maps.

Theorem 2.14. *Let N be an open subset of a complex manifold $\tilde{N}$. Then the following statements are equivalent.*

a) *N is hyperbolically imbedded in $\tilde{N}$.*

b) *$\mathcal{O}(\Delta, N)$ is relatively compact in $\mathcal{O}(\Delta, \tilde{N})$.*

c) *$\mathcal{O}(M, N)$ is relatively compact in $\mathcal{O}(M, \tilde{N})$ for each complex manifold M.*

d) *There exists a Hermitean metric h on $\tilde{N}$ such that for each $f \in \mathcal{O}(\Delta, N)$ and any two points $\zeta', \zeta'' \in \Delta$ the distance in this metric satisfies $d_h(f(\zeta'), f(\zeta'')) \leqq \rho(\zeta', \zeta'')$.*

Property b) constitutes a sharpening of the formulation of our generalized *Montel's theorem* (Theorem 2.12), which lies at the heart of Definition 2.5; in the case $N = \mathbb{C} \setminus \{a, b, c\}$ both formulations coincide. Let us remark that a generalized Montel's theorem, consisting of statement b) in the case $\tilde{N} = \mathbb{CP}^n$, $N = \mathbb{CP}^n \setminus \{2n+1$ hyperplanes in general position$\}$, was found already by Bloch in 1926 and by H. Cartan in 1928. Property d) will be discussed in Sec. 2.6.

Among other results belonging here let us mention that *the manifold $N = \mathbb{CP}^n \setminus \{n+k+1$ hyperplanes in general position$\}$ is hyperbolically imbedded in $\tilde{N} = \mathbb{CP}^n$ modulo A*, where A is the union of finitely many planes in $\mathbb{CP}^n$ of dimension not exceeding k ([40], [23]). This means that $N \cap A$ is hyperbolic modulo A (cf. Definition 2.6) and that for each point $p \in N \setminus A$ and each neighborhood U of p one can find $V \subset U$ such that $k_{\tilde{N}}(V \cap \tilde{N}, \tilde{N} \setminus U) > 0$.

2.5. Generalizations. Let us mention some results on meromorphic continuation of maps. According to Remmert a meromorphic map f from a complex space M into a complex space N is a correspondence such that a) $f(p)$ for each point $p \in M$ is a nonempty compact subset of N; b) the graph $\Gamma_f = \{(p, q) \in M \times N : q \in f(p)\}$ is a complex subspace of dimension equal to $\dim M$; c) there exists a dense subset M_0 of M such that $f(p)$ is a singleton for each p in M_0.

If we in Theorem 2.13 omit the condition about normal selfintersection of A, then only the following can be claimed.

Theorem 2.15 (Kiernan [39]). *If A is an analytic subset of a complex manifold M and N a manifold hyperbolically imbedded in $\tilde{N}$ then each holomorphic map $f: M \setminus A \to N$ extends to a meromorphic map $M \to \tilde{N}$.*

The point is that for the resolution of singularities caused by singularities of A which are not normal one applies the so-called σ-process; this has the effect that the extension becomes meromorphic (cf. the example after Theorem 2.13).

Let us now state some recent results by M.G. Zaĭdenberg [75]. He considered the *locally complete* hyperbolic manifolds introduced by Kobayashi and Kiernan, i.e. submanifolds N of a complex manifold such that for each point

$p \in \bar{N}$ there is a neighborhood $U \subset M$ such that all connected components of hyperbolic manifolds are all algebraic manifolds, analytical polyhedra, domains of holomorphy with sufficiently smooth boundaries etc. An *entire curve, imbedded* in a manifold M with the Hermitean metric h, is a nonconstant holomorphic map $f\colon \Delta \to M$ such that $\|f'(\zeta)\|_h \leqq 1$ for all $\zeta \in \Delta$. Such a *curve* is termed *limiting* on a submanifold $N \subset M$ if on each disk Δ_r one can approximate f by a holomorphic maps $f_r\colon \Delta_r \to N$.

Theorem 2.16 ([75]). *Let N be a connected, relatively compact and locally complete hyperbolic submanifold of a complex manifold M. For N to be complete hyperbolic it is sufficient and to be hyperbolically imbedded in M necessary and sufficient that N does not admit imbedded entire curves and limiting entire curves.*

In particular, a submanifold $N \subset M$, for which there exists a limiting entire curve, cannot be hyperbolically imbedded in M. For instance, the domain $D = \{z \in \mathbb{C}^2\colon 0 < |z_1| < 1,\ |z_1 z_2| < 1\}$, which is biholomorphically equivalent to $\Delta^* \times \Delta$, is completely hyperbolic. But it carries the limiting entire curve $f(\zeta) = (0, \zeta)$ and, consequently, it cannot be hyperbolically imbedded. In [75] one further gives sufficient conditions for complete hyperbolicity and for imbedding of closed subsets as so-called polyhedral domains.

2.6. Geometric Methods. A complex *manifold M* is called *Hermitean* if there is given on the fibers of its tangent bundle a Hermitean bilinear form h: $T(M) \times T(M) \to \mathbb{C}$, which is positive definite and twice smooth on vector fields of class $C^2(M)$. In local coordinates we have in a neighborhood $U \subset M$

$$h_p(u, v) = \sum_{j,k=1}^{n} h_{jk}(p)\, d z_j(u)\, \overline{d z_k(u)}, \quad h_{jk} = h\left(\frac{\partial}{\partial z_j}, \frac{\partial}{\partial z_k}\right). \tag{2.15}$$

where the matrix $h = (h_{jk})$ is Hermitean symmetric ($h_{kj} = \bar{h}_{jk}$) and positive definite (all eigenvalues are positive) and $h_{jk} \in C^2(U)$. The form $\operatorname{Im} h$ is skew-symmetric and $g = \operatorname{Re} h$ is symmetric and defines a Riemannian metric on M; $h(u, u) = g(u, u)$.

On an Hermitean manifold one defines in a natural way the norm of a tangent vector: $\|u\|_h = \sqrt{h_p(u, u)}$, the length of a smooth path $\gamma\colon [0, 1] \to M$:

$$|\gamma| = \int_0^1 \sqrt{h_{\gamma(t)}(\gamma'(t), \gamma'(t))}\, d t,$$

and the distance $d_h(p, q)$ between two points $p, q \in M$ as the least upper bound of all paths connecting them; these are the same as in the Riemannian metric $g = \operatorname{Re} h$. The quantity $h_p(u, u) = H(u, u)$ gives a differential metric on M: it will likewise be denoted by $d s_h^2(u)$.

Let us fix on a Hermitean manifold a point p along with a vector $v \in T_p(M)$ and let us consider a holomorphic curve $f\colon \Delta_r \to M$ such that $f(0) = p$ and

$f'(0)=v$. On the image $S=f(\Delta_r)$ we then have the induced metric

$$h|S=\sum_{j,k=1}^{n} h_{jk}(f(\zeta))f_j'(\zeta)\overline{f_k'(\zeta)}\,d\zeta\,d\bar{\zeta}=H(\zeta)\,d\zeta\,d\bar{\zeta},$$

where $H>0$ too is in $C^2(\Delta_r)$; we take r so small that S lies in a single coordinate neighborhood on M and $f'\neq 0$ on Δ_r. Considering S as a real two dimensional surface we can compute its Gaussian curvature at the point p:

$$K_f(p)=-\frac{1}{2H(0)}\left.\frac{\partial^2 \ln H}{\partial\zeta\,\partial\bar{\zeta}}\right|_{\zeta=0} \tag{2.16}$$

(here $\operatorname{Re}\zeta$ and $\operatorname{Im}\zeta$ are thus isometric coordinates on S). It is not hard to see that the number K_f is invariant under conformal changes of the parameter ζ; it is called the *holomorphic curvature* of S. The least upper bound of $K_f(p)$ over all holomorphic curves f, $f(0)=p$, $f'(0)=v$, is called the *holomorphic sectional curvature* at p in the direction v.

Notice that the holomorphic sectional curvature can be defined for differential metrics that need not be Hermitean (example: the Carathéodory metric or the Kobayashi metric), such that all restrictions to holomorphic curves $f\colon \Delta_r\to M$ have the form $h\,d\zeta\,d\bar{\zeta}$ with a positive function $h\in C^2(\Delta_r)$.

Example 2.7. For $\bar{\mathbb{C}}=\mathbb{CP}^1$ with the spherical metric one has $H(\zeta)=(1+|\zeta|^2)^{-2}$ and the holomorphic curvature equals 1 at all points; for the unit disk Δ with the Poincaré metric one has $H(\zeta)=(1-|\zeta|^2)^{-2}$ and the holomorphic curvature is everywhere -1.

In Sec. 1.1 we have put Schwarz's lemma in the invariant form (1.12), which with the present definition available can be formulated as follows:

Lemma (Ahlfors). *For each holomorphic map $f\colon \Delta\to\Delta$*

$$f^*(ds_\rho^2)\leqq ds_\rho^2,$$

where $ds_\rho^2=|d\zeta|^2/(1-|\zeta|^2)^2$ is the square of the Poincaré metric and $f^*(ds_\rho^2)=|df|^2/(1-|f|^2)$ is its preimage under f (this is only a semimetric, as $df=f'(\zeta)\,d\zeta$ vanishes at the critical points of f).

This lemma extends to maps of manifolds.

Theorem 2.17 ([44]). *Let M be a complex manifold with a Hermitean semimetric h and set $A=\{p\in M\colon ds_h^2(p,v)=0$ for some $0\neq v\in T_p(M)\}$. If the holomorphic sectional curvature does not exceed a negative constant $-k$, then for each holomorphic map $f\colon \Delta\to M$ holds*

$$f^*(ds_h^2)\leq\frac{1}{k}\,ds_\rho^2 \tag{2.17}$$

for all points $\zeta\in\Delta\setminus f^{-1}(A)$.

Thus, the existence of a metric on M with negative holomorphic sectional curvature leads to the contraction property of the Poincaré metric under holomorphic maps from Δ into M (without loss of generality we can take $k=1$ in (2.17)). On this property we can base yet another criterion for the hyperbolicity of a manifold.

Theorem 2.18 ([44]). *If M satisfies the hypotheses of the previous theorem, then M is hyperbolic modulo A and if the metric h is complete it is complete hyperbolic modulo A.*

Hahn and Kim have given hyperbolicity criteria for Hermitean manifolds which are connected with classical theorems by Schottky and Landau. To formulate them we need the following notions.

An Hermitean manifold M satisfies the Schottky condition if for each collection consisting of a point $p \in M$, a relatively compact neighborhood W of p and a number r, $0<r<1$, there exists an $R>0$ such that for each holomorphic map $f\colon \Delta \to M$ holds $d_h(p, f(z)) < R$ for $|z|<r$ and $f(0) \in W$.

If for each relatively compact set $W \subset M$ there exists an R such that for each holomorphic map $f\colon \Delta \to M$ holds $h(f'(0)) \leqq R$ for $f(0) \in W$, we say that M satisfies the Landau condition.

Theorem 2.18′ ([30]). *Let M be any Hermitean manifold. The following conditions are equivalent.*

1) *M is hyperbolic;*
2) *M satisfies the Schottky condition;*
3) *M satisfies the Landau condition.*

Next let us state some results which are related to the continuation theorems of the previous section. For the formulation of the first of them let us associate to the Hermitean form (2.15) the (1, 1)-differential form

$$\omega_h = \frac{i}{2} \sum_{j,k=1}^{n} h_{jk}\, dz_j\, d\bar{z}_k; \tag{2.18}$$

we say that an Hermitean manifold M is a *Kähler manifold* if M carries a metric h such that ω_h is closed, i.e. $d\omega_h = 0$. An equivalent condition is that there exists locally on M real functions $\varphi \in C^2(U)$ such that on U

$$\omega_h = \frac{i}{2}\, \partial \bar{\partial} \varphi = d\, d^c \varphi \tag{2.19}$$

(here $d = \partial + \bar{\partial}$, $d^c = (\partial - \bar{\partial})/4i$ are the differentiation operators).

Example 2.8. The following are Kähler manifolds: 1) The Euclidean metric in $\mathbb{C}^n$, with the Kähler form $\omega = d\,d^c |z|^2$; 2) the spherical metric in $\bar{\mathbb{C}}$ and its generalization, the Fubini-Study metric in $\mathbb{CP}^n$, with $\omega = d\,d^c \ln |z|^2$, $z = (z_0, \ldots, z_n)$ being homogeneous coordinates; 3) the Poincaré metric on Δ, with $\omega = d\,d^c \ln(1-|z|^2)^4$. An Hermitean, non-Kähler manifold is the so-called Hopf manifold, $\mathbb{C}^n \setminus \{0\}$ factored with respect to the relation $w \sim z$, defined by $w = 2^k z\,(k \in \mathbb{Z})$.

Griffiths has proved (cf. [27]).

Theorem 2.19. *Let N be a compact hyperbolic Kähler manifold and A a closed analytic subset of a complex manifold M. Then each holomorphic map $f: M \setminus A \to N$ extends to a meromorphic map $M \to N$.*

Here is the idea of the proof. The general case is first reduced to the case $M \setminus A = \Delta_*^n$. Thereafter one establishes the finiteness of the volume of Δ_*^n in the metric $\omega_f^k \wedge \varphi^{n-k}$, where ω_f is the pullback to M of the Kähler form on N and φ is the Euclidean metric form on $\mathbb{C}^n$, this for any $k = 1, \ldots, n$, and applies Bishop's theorem on continuation of an analytic set to the graph of f in $\Delta_*^n \times N$ (cf. Vol. 7, Part III). The assumption that we have a Kähler manifold is indispensable: the natural projection of $\{z \in \mathbb{C}^n: 0 < |z| < 1\}$ onto the Hopf manifold (which is compact but not Kähler), assigning to each z it equivalence class $\{z\} = \{2^k z: k \in \mathbb{Z}\}$, does not extend meromorphically to the point 0.

In Griffiths [27] one also finds

Theorem 2.20. *If M is an arbitrary complex manifold and N is complete with respect to an Hermitean metric with negative holomorphic sectional curvature, the each meromorphic map $f: M \to N$ is holomorphic.*

Such a result holds also for maps into hyperbolic manifolds. In conclusion let us further refer to a result by S.I. Ivashkovich [32], concerning continuation of locally biholomorphic maps.

2.7. Automorphisms of Hyperbolic Manifolds. It is known in geometry that the isometry group of a Riemannian manifold is a *Lie group*, that is, it can be considered as a manifold on which the group operations are differentiable transformations; its *Lie algebra* consists of vector fields on the manifold with the obvious operations of addition and multiplication by a scalar, and further a multiplication, which is the bracket $[u, v] = u \circ v - v \circ u$ of two vector fields (for details, cf. for instance [45]). In 1935 H. Cartan studied, from this point of view, the group of biholomorphic automorphisms of bounded domains in $\mathbb{C}^n$, considering them as isometries in the Carathéodory metric. Using his own metric, Kobayashi [43] extended Cartan's theory to the case of hyperbolic manifolds. We describe some of his results.

Theorem 2.21. *If M is a hyperbolic manifold of complex dimension n, then:*

(1) *The group* Aut M *of biholomorphic automorphisms is a real Lie group in the topology of uniform convergence on compact sets and its dimension is* $\leqq 2n + n^2$; *if* $\dim \operatorname{Aut} M = 2n + n^2$ *then M is biholomorphically equivalent to a ball in $\mathbb{C}^n$.*

(2) *The stabilizer (isotropy subgroup)* $\{f \in \operatorname{Aut} M: f(p) = p\}$ *of any point $p \in M$ is compact.*

(3) *The Lie algebra* aut M *of* Aut M *consists of globally integrable vector fields on M; if $v \in \operatorname{aut} M$ then iv is not globally integrable, and so, in particular, does not belong to* aut M.

(4) *If M is hyperbolic and compact then* Aut M *must be finite.*

Let us remark that one can deduce from $\dim \operatorname{Aut} M = 2n^2 + n^2$ that $\operatorname{Aut} M$ acts transitively on M (i.e. for any two points p and q in M one can find $f \in \operatorname{Aut} M$ with $f(p) = q$), and this fact is used in the proof that then M must be equivalent to a ball. The compactness of the stabilizers is deduced from Theorem 2.12. Statement (3) is equivalent to the assertion that a complex Lie group which acts effectively on a hyperbolic manifold by necessity is trivial: if the opposite were true, there would act effectively on M a complex one parameter group and this would entail that there exists a nonconstant holomorphic map $\mathbb{C} \to M$, contradicting the hyperbolicity (cf. Sec. 2.3). Let us point out that in view of Theorem 2.18 the propositions (1)–(4) hold also for Hermitean manifolds with the holomorphic sectional curvature bounded from below by a negative constant.

Denoting by $\operatorname{Bim} M$ the group of bimeromorphic maps of a manifold M into itself, we conclude from Theorem 2.20 that for compact hyperbolic manifolds $\operatorname{Bim} M = \operatorname{Aut} M$.

An example of a complete hyperbolic space X such that $\operatorname{Bim} X \neq \operatorname{Aut} X$ can be found in Kodama [46]. A survey of results pertaining to groups of biholomorphic automorphisms of bounded domains may be located in Narasimhan [53].

We end by a result of Kaup's [35], on holomorphic maps of a manifold M into itself (not necessarily automorphisms).

Theorem 2.22. *Let M be compact hyperbolic. Then for each map $f \in \mathcal{O}(M, M)$ one can find an integer $m > 0$ such that for the $2m$-th iterate holds $f^{2m} \equiv f^m$.*

2.8. Variational Problems. Let D be a domain in $\mathbb{C}^n$, z and w two points in D, $v \in \mathbb{C}^n$; consider two variational problems.

I. Find $\sup_f \lambda$ taken over all holomorphic maps $f: \Delta \to D$ such that $f(0) = z$, $f'(0) = \lambda v$ where $\lambda \in \mathbb{R}$.

II. Find $\inf_f \rho(\zeta', \zeta'')$ over all holomorphic maps $f: \Delta \to D$ such that $f(\zeta') = z$, $f(\zeta'') = w$.

The first problem is connected with the Kobayashi seminorm $K_D(z, z)$, the second one with the function $\tilde{K}_D(z, w)$ (cf. Sec. 2.1). In the general case not much is known about neither of them. The following result gives a sufficient condition for the existence of extremals.

Theorem 2.23. *Let D be a domain of the type $\{z \in \mathbb{C}^2 : \varphi(z) < 0\}$, where φ is plurisubharmonic in D and continuous in $\bar{D}$, or one obtained from such a domain upon throwing away an analytic subset A of codimension 1. Then extremal maps exist in both problems.*

This theorem is proved in a similar way as Theorem 2.2. For sets A with $\operatorname{codim} A > 1$ it is not always true: for example, for $D = \{z \in \mathbb{C}^2 : 0 < |z| < 1\}$, $z = (1/2, 0)$, $w = (-1/2, 0)$ and $v = (1, 0)$ neither Problem I or Problem II has extremal maps.

In general the extremals are not unique: for the bidisk $D=\Delta^2$ and $z=(0, 0)$, $w=(1/2, 0)$, $v=(1, 0)$ the extremals of both problems are all maps $f(\zeta)=(\zeta, B(\zeta))$, where B is an arbitrary Blaschke product, with $B(0)=B(1/2)=B'(0)=0$. Lempert [50] established the uniqueness of the extremal map f in both problems for strongly linear convex domains (cf. Sec. 2.1); moreover, he showed that this map is extremal for all $z, w\in f(\Delta)$ and all $v\in f^* T_0(\Delta)$.

The properties of the extremal maps have been studied much less. In the general case one cannot assert that they are proper maps from Δ onto D. For example, if $D=\Delta\setminus\{0\}$ then for the point $z=e^{-1}$ and the vector $v=1$ the extremal for Problem I is the map $f(\zeta)=\exp((\zeta-1)/(\zeta+1))\colon \Delta\to D$, which is not proper. The following result is due to E.A. Poletskiĭ [54].

Theorem 2.24. *Consider the domain* $D=\{z\in\mathbb{C}^n\colon \varphi(z)<0\}$, *where* φ *is plurisubharmonic in* D *and continuous on* $\bar{D}$, *with* $\delta(z, D)>\gamma|\varphi(z)|$ *for some* $\gamma>0$ *and all* $z\in D$ (this condition is fulfilled by all pseudoconvex domains with smooth boundary and by analytic polyhedra). *Then the extremal map in Problem I is almost proper, i.e.* $\lim_{r\to 1} f(e^{i\theta})\in\partial D$ *for a.e.* θ.

In the description of the extremal maps for bounded domains of the form $D=\{z\in\mathbb{C}^n\colon \varphi(z)<0\}$, where $\varphi\in C^1(\bar{D})$ is plurisubharmonic in D and $\nabla\varphi\neq 0$ on ∂D, the following analogue of the Euler-Lagrange variational equations holds (E.A. Poletskiĭ [54]): for the extremal map $f\colon \Delta\to D$ in Problem I there exists a holomorphic vector $g=(g_1, \ldots, g_n)$ consisting of functions $g_j\in H^\infty(\Delta)$, a bounded function $\lambda\colon \partial\Delta\to\mathbb{R}_+$, $\lambda\neq 0$, and a vector $c\in\mathbb{C}^n$ such that for almost all $\zeta\in\partial D$

$$\frac{c}{\zeta}+g(\zeta)=\lambda(\zeta)\,\nabla\varphi(f'(\zeta)). \tag{2.20}$$

It follows from this, in particular, that the function $\zeta\lambda(\zeta)\nabla\varphi(f'(\zeta))=\tilde{f}(\zeta)$ can be continued holomorphically to Δ. In the case of strongly linear convex domains $D\subset\mathbb{C}^n$ this fact was obtained also by Lempert [50], who, using it, proved for such domains: 1) the extremal maps in both problems are proper; 2) $\langle f', \tilde{f}\rangle=\sum_{v=1}^n f'_v\tilde{f}_v=\text{const}>0$ everywhere in Δ and 3) if $\partial D\in C^k$, $k\geqq 3$, then $f, \tilde{f}\in C^{k-2}(\bar{D})$. Relying on these properties of the extremals, he established the following result:

Theorem 2.25 (Lempert). *If* $D\subset\mathbb{C}^n$ *is strongly linear convex then* a) *the three metrics coincide in* D:

$$c_D(z, w)=k_D(z, w)=\tilde{k}_D(z, w) \quad \textit{for all } z, w\in D,$$

and b) *if* $\partial D\in C^k$, $k\geqq 6$, *then* $k_D\in C^{k-4}(D\times D\colon z\neq w)$.

Let us also mention an interesting connection between the variational Problem II and the Monge-Ampère equation displayed by Lempert. Let $D\subset\mathbb{C}^n$ be a strictly linearly convex domain with boundary of class C^6; fix a point $w\in D$. For each $z\in D$ let us construct the solution f of this problem with $f(\zeta)=z$, $f(0)=w$, writing $\Phi_w(z)=\zeta f'(0)/\|f'(0)\|$. It turns out that Φ_w is in

$C^2(D\setminus\{w\})$ and maps D homeomorphically onto the unit ball B in $\mathbb{C}^n$. The function $u(z)=\ln|\Phi_w(z)|$ is plurisubharmonic in D, vanishes on ∂D and for $z \neq w$ satisfies the complex *Monge-Ampère equation*

$$\det\left(\frac{\partial^2 u}{\partial z_j \partial \bar{z}_k}\right)=0; \tag{2.21}$$

also $u(z)-\ln|z-w|=O(1)$ so that u is a fundamental solution of this equation (cf. the following section).

Concerning the coincidence of the Carathéodory and the Kobayashi norm there is further the following result by Stanton [67]: Let M be a connected hyperbolic manifold. Assume that there exists a point $p \in M$ such that $C_M(p, v)=K_M(p, v)$ and that at least one of these norms is Hermitean and of class C^∞. Then M is biholomorphically equivalent to a ball. The assumption of Hermiteness is here essential: by Theorem 2.25 the norms K and C coincide on every ellipsoid in $\mathbb{C}^n$, but not all such ellipsoids are biholomorphically equivalent to a ball.

The hypothesis of a differential metric $H(p, q)$ being Hermitean, apparently, expresses the fact that in this metric the unit ball in the tangent space $\{v \in T_p(M): \sum_{j,k=1}^n h_{jk} v_j \bar{v}_k < 1\}$ is biholomorphically (or even linearly) equivalent to a ball. In this connection, let us mention an earlier result by Stanton: if, in the hypothesis of the preceding theorem, there exists a point $p \in M$ with $C_M(p, v)=K_M(p, v)$ such that the unit ball in $T_p(M)$ is a polydisk then M is biholomorphically equivalent to a polydisk.

Suzuki [68] proved for arbitrary hyperbolic manifolds, without the Hermiteness assumption, that the holomorphic sectional curvature in the Carathéodory metric does not exceed -1 and in the Kobayashi metric is greater or equal to -1, so that if the two metrics coincide the curvature must be -1. In the same paper it is also shown that the metrics (and norms) of Kobayashi and Carathéodory coincide on every bounded symmetric domain. (A *domain* $D \subset \mathbb{C}^n$ is called *symmetric* if the group $\operatorname{Aut} D$ acts transitively on D and if there exists an automorphism with a single fixed point in D, whose square is the identity map.)

2.9. The Invariant Green's Function. [2] In conclusion, let us consider a notion which in a natural way is the multidimensional analogue of the classical Green's function. In the domain $D \subset \mathbb{C}^n$ let us fix two points z and w. If f is a holomorphic map of the unit disk $\Delta \subset \mathbb{C}$ into D such that $f(0)=z$ write

$$u_f(z, w)=\sum_{f^{-1}(w)} k_\nu \ln|\zeta_\nu|,$$

where the sum (possibly equal to $-\infty$) is extended over all preimages $\zeta_\nu \in \Delta$ of the point w, k_ν being the multiplicity of ζ_ν. The *invariant Green's function*

[2] The results of this Section are due to E.A. Poletskiĭ.

of D is the function

$$g_D(z, w) = \inf_f u_f(z, w), \qquad (2.22)$$

where the greatest lower bound runs over all holomorphic maps $f\colon \Delta \to D$ with $f(0) = z$.

It follows from this definition that g_D is invariant for biholomorphic maps of D. It is clear that $g_D(z, w) \leqq 0$ and, for bounded domains D, we have, in a neighborhood of w, $g_D(z, w) = \ln|z-w| + O(1)$, where $O(1)$ is a bounded function.

Let us mention some results which justify our definition.

Theorem 2.26. *g_D is plurisubharmonic in z for $w \in D$ fixed.*

Let us sketch the proof. The semicontinuity is established in the usual way so that it remains to show that for each point $z \in D \setminus \{w\}$ and each vector $v \in \mathbb{C}^n$ of sufficiently small modulus holds

$$g_D(z, w) \leqq \frac{1}{2\pi} \int_0^{2\pi} g_D(z + e^{i\varphi} v, w)\, d\varphi. \qquad (2.23)$$

It is not hard to see that for each $\varepsilon > 0$ there exists a map $f(\zeta, \lambda)\colon \Delta \times \partial\Delta \to D$, holomorphic in ζ and continuous in λ, such that $f(0, \zeta) = z + \lambda v$

$$\int_0^{2\pi} u_f(z + e^{i\varphi} v, w)\, d\varphi \leqq \frac{1}{2\pi} \int_0^{2\pi} g_D(z + e^{i\varphi} v, w)\, d\varphi + \varepsilon. \qquad (2.24)$$

The map f can be approximated by a map $f_1\colon \Delta \times \partial D \to D$ such that $f_1(\zeta, \lambda) = \sum_{k=-N}^{N} c_k(\zeta)\, \lambda^k$ with coefficients c_k that are holomorphic in Δ, as before, with $f_1(0, \lambda) = z + \lambda v$, inequality (2.24) too being fulfilled (with a different ε). As $c_k(0) = 0$ for $k < 0$, we see that for sufficiently large integers m the map $f_1(\zeta \lambda^m, \lambda)$, which we anew denote by $f(\zeta, \lambda)$, is holomorphic in the bidisk $\Delta^2 = \Delta \times \Delta$, mapping it into D, with $f(0, \lambda) = z + \lambda v$ for every $\lambda \in \Delta$ and (2.24) being satisfied.

The set $\{(\zeta, \lambda) \in \Delta^2\colon f(\zeta, \lambda) = w\}$ of the preimages of the point $w \in D$ taken with their multiplicities in ζ defines a divisor in Δ^2 [3]; let it be the divisor of the function $h(\zeta, \lambda) \in \mathcal{O}(\Delta^2)$. The function $v(\zeta, \lambda) = \ln|h(\zeta, \lambda)|$ is then plurisubharmonic in Δ^2 and its generalized Laplacean $\Delta_\zeta v$ is for λ fixed a measure supported by those points $\zeta \in \Delta$ for which $h(\zeta, \lambda) = 0$, i.e. on the set of preimages of w taken with their multiplicities. Therefore

$$u_f(z + \lambda v, w) = \frac{1}{2\pi} \int_\Delta \ln|\zeta|\, \Delta_\zeta v(\zeta, \lambda)\, d\sigma, \qquad (2.25)$$

where $d\sigma$ is the area element. For fixed $\alpha \in [0, 2\pi]$, $v_\alpha(\zeta) = v(e^{i\alpha}\zeta, \zeta)$ is subharmonic and so, by the Riesz representation,

$$v_\alpha(0) = \frac{1}{2\pi} \int_\Delta \ln|\zeta|\, \Delta_\alpha v(\zeta, \lambda)\, d\sigma + \frac{1}{2\pi} \int_0^{2\pi} v_\alpha(e^{i\varphi})\, d\varphi,$$

[3] It is clear that one can neglect isolated points of this set.

or, as $v_\alpha(0)=v(0,0)$ does not depend on α,

$$v(0,0)=\frac{1}{4\pi^2}\int_0^{2\pi} d\alpha\left(\int_\Delta \ln|\zeta|\,\Delta v_\alpha\,d\sigma+\int_0^{2\pi} v_\alpha(e^{i\varphi})\,d\varphi\right).$$

But by the plurisubharmonicity of v

$$v(0,0)\leqq\frac{1}{4\pi^2}\int_0^{2\pi} d\alpha\left(\int_\Delta \ln|\zeta|\,\Delta_\zeta v(\zeta,e^{i\varphi})\,d\sigma+\int_0^{2\pi} v(e^{i\alpha},e^{i\varphi})\,d\alpha\right),$$

and $v(e^{i(\varphi+\alpha)},e^{i\varphi})=v_\alpha(e^{i\varphi})$, so that in the last two formulae the integrals over $[0,2\pi]\times[0,2\pi]$ are equal and, consequently,

$$\int_0^{2\pi} d\alpha\int_\Delta \ln|\zeta|\,\Delta v_\alpha\,d\sigma\leqq\int_0^{2\pi} d\varphi\int_\Delta \ln|\zeta|\,\Delta_\zeta v(\zeta,e^{i\varphi})\,d\sigma.$$

Hence, there exists $\alpha\in[0,2\pi]$ such that

$$\int_\Delta \ln|\zeta|\,\Delta v_\alpha\,d\sigma\leqq\frac{1}{2\pi}\int_0^{2\pi} d\varphi\int_\Delta \ln|\zeta|\,\Delta_\zeta v(\zeta,e^{i\varphi})\,d\sigma=\int_0^{2\pi} u_f(z+e^{i\varphi}v,w)\,d\varphi.$$

in view of (2.25). Using (2.24) once more we get

$$\int_\Delta \ln|\zeta|\,\Delta v_\alpha\,d\sigma\leqq\frac{1}{2\pi}\int_0^{2\pi} g_D(z+e^{i\varphi}v,w)\,d\varphi+\varepsilon.$$

The function $v_\alpha(\zeta)=\ln|h(e^{i\alpha}\zeta,\zeta)|$ has its singularities at the preimages of the point w under the holomorphic map $F(\zeta)=f(e^{i\alpha}\zeta,\zeta)\colon \Delta\to D$, $F(0)=z$. Therefore, the left hand side of the last inequality equals $u_F(z,w)$. As by definition $g_D(z,w)\leqq u_F(z,w)$ this establishes (2.23).

Theorem 2.27. *The invariant Green's function of a domain $D\subset\mathbb{C}^n$ can also be defined by the relation*

$$g_D(z,w)=\sup v_w(z), \tag{2.26}$$

where the least upper bound is taken over all non-positive plurisubharmonic functions v_w which near the point $w\in D$ have the form $v_w(z)=\ln|z-w|+O(1)$, $O(1)$ bounded.

Indeed, let us denote by $g_1(z,w)$ the right hand side of (2.26). As by Theorem 2.26 $g_D(z,w)$ is nonpositive, plurisubharmonic and has at the point w a singularity of the desired kind, we see that $g_D(z,w)\leqq g_1(z,w)$. On the other hand, for every $\varepsilon>0$ there exists a holomorphic map $f\colon D\to\Delta$, $f(0)=z$, such that $g_D(z,w)\geqq u_f(z,w)+\varepsilon$. If $v_w(z)$ is an arbitrary plurisubharmonic function in D with the right type of singularity at w then $v_w\circ f(\zeta)$ is a subharmonic function on Δ which at each preimage $\zeta_v\in f^{-1}(w)$ of multiplicity k_v has a singularity of type $k_v\cdot\ln|\zeta-\zeta_v|$. Consequently, by the Riesz representation formula, we get

$$v_w \circ f(0) = v_w(w) = \frac{1}{2\pi} \int_{\Delta} \ln|\zeta| \, \Delta v_w \circ f(\zeta) \, d\sigma + \frac{1}{2\pi} \int_0^{2\pi} v_w \circ f(e^{i\varphi}) \, d\varphi \leqq u_f(z, w),$$

where we have exploited also the fact that v_w is nonpositive. But $g_D(z, w) \geqq v_w(z) + \varepsilon$ for every admissible function v_w. Therefore also $g_D(z, w) \geqq g_1(z, w) + \varepsilon$. As ε is arbitrary, this proves everything.

This shows that if $N=1$, i.e. for a domain D in $\mathbb{C}$, g_D coincides with the classical Green's function.

Theorem 2.28. a) *If $D \subset \mathbb{C}^n$ is a bounded domain then $g_D(z, w)$ is continuous in w in $D \setminus \{z\}$*; b) *if $D = \{z \in \mathbb{C}^n : \varphi(z) < 0\}$, where φ is a plurisubharmonic function, continuous in a neighborhood of $\bar{D}$*, then *$g_D(z, w)$ is continuous in both variables in $(D \times D) \setminus \{z = w\}$.*

The first part of this theorem is elementary. The proof of the second part is based on some properties of plurisubharmonic functions (cf. A. Sadullaev [61]).

A plurisubharmonic *function* v in a domain $D \subset C^n$ is called *maximal* if for each domain $G \Subset D$ the inequality $u \mid \partial G \leqq v \mid \partial G$ implies that $u(z) \leqq v(z)$ for all $z \in G$.

Theorem 2.29. *$g_D(z, w)$ is maximal in z in $D \setminus \{w\}$.*

Indeed, let $G \Subset D$ and let u be plurisubharmonic in D with $u \mid \partial G \leqq g_D \mid \partial G$. For fixed $z \in G$ there exists a holomorphic map $f: \Delta \to D$ such that $f(0) = z$ and $u_f(z, w) \leqq g_D(z, w) - \varepsilon$. Let us further consider the subharmonic (in Δ) functions $u_1 = u \circ f$, $g_1 = g_D \circ f$ as well as the function

$$h(\zeta) = \sum k_\nu \ln \left| \frac{\zeta - \zeta_\nu}{1 - \bar{\zeta}_\nu \zeta} \right|,$$

with the summation over all preimages $\zeta_\nu \in f^{-1}(w)$, k_ν being the multiplicities of f at the point ζ. Apparently $h(0) = u_f(z, w)$ so, repeating the argument in the proof of Theorem 2.27, we see that $g_1(\zeta) \leqq h(\zeta)$ everywhere in Δ.

None of the points ζ_ν belongs to the open set $G_1 = f^{-1}(G)$, as $w \in G$. Consequently, h is harmonic in G_1. On the boundary of G_1 we have $u_1 \mid \partial G_1 \leqq g_1 \mid \partial G_1 \leqq h \mid \partial G_1$, whence, by the harmonicity of h, $u_1(0) \leqq h(0)$ or $u(z) \leqq u_f(z, w) \leqq g_D(z, w) - \varepsilon$. As ε is arbitrary, we conclude that $u(z) \leqq g_D(z, w)$ and the proof is complete.

For $n=1$ the only maximal functions are precisely the harmonic functions. If $n>1$ the maximal functions satisfy in a generalized sense the Monge-Ampère equation (cf. Vol. 8, Part III)

$$\det\left(\frac{\partial^2 u}{\partial z_j \, \partial \bar{z}_k}\right) = 0, \tag{2.27}$$

which is a multidimensional complex analogue of the Laplace equation. Therefore the Green's function $g_D(z, w)$ in $D \setminus \{w\}$ is a generalized solution of the Monge-Ampère equation with a singularity of type $\ln|z-w|$ at the point w. If D is strictly pseudoconvex, then this is the only solution of this kind with vanishing boundary values.

Consider two more functions which are variants of the Carathéodory and the Kobayashi distances (cf. Sec. 1.1 and 2.1). The first of these, the *Carathéodory function*, is defined for a domain $D \subset \mathbb{C}^n$ by

$$c_1(z, w) = \inf_{f_w} \ln \frac{1}{|f_w(z)|}, \tag{2.28}$$

with the greatest lower bound taken over all holomorphic maps $f_w : D \to \Delta$, $f_w(w) = 0$. It is connected with the Carathéodory distance by the evident relation

$$c_1(z, w) = \ln \frac{e^{2c(z, w)} + 1}{e^{2c(z, w)} - 1}.$$

The second one, the *Kobayashi function*, is

$$k_1(z, w) = \sup_f \ln\left(\frac{1}{|\zeta|}\right), \tag{2.29}$$

where now the least upper bound runs over all holomorphic maps $f : \Delta \to D$, $f(0) = z$, ζ being the preimage of w of the least modulus. Apparently

$$k_1(z, w) = \ln \frac{e^{2\tilde{k}(z, w)} + 1}{e^{2\tilde{k}(z, w)} - 1} = -\ln|\Phi_w(z)|,$$

where $\tilde{k}$ is defined in Sec. 2.1 and Φ_w in Sec. 2.8.

Both functions c_1 and k_1 are symmetric in their arguments and lower semicontinuous in each of them. For a bounded domain D both have near $w \in D$ the form $\ln|z-w| + O(1)$, $O(1)$ a bounded function. They are connected with Green's function via the inequalities

$$k_1(z, w) \leqq -g_D(z, w) \leqq c_1(z, w). \tag{2.30}$$

The Carathéodory function $c_1(z, w)$ is plurisuperharmonic in each variable when the other is fixed, while the Kobayashi function $k_1(z, w)$ does not always enjoy this property. If the latter is plurisuperharmonic, then by Theorem 2.27 $-k_1(z, w) = g_D(z, w) = \ln|\Phi_w(z)|$ is a generalized solution of the Monge-Ampère equation; this complements Lempert's results mentioned in Sec. 2.8.

It follows from the inequalities (2.30) that $k_1(z, w)$ coincides with $-g_D(z, w)$ and is, in particular, plurisubharmonic for all domains D such that $\tilde{k}_D \equiv c_D$ (which, again, entails $k_1 \equiv c_1$). It follows from Lempert's results (cf. Sec. 2.8) that all three functions k_1, c_1 and $-g_D$ coincide for strongly linear convex domains in $\mathbb{C}^n$. But, as is seen from the definitions, they differ for plane annuli.

§ 3. The Bergman Metric

3.1. Definition and Basic Properties. This metric was introduced by Stefan Bergman in 1929 for domains D in $\mathbb{C}^2$; we give here the general definition (cf., for example, A. Weil [73]). Let M be a complex n-dimensional manifold and $B(M)$ the Hilbert space of holomorphic forms of bidegree $(n, 0)$ with the inner product

$$(\alpha, \beta)=\frac{i^{n^2}}{2^n}\int\limits_M \alpha\wedge\bar{\beta} \tag{3.1}$$

(only those forms α for which $\|\alpha\|^2=(\alpha, \alpha)<\infty$ are members of $B(M)$).

Let us assume that $B(M)$ is nontrivial and let $\{\omega_j\}$ be an orthonormal basis there. One can show that the series

$$\beta_M(p, q)=\frac{i^{n^2}}{2^n}\sum_{j=1}^{\infty}\omega_j(p)\wedge\overline{\omega_j(q)}, \quad p, q\in M, \tag{3.2}$$

converges uniformly on compact subsets of M and that its sum does not depend on the choice of the basis $\{\omega_j\}$. Here the convergence is understood in the following sense: if $z=(z_1, \ldots, z_n)$ are local coordinates in M in a neighborhood of a point p and $w=(w_1, \ldots, w_n)$ local coordinates in a neighborhood of a point q, then locally $\omega_j(p)=\varphi_j(z)\,dz$ and $\omega_j(q)=\varphi_j(w)\,dw$, where $dz=dz_1\ldots dz_n$, dw being defined analogously. Then

$$\omega_j(p)\wedge\overline{\omega_j(q)}=\varphi_j(z)\,\overline{\varphi_j(w)}\,dz\wedge d\bar{w}$$

and the convergence of (3.2) means the convergence of the series with general term $\varphi_j(z)\,\overline{\varphi_j(w)}$.

It turns out that this series converges uniformly on compact sets together with all derivatives with respect to the local coordinates, so that $\beta_M(p, q)$ is a form of bidegree (n, n) on M with coefficients of class C^∞ (even real analytic).

The following (n, n)-form has a special interest:

$$\beta_M(p)=\beta_M(p, p)=\frac{i^{n^2}}{2^n}\sum_{j=1}^{\infty}\omega_j(p)\wedge\overline{\omega_j(p)}. \tag{3.3}$$

Theorem 3.1 ([73]). *If $B(M)$ is nontrivial on the complex manifold M, then the (n, n)-form β_M is real analytic on M, everywhere nonnegative, does not depend on the choice of the orthonormal basis $\{\omega_j\}$ and invariant under biholomorphic automorphisms of M.*

The form $\beta_M(p, q)$ is called the *Bergman kernel*, and $\beta_M(p)$ the *Bergman form* of M. In view of the orthonormality of $\{\omega_j\}$ the Bergman kernel has the reproducing property: for each $\omega\in B(M)$

$$\omega(p)=\int\limits_M \beta_M(p, q)\wedge\omega(q). \tag{3.4}$$

If $\omega_j(p)=\varphi_j(z)\,dz$, $\omega_j(q)=\varphi_j(w)\,dw$ in terms of local coordinates z and w in neighborhoods of the points p and q then

$$\beta_M(p, q)=B_M(z, w)\frac{i^{n^2}}{2^n}\,dz\wedge d\bar{w},$$

where

$$B_M(z, w)=\sum_{j=1}^{\infty}\varphi_j(z)\,\overline{\varphi_j(w)}, \tag{3.5}$$

and, in particular,

$$\beta_M(z)=B_M(z)\frac{i^{n^2}}{2^n}\,dz\wedge d\bar{z},$$

where

$$B_M(z)=\sum_{j=1}^{\infty}|\varphi_j(z)|^2. \tag{3.6}$$

Definition 3.1. Let M be a complex n-dimension manifold with nontrivial $B(M)$, set $E=\{p\in M: \omega(p)=0$ for all $\omega\in B(M)\}$ and pick an orthonormal basis $\{\omega_j\}$ in $B(M)$. If $\omega_j(z)=\varphi_j(z)\,dz$ in terms of local coordinates z in a neighborhood of the point $z\in M$ and if this neighborhood is contained in $M\setminus E$ then the (1, 1)-form

$$dd^c \ln B_M(z)=\frac{i}{2}\sum_{k=1}^{n}\frac{\partial^2 \ln B_M(z)}{\partial z_j\,\partial\bar{z}_k}\,dz_j\wedge d\bar{z}_k \tag{3.7}$$

is defined there; the corresponding bilinear form

$$ds_B^2(v)=B_M(p, v)=\sum_{j,k=1}^{n}\frac{\partial^2 \ln B_M(z)}{\partial z_j\,\partial\bar{z}_k}\,dz_j(v)\,d\bar{z}_k(v) \tag{3.8}$$

is called the metric *Bergman form* of M.

This form does not depend on the choice of the basis or on the local coordinates. It defines an Hermitean and, even more, a Kählerian metric on $M\setminus E$, called the *Bergman metric*; this metric degenerates on E. It is easy to see that the Bergman metric is invariant under biholomorphic maps of M, but it does not enjoy the contraction property for holomorphic maps.

In fact, by Stanton's theorem in Sec. 2.8, in every domain, which is not biholomorphically equivalent to the ball, in which the metrics of Carathéodory and Kobayashi coincide, these metrics can not be Hermitean, and consequently not Bergmanian. If the Bergman metric of such a domain D had the contraction property, then, in view of the extremal properties of the Carathéodory and the Kobayashi metrics (cf. Sec. 2.1), it would have been enclosed between them, and as these metrices coincide, then it would indeed coincide with them. But the Bergman metric in D is different from c_D and k_D. Consequently, there exists a holomorphic transformation of D which does not everywhere decrease the Bergman metric. By Lempert's Theorem 2.25 the class under consideration includes all strongly linear convex domains, not equivalent to the ball; in particular, there are such domains in arbitrary small C^2-neighborhoods of the ball.

The following remark is useful. Pick an orthonormal basis $\{\omega_j\}$ in $B(M)$ such that at a fixed point $p\in M\omega_1(p)\neq 0$ but $\omega_j(p)=0$ for all $j\geqq 2$. A simple calculation reveals that for each vector $v\in T_p(M)$ holds

$$B_M(p, v)=\frac{1}{B_M(z)}\sum_{j=2}^{\infty}|d\varphi_j(v)|^2, \tag{3.9}$$

where the coefficients φ_j are the local expressions of the ω_j, B_M being defined by (3.6). From this we may, in particular, conclude that $B_M(p, v)>0$ iff there exists a form $\omega\in B(M)$ such that $\omega(p)=0$ while $\partial\varphi/\partial v(p)\neq 0$.

In the case of domains D in $\mathbb{C}^n$ we may take for local coordinates the standard coordinates in $\mathbb{C}^n$. Therefore it is no point to consider forms ω_j and one may restrict oneself to the coefficients φ_j; instead of $\beta_D(p, q)$ and $\beta_D(p)$ it suffices to consider just the functions $B_D(z, w)$ and $B_D(z)$. Similarly, we may regard $B(D)$ as a Hilbert space of functions with the inner product

$$(\varphi, \psi)=\int_D \varphi\bar{\psi}\, dV$$

(dV is the Euclidean volume element in $\mathbb{C}^n$).

Example 3.1. In certain domains in $\mathbb{C}^n$ one can use as an orthonormal basis the functions $\varphi_k(z)=\lambda_k z^k$, where $k=(k_1, \ldots, k_n)$ is a multi-index and the λ_k are normalization coefficients >0 depending on the domain. In particular, for the unit ball Δ^n and the ball B^n a computation leads to the following expressions for the Bergman kernel:

$$B_{\Delta^n}(z)=\frac{1}{\pi^n}\prod_{k=1}^{n}\frac{1}{(1-|z_k|^2)^2}, \qquad B_{B^n}(z)=\frac{n!}{\pi^n(1-|z|^2)^{n+1}}.$$

3.2. Variational Properties. At the basis of the definition of the Bergman metric on manifolds M with nontrivial $B(M)$ one can put its variational properties. In domains in $\mathbb{C}^n$ one can, for instance, fixing a point $w\in D$, minimize the norm $(\varphi, \varphi)=\|\varphi\|^2$ within the class of all functions $\varphi\in B(D)$ such that $\varphi(w)=1$. Such an approach is carried out, for example, in [62].

On manifolds, when the coefficients of the forms depend on the local coordinates, it is better to proceed as follows. The forms $\omega\wedge\bar{\omega}$ on M have maximal degree (n, n) and, consequently, must be proportional to a (real) volume form on M with a real proportionality coefficient (therefore, they can be compared in magnitude). We then search, at a fixed point $q\in M$, for the maximum of this coefficient on the unit sphere of $B(M)$, i.e. the set $\{\omega\in B(M)\colon \|\omega\|=1\}$, where

$$\|\omega\|^2 = \frac{i^{n^2}}{2^n} \int_M \omega \wedge \bar{\omega}. \tag{3.10}$$

Using standard facts from functional analysis, one can prove that the solution of this problem exists for $q \notin E$ and is uniquely defined up to a constant factor of modulus 1. This solution, let it be denoted $\omega_0(p, q)$, is orthogonal in the sense of the inner product in $B(M)$ to the kernel of the maximized functional, i.e. the set $\{\omega \in B(M) \colon \omega(q) = 0\}$. From this it is easy to see that the Bergman kernel (3.3) can be expressed in a simple way in terms of the extremal function of our problem:

$$\beta_M(p) = \omega_0(p, p) \wedge \overline{\omega_0(p, p)}. \tag{3.11}$$

Let us consider another variational problem. For the form $\omega \in B(M)$ with the local representation $\omega(p) = \varphi(z)\, dz$ and the tangent vector $v \in T_p(M)$ set $\partial\omega/\partial v(p) = \partial\varphi/\partial v\, dz$. At a fixed point $q \in M$ let us maximize $\partial\omega/\partial v \wedge \partial\bar{\omega}/\partial v$ over all $\omega \in B(M)$, $\|\omega\| = 1$, such that $\omega(q) = 0$. The solution $\omega_1(p, q)$ of this problem is orthogonal to all $\omega \in B(M)$ with $\partial\omega/\partial(q) = 0$. It is seen from (3.9) that if we in $B(M)$ choose an orthonormal basis consisting of $\omega_0(p) = \omega_0(p, q)$, $\omega_1(p) = \omega_1(p, p)$ and forms such that $\omega(p) = \partial\omega/\partial v(p) = 0$, then in this basis the Bergman form (3.8) takes a very simple form:

$$B_M(p, v) = \frac{1}{B_M(z)} \left| \frac{\partial \varphi_1(z)}{\partial v} \right|^2. \tag{3.12}$$

where B_M is defined in (3.6) and φ_1 is the local expression of the form $\omega_1(p)$.

With the aid of this technique one can compare the Bergman and the Carathéodory metric.

Theorem 3.2 (Hahn [28]). *On every complex manifold M with $B(M)$ nontrivial for arbitrary $p \in M$ and $v \in T_p(M)$ holds*

$$B_M(p, v) \geqq [C_M(p, v)]^2. \tag{3.13}$$

It is very easy to prove this theorem: for each holomorphic function $f \colon M \to \Delta$, $f(p) = 0$, we have $f\omega_0(p) = 0$ and $\| f\omega_0 \| \leqq 1$, where $\omega_0 = \varphi_0\, dz$ is the extremal form in the first variational problem. Therefore, if $\omega_1(p) = \varphi_1(p)\, dz$ is the solution of the second one, we obtain for each $v \in T_p(M)$

$$\left| \frac{\partial (f\varphi_0)}{\partial v} \right|_z^2 = \left| \frac{\partial f}{\partial v} \right|_z^2 |\varphi_0(z)|^2 \leqq \left| \frac{\partial \varphi_1}{\partial v} \right|_z^2,$$

and as $|\varphi_0(z)|^2 = B_M(z)$ in the basis chosen, then by (3.12) $|\partial f/\partial v|^2 \leqq B_M(p, v)$.

One can construct a nondegenerate Bergman metric on some compact manifolds, on which the Carathéodory metric always is degenerate. As an example [43] we cite algebraic hypersurfaces in $\mathbb{CP}^n$ of degree $\geqq n+2$. Let us also mention that Diederich and Sibony [18] displayed a domain in $\mathbb{C}^n$ such that $C_D(z, v)=0$ while $B_D(z, v)>0$ for $v \neq 0$.

3.3. Other Properties. The Bergman metric (3.8) is positive definite on manifolds with nontrivial $B(M)$ and, therefore, one can define the distance $b_M(p, q)$ between two points p and q in the usual way. In domains in $\mathbb{C}^n$ the positive definiteness of this form means that $\ln B_M(z)$ is strictly plurisubharmonic. From this one readily concludes (Bremerman) that if the domain is complete in the Bergman metric then it is a domain of holomorphy (cf. [44]).

Kobayashi proved that the Bergman metric is complete on generalized analytic polyhedra, and Wu, Green and Kerzman established the same fact for strictly pseudoconvex domains (with twice smooth boundaries); cf. [44].

There are many papers devoted to the study of the boundary behavior of the Bergman metric. Let us mention a result by Diederich [16].

Theorem 3.3. *Let the domain $D \subset \mathbb{C}^n$ be strictly pseudoconvex at a boundary point a, which we without loss of generality may take to be the origin 0. Pick in a neighborhood $U \ni 0$ a system of coordinates z and a function $\varphi \in C^2(U)$ such that $D \cap U=\{z \in U: \varphi(z)<0\}$ and $\partial\varphi/\partial z_1(0)=-1$, $\partial\varphi/\partial z_k(0)=0$ for $k=2, \ldots, n$; set $K_\lambda=\{z \in D: |z|<\lambda|z_1+\bar{z}_1|\}$. Then for each $\lambda>1/2$ we have the following limit relations as $z \to 0$ through points in K_λ:*

$$\lim\,(z_1+\bar{z}_1)^2 \frac{\partial^2 \ln B_D(z)}{\partial z_1\,\partial \bar{z}_1}=n+1,$$

$$\lim\,(z_1+\bar{z}_1)^{3/2} \frac{\partial^2 \ln B_D(z)}{\partial z_j\,\partial \bar{z}_k}=0 \quad (k=2, \ldots, n),$$

$$\lim\,(z_1+\bar{z}_1) \frac{\partial^2 \ln B_D(z)}{\partial z_j\,\partial \bar{z}_k}=(n+1) \frac{\partial^2 \varphi}{\partial z_j\,\partial \bar{z}_k}(0) \quad (j, k=2, \ldots, n).$$

If we compare this to Graham's result mentioned in Sec. 2.2, we convince ourselves that the metrics of Carathéodory, Kobayashi and Bergman must be equivalent for strictly pseudoconvex domains. However, Diederich and Fornaess [17] gave an example of a pseudoconvex domain $D \subset \mathbb{C}^3$ with infinitely smooth boundary such that the ratio $B(z, v)/K(z, v)$ is not bounded from above on $D \times \mathbb{C}^3$.

Fefferman used result on the boundary behavior of the Bergman metric to prove that biholomorphic maps of strictly pseudoconvex domains $D \subset \mathbb{C}^n$ with smooth boundaries have smooth extensions up to the boundary. In this connection he established, in particular, the following result. Let z^0 be a point in D, v^0 a point in the unit sphere $S \subset \mathbb{C}^n$ and $\gamma(t, z^0, v^0)$ the geodesic in the Bergman metric on D defined by the initial conditions $\gamma(0, z^0, v^0)=z^0$, $(\partial\gamma/\partial t)(0, z^0, v^0)=v^0$. Then either this geodesic never leaves a certain compact set of D or else the limit $\lim_{t\to\infty} \gamma(t, z^0, v)=\pi_{z^0}(v)$ exists for all $v \in S$ sufficiently closed to v^0 and belongs to the boundary of the domain ∂D; in the second case π_{z^0} establishes a diffeomorphism between a neighborhood of v^0 on S and a neighborhood of $\pi_{z^0}(v^0)$ on ∂D.

Herbort [31] has studied the geodesics which do not reach out to the boundary. He showed that for a strictly pseudoconvex domain $D \subset \mathbb{C}^n$ with C^∞-boundary each nontrivial homotopy class of loops contains a closed geodesic in the Bergman metric and if the fundamental group D is infinite, then through each point, not lying on a closed geodesic, there passes a geodesic spiral which does not go out to the boundary.

In [22] Fefferman established an interesting connection between the Bergman function $B_D(z)$ and the complex Monge-Ampère equation, and using this function, he constructed a Lorentz metric on the boundary of the domain, where the light rays are connected with the Moser chains (cf. S.I. Pinchuk's Part in this Volume).

§ 4. Invariant Volume Forms

4.1. Definition and Simplest Properties. In [43] Kobayashi introduced a volume analogue of the metrics c_M and k_M considered in §§ 1 and 2. Let B^n be the unit ball in $\mathbb{C}^n$, denoting by $\operatorname{vol} E$ the volume of a set $E \subset B^n$, computed in the Bergman metric of B^n.

Definition 4.1. Let M be a complex n-dimensional manifold and $Q \subset M$ any subset of M. The *invariant C-volume* (or Carathéodory volume) of Q is defined as

$$V_M^c(Q) = \sup \operatorname{vol} f(Q), \tag{4.1}$$

where the least upper bound is taken over all holomorphic maps $f\colon M \to B^n$. The *invariant K-volume* (or Kobayashi volume) of Q is defined as

$$V_M^k(Q) = \inf \sum_{j=1}^{m} \operatorname{vol} E_j, \tag{4.2}$$

where the greatest lower bound is taken over all possible families of Borel measurable sets $E_j \subset B^n$ and holomorphic maps $f^j\colon B^n \to M$ such that $\bigcup_{j=1}^{n} f^j(E_j) \supset Q$.

Instead of the unit ball B^n one can also take the unit polydisk Δ^n with the Bergman volume

$$\beta(\zeta) = \left(\frac{i}{2}\right)^n \prod_{k=1}^{n} \frac{d\zeta_k \wedge d\bar{\zeta}_k}{(1-|\zeta_k|^2)^2}. \tag{4.3}$$

Eisenman and Kobayashi [21], [22] (cf. further M.N. L'vovskiĭ [51]) gave infinitesimal analogues of these invariant volumes.

Definition 4.2. Let M be a complex n-dimensional manifold; the *invariant Carathéodory pseudo-volume form of* M at the point $p \in M$ is defined as

$$\Omega_M^c(p) = \sup f^* \beta|_p, \tag{4.4}$$

where the least upper bound extends over all holomorphic maps $f: M \to \Delta^n$, $f(p)=0$, while the *Kobayashi pseudo-volume form* is

$$\Omega_M^k(p) = \inf (f^{-1})^* \beta|_0, \tag{4.5}$$

where the greatest lower bound extends over all holomorphic maps $f: \Delta^n \to M$, $f(0)=p$, nondegenerate at the point $\zeta=0$.

Let us make explicit these definitions. Let $z=(z_1, \ldots, z_n)$ be local coordinates near the point p and let $\Phi=(i/2)^n \prod_{k=1}^n dz_k \wedge d\bar{z}_k$ be the Euclidean volume form. In the first case f is locally of the form $\zeta_j = f_j(z)$, $j=1, \ldots, n$, and as $f(p)=0$ then, taking account of (4.3), we get $f^*\beta|_p = |J_f(p)|^2 \Phi$, where $J_f(p) = \det f'(p)$ is the Jacobian of the map f. Therefore

$$\Omega_M^c(p) = \sup_f |J_f(p)|^2 \Phi = H^c(p)\,\Phi; \tag{4.6}$$

it is clear that this expression does not depend on the local coordinates.

In the second case we must consider, instead of f, the inverse f^{-1} in a neighborhood of $\zeta=0$. As $\det(f^{-1})' = (\det f')^{-1}$, we get

$$\Omega_M^k(p) = \sup_f \frac{1}{|J_f(0)|^2}\,\Phi = H^k(p)\,\Phi. \tag{4.7}$$

The coefficients H^c and H^k in the local expressions for Ω_M^c and Ω_M^k are nonnegative but may vanish at some points of M (this is independent of the choice of local coordinates) so that in the general case Ω_M^c and Ω_M^k are only pseudo-volume forms.

The form Ω_M^c is continuous on M, Ω_M^k only semicontinuous. Both forms are invariant under biholomorphic maps, and under holomorphic maps $f: M \to N$ they do not grow in the sense that

$$f^*\Omega_N(p) \leqq \Omega_M(p) \quad \text{for all } p \in M \tag{4.8}$$

(the pullback $f^*\Omega_N$ and the form Ω_M are compared by their coefficients as forms of maximal degree).

We have also

Theorem 4.1. a) *The form Ω_M^k is the largest among the pseudo-volume forms on M which do not increase under holomorphic maps $\Delta^n \to M$, whereas Ω_M^c is the least among those which enjoy the same property for maps $M \to \Delta^n$;*

b) *On every complex manifold $\Omega_M^c(p) \leqq \Omega_M^k(p)$.*

The Kobayashi form agrees with passage to holomorphic coverings: if $\pi: \tilde{M} \to M$ is such a covering, then $\Omega_{\tilde{M}}^k(p) = \pi^* \Omega_M^k(p)$ at each point $p \in \tilde{M}$. The Carathéodory form does not have this property.

Finally, let us mention that Yau [74] has for each $m \leqq n = \dim M$ introduced an m-form generalizing the Kobayashi form:

Definition 4.3. Fix a point $p \in M$ and an m-vector $v = v^1 \wedge \ldots \wedge v^m$ in the m-th power $(T_p(M))^m$ of $T_p(M)$. The *invariant m-measure* is defined as

$$K_M^{(m)}(p, v) = \inf r, \tag{4.9}$$

where the greatest lower bound extends over all maps meromorphic in the sense of Remmert (cf. Sec. 2.5) $f: \Delta^m \to M$ such that $f(0) = p$, $f^*(u) = v$, and all m-vectors $u \in (T_0(\Delta^m))^m$ of length r in the Poincaré-Bergman metric of the polydisk Δ^m.

For $m = 1$ this definition gives a norm majorized by the Kobayashi norm $K_M(p, v)$ and for $m = n$ it leads to a pseudo-volume form which differs from (4.5) only in the respect that we allow meromorphic maps, not just holomorphic ones. Exactly as Ω_M^k the latter is semicontinuous, does not increase under meromorphic maps $f: M \to N$, agrees with transition to holomorphic coverings. On every complex manifold holds $\Omega_M^c \leqq \tilde{\Omega}_M^k \leqq \Omega_M^k$.

4.2. Measure Hyperbolicity. Let us call a subset E of a complex m-dimensional manifold M a set of zero measure if locally, in the neighborhood of each point, it has zero $2n$-dimensional Hausdorff measure in a coordinate neighborhood of the point.

Definition 4.4. The *manifold M* is called *measure hyperbolic* (respectively *C-measure hyperbolic*) if $\Omega_M^k(p) \neq 0$ (respectively $\Omega_M^c(p) \neq 0$) outside a set of zero measure.

It is interesting to compare measure hyperbolicity to hyperbolicity with respect to invariant metrics. It is not hard to see that if $C_M(p, v) \neq 0$ for all $v \neq 0$, then the coefficient $H^c(p)$ in the expression (4.6) for the form $\Omega_M^c(p)$ in local coordinates z near the point p satisfies $H^c(p) \leqq \pi^n / \mathrm{vol}\, B^c(p)$, where $\mathrm{vol}\, B^c(p)$ is the volume of the Carathéodory unit ball in $T_p(M)$ evaluated with the aid of the Euclidean volume form Φ. Analogously, if $K_M(p, v) \neq 0$ then in the same coordinates $H^k(p) \geqq \pi^n / \mathrm{vol}\, B^k(p)$. As always $B^k(p) \subset B^c(p)$, it follows that

$$H^c(p) \leqq \frac{\pi^n}{\mathrm{vol}\, B^c(p)} \leqq \frac{\pi^n}{\mathrm{vol}\, B^k(p)} \leqq H^k(p). \tag{4.10}$$

M.N. L'vovskiĭ [51] proved that the class of C-volume hyperbolic manifolds coincides with the class of complex manifolds M for which the Carathéodory distance satisfies $c_M(p, q) > 0$ for all points $p \in M$ and all points $q \in M \setminus A$ disjoint from p, where A is a certain analytic subset of M. It is likewise easy to see that $C_M(p, v) = 0$ for some vector $v \neq 0$ iff $\Omega_M^c(p) \neq 0$. The analogous statement for the Kobayashi metric is not true.

Example 4.1. Let $D=\{z\in\mathbb{C}^n: |z_1|<1, |z_2|<|z_1|^{-1/2}\}$; then $K_D(0, v)=0$ for $v=(1, 0)$, while $\Omega_D^k(0)\neq 0$, because vol $B^k(0)<\infty$ and $H^k(0)>0$, this in view of (4.10).

Furthermore, if a manifold M is hyperbolic modulo a closed set of measure zero, then it must be measure hyperbolic. But the Kobayashi metric of measure hyperbolic manifold can vanish on rather complicated sets:

Example 4.2. Let $E\subset\Delta$ be an arbitrary polar set and $u\not\equiv\infty$ a superharmonic function such that $E=\{\zeta\in\Delta: u(\zeta)=\infty\}$. Consider the domain $D=\{z\in C^2: |z_1|<1, |z_2|<u(z_1)\}$. Arguing as in Example 2.3, one can show that $K_D(z, v)\geqq c|v|/u(z_1)$ for each point $z\in D$ and each $v\in\mathbb{C}^2$, with a suitable $c>0$. From this it follows that vol $B^k(z)<\infty$ for all $z=(z_1, z_2)$: $z_1\notin E$ so that $H^k(z)>0$ for such a z, i.e. D is measure hyperbolic. However, the Kobayashi distance between any two points (z_1, z_2') and (z_1, z_2'') with $z_1\in E$ equals 0.

For pseudo-volume forms, as well as for metrics, one can introduce the notion of negligible set, and for the Carathéodory form one can establish the negligibility of analytic subsets: if A is an analytic subset of a complex manifold M, with codim $A\geqq 1$, then $\Omega_{M\setminus A}^c\equiv\Omega_M^c$. For the Kobayashi form one does not have such a result:

Example 4.3 (V.V. Rabotin). Let $M=B^n$, $A=\{0\}$; then by an old result of Carathéodory's [13] $\Omega_M^k(0)=n^{n/2}\,\Phi$, where Φ is the Euclidean volume form; the extremal map realizing $\Omega_M^k(0)$ is $f(\zeta)=n^{-1/2}\,U\zeta$, where U is a unitary $n\times n$ matrix. From this one readily concludes that the omitting of points decreases the Kobayashi form: $\Omega_{M\setminus A}^k(p)>\Omega_M^k(p)$ for all $p\in M\setminus A$.

4.3. Applications. The monotonicity property of volumes and forms makes it possible to obtain generalizations of the classical theorems of Picard and Schottky-Landau. Here is one result of this nature.

Theorem 4.3 ([44]). *Let M be a complex manifold such that $\Omega_M^k(p)=0$ on a nonempty open subset and let N be measure hyperbolic. Then each holomorphic map f: $M\to N$ is degenerate at all points of M.*

It is well-known that every nonconstant holomorphic map from $\bar{D}\setminus\{a, b, c\}$ (the points a, b, c being disjoint) onto itself is by necessity biholomorphic and there are only 6 such maps. Using the notion of measure hyperbolicity, V.V. Rabotin [56] proved that this fact generalizes to proper holomorphic maps from $M\setminus D$ into itself, where M is a smooth projective manifold and D a divisor of M with normal selfintersections satisfying a certain condition on its size: each such map is biholomorphic and their number is finite.

Rosay [58] proved that if the coefficients $H^c(z)$ and $H^k(z)$ of the volume forms of Carathéodory and Kobayashi of a bounded domain in $\mathbb{C}^n$ coincide at some point then the domain must be equivalent to a ball (here we have to take B^n in place of Δ^n in the definition). I.M. Dektyarev [15] obtained a generalization of this result: let M and N be hyperbolic manifolds of the same dimension and set (in local coordinates) $J_{pq}=\sup_f |J_f(p)|$, where the least upper bound runs over all holomorphic maps $f\colon M\to N$ with $f(p)=f(q)$; then if $J_{pq}\cdot J_{qp}=1$ for some pair of points, then there exists a biholomorphic extremal map from M onto N and is defined uniquely up to automorphisms of the manifolds fixing the given points.

It is interesting to consider such geometric characteristics of a complex manifold M as the full volume computed with the aid of the Kobayashi form:

$$V(M)=\int_M \Omega_M^k. \tag{4.11}$$

It follows from the Gauss-Bonnet formula that if M is a compact Riemann surface of genus g then its total Kobayashi volume is $V(M)=2\pi(g-2)$. Using the formula for the degree $\deg f$ of a proper holomorphic map $f\colon M\to N$ of complex manifolds of the same dimension in the case when the total Kobayashi volume of N is positive and finite:

$$\deg f=\int_M f^*\Omega_N^k/\int_N \Omega_N^k,$$

and further the monotonicity property for volumes, Yau [74) obtained the following result.

Theorem 4.4. *Let M and N be complex n-dimension manifold and assume that the total Kobayashi volume of N satisfies* $0<V(N)<\infty$. *Then*

(1) *if* $V(M)<V(N)$ *then there exists no nondegenerate proper holomorphic maps* $f\colon M\to N$;

(2) *if* $V(M)<2V(N)$ *then each nondegenerate proper holomorphic map* $f\colon M\to N$ *is biholomorphic*

4.4. Ricci Curvature

Definition 4.5. Let Ω be an arbitrary twice smooth volume form. i.e. a positive (n, n)-form, on an n-dimensional complex manifold M. If in local coordinates on M this form has the representation $\Omega(p)=h(z)\,\Phi$, where $h(z)>0$ and Φ is the Euclidean volume form then the *Ricci form* of Ω is defined as the Hermitean (1, 1)-form with the local representation:

$$\operatorname{Ric}\Omega=dd^c\ln h=\frac{i}{2}\sum_{j,k=1}^{n}\frac{\partial^2\ln h}{\partial z_j\,\partial\bar z_k}\,dz_j\wedge d\bar z_k. \tag{4.12}$$

It is clear that this form does not depend on the choice of local coordinates.

We say that the *Ricci curvature* of M does not surpass a negative constant $-c$, if there exists a volume form Ω on M such that all points $p\in M$ its Ricci form $\operatorname{Ric}\Omega$ is positive definite and its n-th exterior power satisfies

$$(\operatorname{Ric}\Omega)^n\geqq c\Omega. \tag{4.13}$$

Example 4.4. On a complex 1-dimensional manifold S with the form $\Omega=i/2h(z)\,dz\wedge d\bar z$, the Ricci form $\operatorname{Ric}\Omega$ is positive definite iff its Gaussian curvature is negative. On the polydisc Δ^n the proportionality coefficient between the invariant and the Euclidean volume forms equals, in view of (4.3), $h=\prod_{k=1}^{n}(1-|z_k|^2)^{-2}$ and, consequently, we have $\operatorname{Ric}\Omega_{\Delta}n=2\beta(z)$; thus the Ricci curvature on Δ^n does not surpass $-2^n n!$. For the Fubini-Study metric ω on $\mathbb{CP}^n$, $\operatorname{Ric}\Omega=-(n+1)\,\omega$ is negative definite and, consequently, the Ricci curvature is positive.

The following generalization of the Schwarz lemma in invariant form is due to Kobayashi and Griffiths:

Theorem 4.5. *If M is a complex n-dimensional manifold whose Ricci curvature does not surpass a negative constant $-c$, then for each holomorphic map $f\colon \Delta^n \to M$ holds*

$$f^*\Omega \leqq c'\beta, \tag{4.14}$$

where $c'>0$ is a constant depending only on c and n, while β is the invariant volume form (4.3) *for the polydisk Δ^n.*

One can always normalize Ω in such a way that $c=c'=1$. Then, by (4.14), Ω is not increased by holomorphic maps $f\colon \Delta^n\to M$ and so, by Theorem 4.1, everywhere on M

$$\Omega_M^k \geqq \Omega .$$

Thus, the existence of an arbitrary, not necessarily invariant metric with negative Ricci curvature on a complex manifold secures that it is measure hyperbolic. This approach was developed in papers by Griffiths, Carlson and others (cf. Part II).

Concluding Remarks

1. We have included in this survey only the most useful and most well-known of the invariant metrics. The list of such metrics is, of course, not exhausted by this. For example, one can consider the following generalization of the Kobayashi norm:

$$\tilde{K}^{(m)}(p, v)=\inf r^{-1/m}$$

where the greatest lower bound is taken over all holomorphic maps such that

$$f(0)=p,\ f'(0)=\ldots=f^{(m-1)}(0)=0,\ f^{(m)}(0)=rv;$$

apparently

$$\tilde{K}_M^{(m)}(p, v) \leqq \sqrt[m]{m!}\, K_M(p, v).$$

Or one can take $\inf r^{-1}$ just over holomorphic *imbeddings* $\Delta\to M$ with $f(0)=p$, $f'(0)=v$. Then one gets an invariant norm majorizing the Kobayashi norm but which does increase only under holomorphic imbeddings of the manifold [29].

Let us further mention a variant of the Carathéodory pseudometric suggested by Sibony [65]: this time bounded holomorphic functions are replaced by bounded plurisubharmonic ones. In that paper the connection of this notion with hyperbolicity in the sense of Carathéodory and Kobayashi and with Montel manifolds is investigated. A related circle of ideas is touched upon in Berg [6] and in Kerzman-Rosay [36].

Generalizing the biholomorphically invariant norms on cohomology groups, introduced by Chern, Levine and Nirenberg, Kalka [34] defined an invariant outer measure on submanifolds of complex manifolds. This measure, which is reminiscent of the Bergman metric, behaves in an uncontrolled way under holomorphic maps. Bedford [5] considered norms on homology groups which do not increase under homomorphisms of these groups induced by injective holomorphic maps and gave several applications of this.

2. There is a long series of papers by Grauert, Chern, Kobayashi, Griffiths, Shiffman, and others, where problems of analysis in several complex variables, similar to those considered above, are studied by other geometric methods. It is question about continuation theorems of the type of the Picard theorems, theorems of Schottky-Landau type etc. The methods which we are speaking of are related to items such as Chern classes of holomorphic vector bundles, various curvatures etc. These methods are widely used notably in the contemporary value distribution theory for holomorphic maps. Some ideas about this can be found in the book [63], cf. also the paper [2] by V.A. Babets and Part II.

3. Invariant metrics have important applications in the theory of Teichmüller spaces; let us outline some of them. The Teichmüller space $T=T(g, n)$ consists of the collection of Riemann surfaces of genus g with n punctures and a distinguished system of generators of the fundamental group, equipped with the following metric: the distance between two surfaces $X, Y \in T$ is defined to be

$$\tau(X, Y)=\tfrac{1}{2}\inf \ln Q_f,$$

where the greatest lower bound is taken over all quasiconformal maps $f: X \to Y$ such that $f_*: \pi_1(X) \to \pi_1(Y)$ fixes the distinguished generators and Q_f is the quasiconformal coefficient of f (hence, the distance between conformally equivalent distinguished surfaces is zero). One can define a complex structure on $T(g, n)$ in such a way that the space can be realized as a domain in $\mathbb{C}^{3g-3+n}$. Thus one can consider in T the invariant metrics of Carathéodory, Kobayashi and Bergman.

As first shown by Royden for $n=0$ and then by Earl and Kra [19] for arbitrary g and n, the metric τ coincides with $\tilde{k}_T$ and thus also with the Kobayashi metric k_T. However the Carathéodory distance c_T is strictly less than τ for some pairs of points in $T(g, n)$, provided $3g+n>5$ and for $g>2$ also $n \geqq 1$ (S.L. Krushkal [48]). In [47] it is proved that $T(g, n)$ is complete in the sense of Carathéodory. Hence its realization is a domain of holomorphy.

Earl and Kra applied their result to determine the group Aut T: they proved that a point $X \in T$ can be mapped onto $Y \in T$ iff the surfaces X and Y are conformally equivalent. For the proof one uses the fact that the set of failure of smoothness of K_T allows a reconstruction of the conformal class of the surface X. Let us also remark that Royden computed the Bergman metric of $T(g, 0)$, showing that it is induced by the same metric for Siegel's upper halfplane under the map defined by integration of holomorphic differentials along the distinguished generators of $\pi_1(X)$.

If one introduces on a surface $X \in T$ the metric $t_X(p, q) = 1/2 \cdot \inf\{\ln Q_f : f$ a quasiconformal automorphism of X homotopic to the identity, with $f(p) = q\}$ then, as shown by S.L. Krushkal, $dt_X(p, v) \leqq K_X(p, v)$ for each point $p \in X$ and $v \in T_P(X)$. This result was used by him to prove a distortion theorem for quasiconformal maps of Riemann surfaces. Cf. [48].

4. A variety of deep applications of invariant metrics can be found in algebraic geometry; it is not possible to describe even the main features of these applications. For example, let us mention Mumford's result [40] which generalizes Hirzebruch's proportionality theorem to quotients of bounded symmetric domains of finite volume: it is based on the use of invariant volumes in products of complete punctured disks by which one covers noncompact algebraic varieties. This result has important applications.

As a second application of hyperbolic analysis in algebraic geometry we mention the study of families of algebraic varieties with the aid of the period map. From the contraction property of invariant metrics under such maps one derives deep results on the monodromy of families of algebraic varieties, degeneracy etc. In particular, by similar considerations Deligne proved a theorem to the effect of the finiteness of the number of families of Abelian varieties over a hyperbolic basis, which for Faltings served as the prototype for an analogous arithmetical theorem, forming the foundation for his proof of the Mordell conjecture.

References

For the convenience of the reader, references to reviews in Zentralblatt für Mathematik (Zbl.), compiled using the MATH database, and Jahrbuch über die Fortschritte der Mathematik (Jrb.) have been included as far as possible.

1. Ahern, P., Schneider, R.: Isometries of H^∞. Duke Math. J. *42*, 321–326 (1975). Zbl. 354.32023
2. Babets, V.A.: The geometry of logarithmically tangent bundles and hyperbolic manifolds. Funkts. Anal. Prilozh. *16*, No. 2, 64–65 (1982) [Russian]. English transl.: Funct. Anal. Appl. *16*, 127–128 (1982). Zbl. 523.32019
3. Barth, T.: The Kobayashi distance induces the standard topology. Proc. Am. Math. Soc. *35*, 439–441 (1972). Zbl. 259.32007
4. Barth, T.: The Kobayashi indicatrix at the center of a circular domain. Proc. Am. Math. Soc. *88*, 527–530 (1983). Zbl. 494.32008
5. Bedford, E.: Invariant forms on complex manifolds with applications to holomorphic mappings. Math. Ann. *265*, 377–397 (1983). Zbl. 532.32015

6. Berg, G.: Bounded holomorphic functions of several variables. Ark. Mat. *20*, 249–270 (1982). Zbl. 535.32007
7. Borel, A., Narasimhan, R.: Uniqueness conditions for certain holomorphic mappings. Invent. Math. *2*, 247–255 (1967). Zbl. 145, 318
8. Brody, R.: Compact manifolds and hyperbolicity. Trans. Am. Math. Soc. *235*, 213–219 (1978). Zbl. 416.32013
9. Campbell, L., Howard, A., Ochiai, T.: Moving holomorphic discs off analytic subsets. Proc. Am. Math. Soc. *60*, 106–107 (1976). Zbl. 314.32014
10. Campbell, L., Ogawa, R.: On preserving the Kobayashi pseudometric. Nagoya Math. J. *57*, 37–47 (1975). Zbl. 312.32014
11. Carathéodory, C.: Über das Schwarzsche Lemma bei analytischen Funktionen von zwei komplexen Veränderlichen. Math. Ann. *97*, 76–98 (1926). Jrb. 52, 345. Also: Gesammelte mathematische Schriften IV, München, Beck (1956; Zbl. 73, 1) 132–159
12. Carathéodory, C.: Über die Geometrie der Abbildungen, die durch analytische Funktionen von zwei Veränderlichen vermittelt werden. Abh. Math. Semin. Univ. Hamb. *6*, 96–145 (1928) Jrb. 54, 372. Also: Gesammelte mathematische Schriften IV, München, Beck (1956; Zbl. 73, 1) 167–227
13. Carathéodory, C.: Über die Abbildungen, die durch Systeme von analytischen Funktionen von mehreren Veränderlichen erzeugt werden. Math. Z. *34*, 758–792 (1932). Also: Gesammelte mathematische Schriften III, München, Beck (1955; Zbl. 58, 241) 406–448
14. Chirka, E.M.: The theorems of Lindelöf and Fatou in $\mathbb{C}^n$. Mat. Sb., Nov. Ser. *92* (134), 622–644 (1973) [Russian]. Zbl. 285.32005. English transl.: Math. USSR, Sb. *21*, 619–639 (1973)
15. Dektyarev, I.M.: An equivalence criterion for hyperbolic manifolds. Funkts. Anal. Prilozh. *15*, No. 4, 73–74 (1981) [Russian]. Zbl. 506.32011. English transl.: Funct. Anal. Appl. *15*, 292–293 (1982)
16. Diederich, K.: Über die 1. und 2. Abteilung der Bergmanschen Kernfunktion und ihr Randverhalten. Math. Ann. *203*, 129–170 (1973). Zbl. 253.32011
17. Diederich, K., Fornaess, J.: Comparison of the Bergman and the Kobayashi metric. Math. Ann. *254*, 257–262 (1980). Zbl. 429.32031
18. Diederich, K., Sibony, N.: Strange complex structures on Euclidean space. J. Reine Angew. Math. *311/312*, 397–407 (1979). Zbl. 413.32001
19. Earle, C., Kra, I.: On isometries between Teichmüller spaces. Duke Math. J. *41*, 583–591 (1974). Zbl. 293.32020
20. Eastwood, A.: A propos des variétés hyperboliques completes. C. R. Acad. Sci., Paris, Sér. A, *280*, 1071–1075 (1975). Zbl. 301.32021
21. Eisenman, D. (=Pelles, D.): Intrinsic measures on complex manifolds and holomorphic mappings. Mem. Am. Math. Soc. *96* (1970). Zbl. 197, 50
22. Fefferman, C.: Monge-Ampère equations, the Bergman kernel and geometry of pseudoconvex domains. Ann. Math., II. Ser. *103*, 395–416 (1976). Zbl. 322.32012
23. Fujimoto, H.: Families of holomorphic maps into projective space omitting some hyperplanes. J. Math. Soc. Japan *25*, 235–249 (1973). Zbl. 253.32012
24. Graham, I.: Boundary behavior of the Carathéodory and Kobayashi metrics on strongly pseudoconvex domains in $\mathbb{C}^n$ with smooth boundary. Trans. Am. Math. Soc. *207*, 219–240 (1975). Zbl. 305.32011
25. Green, M.: Holomorphic maps into complex projective space omitting hyperplanes. Trans. Am. Math. Soc. *169*, 89–103 (1972). Zbl. 256.32015
26. Green, M.: The hyperbolicity of the complement of $2n+1$ hyperplanes in general position in $\mathbb{P}^n$ and related results. Proc. Am. Math. Soc. *66*, 109–113 (1977). Zbl. 366.32013
27. Griffiths, P.: Two theorems on extensions of holomorphic mappings. Invent. Math. *14*, 27–62 (1971). Zbl. 223.32016
28. Hahn, K.: Inequality between the Bergman and Carathéodory differential metrics. Proc. Am. Math. Soc. *68*, 193–194 (1978). Zbl. 376.32020
29. Hahn, K.: Some remarks on a new pseudo-differential metric. Ann. Pol. Math. *39*, 71–81 (1981). Zbl. 476.32031

30. Hahn, K., Kim, K.T.: Hyperbolicity of a complex manifold and other equivalent properties. Proc. Am. Math. Soc. *91*, 49–53 (1984). Zbl. 518.32017
31. Herbort, G.: On the geodesics of the Bergman metric. Math. Ann. *264*, 39–51 (1983). Zbl. 497.32022
32. Ivashkovich, S.M.: Continuation of locally biholomorphic maps of domains in complex projective space. Izv. Akad. Nauk SSSR, Ser. Mat. *47*, No. 1, 197–206 (1983) [Russian]. Zbl. 523.32009. English transl.: Math. USSR, Izv. *22*, 181–189 (1984)
33. Illarionov, M.A.: The Kobayashi metric on a fiber space. Vestn. Mosk. Univ, Ser. I 1980, No. 2, 35–36 (1980) [Russian]. Zbl. 436.32021. English transl.: Mosc. Univ. Math. Bull. *35*, No. 2, 39–40 (1980)
34. Kalka, M.: Measures associated to Chern, Levine and Nirenberg norms. Proc. Am. Math. Soc. *88*, 404–406 (1983). Zbl. 578.32043
35. Kaup, W.: Reelle Transformationsgruppen und invariante Metriken auf komplexen Räumen. Invent. Math. *3*, 43–70 (1967). Zbl. 157, 134
36. Kerzman, N., Rosay, J.-P.: Fonctions plurisousharmoniques d'exhaustion bornées et domaines taut. Math. Ann. *257*, 171–184 (1981). Zbl. 451.32012
37. Khenkin, G.M.: An analytic polyhedron holomorphically nonequivalent to a strictly pseudoconvex domain. Dokl. Akad. Nauk SSSR *210*, 1026–1029 (1973) [Russian]. Zbl. 288.32015. English transl.: Sov. Math., Dokl. *14*, 858–862 (1973)
38. Kiernan, P.: Some results concerning hyperbolic manifolds. Proc. Am. Math. Soc. *25*, 588–592 (1970). Zbl. 198, 424
39. Kiernan, P.: Extensions of holomorphic mappings. Trans. Am. Math. Soc. *172*, 347–355 (1972). Zbl. 255.32014
40. Kiernan, P., Kobayashi, S.: Holomorphic mappings into projective spaces with lacunary hyperplanes. Nagoya Math. J. *50*, 199–216 (1973). Zbl. 262.32010
41. Kim, D.: Complete domains with respect to the Carathéodory distance. II. Proc. Am. Math. Soc. *53*, 141–142 (1975). Zbl. 309.32010
42. Kobayashi, S.: Invariant distances on complex manifolds and holomorphic mappings. J. Math. Soc. Japan *19*, 460–480 (1967). Zbl. 158, 332
43. Kobayashi, S.: Hyperbolic manifolds and holomorphic mappings. New York: Dekker 1970. Zbl. 207, 379
44. Kobayashi, S.: Intrinsic distances, measures and geometric function theory. Bull. Am. Math. Soc. *82*, 357–416 (1976). Zbl. 346.32031
45. Kobayashi, S., Nomizu, K.: Foundations of differential geometry, I, II. New York – London: Interscience 1963, 1969. Zbl. 119, 375. Zbl. 175, 485
46. Kodama, A.: On bimeromorphic automorphisms of hyperbolic complex spaces. Nagoya Math. J. *73*, 1–5 (1979). Zbl. 367.32002
47. Krushkal', S.L.: Two theorems on Teichmüller spaces. Dokl. Akad. Nauk. SSSR *228*, 290–292 (1976) [Russian]. Zbl. 347.32011. English transl.: Sov. Math., Dokl. *17*, 704–707 (1976)
48. Krushkal', S.L., Kühnau, R.: Quasikonforme Abbildungen – neue Methoden und Anwendungen. Leipzig: Teubner-Texte zur Mathematik, Bd. 54 (1983). Zbl. 539.30001. Novosibirsk: Nauka 1984 [Russian]. Zbl. 543.30001
49. Kwack, M.: Generalizations of the big Picard theorem. Ann. Math., II. Ser. *90*, 9–22 (1969). Zbl. 179, 121
50. Lempert, L.: La métrique de Kobayashi et la représentation des domaines sur la boule. Bull. Soc. Math. Fr. *109*, 427–474 (1981). Zbl. 492.32025
51. L'vovskiĭ, M.N.: Integral representations of hyperbolic volumes. Vestn. Mosk. Univ., Ser. I *31*, *2*, 21–27 (1976) [Russian]. Zbl. 337.32007 English transl.: Mosc. Univ. Math. Bull. *31*, *1/2* 76–81 (1976)
52. Mumford, D.: Hirzebruch's proportionality theorem in the non-compact case. Invent. Math. *42*, 239–272 (1977). Zbl. 365.14012
53. Narasimhan, R.: Automorphisms of bounded domains in $\mathbb{C}^n$. Astérisque *32/33*, 213–224 (1976). Zbl. 336.32020
54. Poletski, E.A.: The Euler-Lagrange equations for extremal holomorphic mappings of unit disk. Mich. Math. J. *30*, 317–333 (1983). Zbl. 577.32022

55. Pinchuk, S.I.: Holomorphic nonequivalence of some classes of domains in $\mathbb{C}^n$. Mat. Sb., Nov. Ser. *111* (153), 67–94 (1980) [Russian]. Zbl. 442.32005 English transl.: Math. USSR, Sb. *39*, 61–86 (1981)
56. Rabotin, V.V.: The finiteness of the number of holomorphic maps of some algebraic varieties. In: Some questions of multivariate complex analysis, pp. 129–138. Krasnoyarsk: Akad. Nauk SSSR, Sibirsk. Otdel., Inst. Fiz. 1980 [Russian]. Zbl. 486.32012
57. Rabotin, V.V.: Some remarks on the Borel-Narasimhan theorem. Sb. Nauchn. Tr., Tashk. Gos. Univ. 576, 77–80 (1979) [Russian]. Not reviewed in Zbl. (MR). R. Zh. Mat. 1980, 6A664
58. Rosay, J.-P.: Sur une caractérization de la boule parmi les domaines de $\mathbb{C}^n$ par son groupe d'automorphismes. Ann. Inst. Fourier 29, 91–97 (1979). Zbl. 402.32001
59. Rosay, J.-P.: Un exemple d'ouvert borné de $\mathbb{C}^3$ "taut" mais non hyperbolic complet. Pac. J. Math. 98, 153–156 (1982). Zbl. 485.32013
60. Royden, H.L.: Remarks on the Kobayashi metric. In: Several complex variables, II, Maryland, 1970. Lect. Notes Math. 185, 125–137. Berlin etc.: Springer 1971. Zbl. 218.32012
61. Sadullaev, A.: Plurisubharmonic measures and capacities on complex manifolds. Usp. Mat. Nauk *36*, No. 4, 53–105 (1981) [Russian]. Zbl. 475.31006. English transl.: Russ. Math. Surv. *36*, No. 4, 61–119 (1981)
62. Shabat, B.V.: Introduction to complex analysis, II. Moscow: Nauka 1985 [Russian]. Zbl. 578.32001
63. Shabat, B.V.: Distribution of values of holomorphic maps. Moscow: Nauka 1982 [Russian]. Zbl. 537.32008. English transl.: Transl. Math. Monogr. 61, Providence (1985)
64. Sibony, N.: Prolongement des fonctions holomorphes bornées et métrique de Carathéodory. Invent. Math. *29*, 205–230 (1975). Zbl. 333.32011
65. Sibony, N.: A class of hyperbolic manifolds. Ann. Math. Stud. *100*, 357–372 (1981). Zbl. 476.32033
66. Simha, R.: The Carathéodory metric of the annulus. Proc. Am. Math. Soc. *50*, 162–164 (1975). Zbl. 281.30010
67. Stanton, C.M.: A characterization of the ball by its intrinsic metrics. Math. Ann. *264*, 271–275 (1983). Zbl. 501.32002
68. Suzuki, M.: The intrinsic metrics of the domains in $\mathbb{C}^n$. Math. Rep., Toyama Univ. *6*, 143–177 (1983). Zbl. 522.32020
69. Urata, T.: Holomorphic mappings onto a certain compact complex analytic space. Tôhoku Math. J., II. Ser. *33*, 573–585 (1981). Zbl. 477.32025
70. Vesentini, E.: Complex geodesics and holomorphic maps. Symp. Math. *26*, 211–230 (1982). Zbl. 506.32008
71. Vigué, J.P.: La distance de Carathéodory n'est pas intérieure. Result. Math. *6*, 100–104 (1983). Zbl. 552.32022
72. Vigué, J.P.: The Carathéodory distance does not define the topology. Proc. Am. Math. Soc. *91*, 223–224 (1984). Zbl. 555.32016
73. Weil, A.: Introduction à l'étude des variétés kähleriennes. Paris: Hermann 1958. Zbl. 137, 411
74. Yau, S.-T.: Intrinsic measures on compact complex manifolds. Math. Ann. *212*, 317–329 (1975). Zbl. 313.32031
75. Zaĭdenberg, M.G.: Picard's theorem and hyperbolicity. Sib. Mat. Zh. *24*, 44–55 (1983) [Russian]. Zbl. 579.32039. English transl.: Sib. Math. J. *24*, 858–867 (1983)

IV. Finiteness Theorems for Holomorphic Maps

M.G. Zaĭdenberg, V.Ya. Lin

Translated from the Russian
by J. Peetre

Contents

Introduction

This Part is devoted to a direction of complex analysis which has its roots in the theorem of Liouville (Liouville (1844) for doubly periodic functions; Cauchy (1844) in the contemporary formulation) and Picard (1879) on the nonexistence of nonconstant holomorphic functions $f\colon \mathbb{C}\to D=\{z\in\mathbb{C}: |z|<1\}$ and $f\colon \mathbb{C}\to\mathbb{C}\setminus\{0, 1\}$. The first results on the finiteness of sets of holomorphic maps were obtained in the second half of the past century within the framework of Riemann surfaces, which subject then began to take shape. Schwarz (1879) and Poincaré (1885) proved the finiteness of the group $\operatorname{Aut} R_g$ of the automorphism of compact Riemann surfaces of genus $g>1$. Hurwitz (1893) completed this result by establishing the explicit bound $\#\operatorname{Aut} R_g \leqq 84(g-1)$; we owe to him several other remarkable results on maps of Riemann surfaces (some of these will be set forth in § 2 and § 3 of Chap. 1). De Franchis (1913) and Severi (1926) proved the finiteness of the set $\operatorname{Hol}_*(R_{g_1}, R_{g_2})$ of nonconstant holomorphic maps of compact Riemann surfaces $R_{g_1}\to R_{g_2}$ in the hypothesis $g_2>1$. Moreover, they established that, for a fixed Riemann surface R_{g_1}, the number of all pairs (f, R_{g_2}), where R_{g_2} is a compact Riemann surface of genus $g_2>1$ and $f\in\operatorname{Hol}_*(R_{g_1}, R_{g_2})$, is finite (and admits estimates depending only on g_1).

In the sequel it became clear that analogous statements likewise hold in a noncompact situation; it is only essential to make sure that one deals with hyperbolic Riemann surfaces of finite type (or, what is the same, smooth algebraic curves with negative Euler characteristic). In contrast to compactness, hyperbolicity is essential. The ("exceptional") nonhyperbolic Riemann surfaces are $\mathbb{CP}^1$, $\mathbb{C}=\mathbb{CP}^1\setminus\{\infty\}$, $\mathbb{C}^*=\mathbb{C}\setminus\{0\}$ and surfaces homeomorphic to a torus. They are all complex-homogeneous: their automorphism groups are transitive complex Lie groups of biholomorphic maps. Therefore each holomorphic transformation $f: X \to R$ into such a surface R "propagates" in a family of maps $a \circ f: X \to R$, $a \in \operatorname{Aut} R$.

It is well-known that the homotopy classification of holomorphic maps of a Stein space X into a complex-homogeneous manifold Y (i.e. into a homogeneous space of a complex Lie group) coincides with the homotopy classification of continuous maps from X into Y. This is a manifestation of the so-called Oka-Cartan-Grauert principle (cf. Vol. 10, Part II). Spaces distinguished by various hyperbolicity conditions enjoy properties very different from those of complex homogeneous spaces. In many cases holomorphic maps into such spaces display a remarkable rigidity. Apparently Carathéodory (1932; cf. Chap. 1, Sec. 4.2) first understood that holomorphic maps even of topologically trivial multidimensional domains enjoy such a rigidity. What exactly this rigidity amounts to, depends on the case at hand. Sometimes it is the discreteness or finiteness of some classes of maps (which comprises expliciting their complete description); sometimes it is the coincidence of holomorphic maps normalized by suitable conditions, or the nonexistence of such maps. The goal of the Part is to set forth results of this type (classical as well as results obtained within the last decade) and to clarify the connections between them.

In Chap. 1 we mainly collect results where the topological properties of the spaces and the maps are essential. Often these are properties of the fundamental groups or the (co)homology groups and their homomorphisms induced by holomorphic maps. A typical example is Hurwitz's theorem on the coincidence of automorphisms of surfaces R_g $(g > 1)$ inducing one and the same automorphism of the one dimensional homology group $H_1(R_g, \mathbb{Z})$ (we reproduce a simple proof of this theorem based on Lefschetz's formula for the number of fixed points). Another example is a beautiful theorem, most useful in the applications, by Borel and Narasimhan (Sec. 3.1). In the same vein, we have the results of Secs. 1.1, 1.2 on Liouville and Picard spaces, and likewise the multidimensional generalizations of Hurwitz's inequality on the degree of maps of compact Riemann surfaces (§ 2). Sec. 4.3 (and in part 1.2, Example 2) is devoted to holomorphic maps of some concrete spaces connected with polynomials without multiple roots (here the algebraic properties of the Artin braid groups and their homomorphisms are of importance); this circle of ideas arose in connection with algebraic equations over function rings and problems on superposition of algebraic functions of several variables. Finally, Sections 3.3 and 4.2 deal with results of a somewhat different nature (the coincidence theorem of Pólya-Nevanlinna-Cartan and its generalizations, and likewise some rigidity properties of bounded multidimensional domains). The manifolds considered there are topologically trivial and in their study purely analytic aspects prevail.

Almost all results of Chap. 2 may be considered as multidimensional analogues of the theorems of Hurwitz and de Franchis-Severi. Here such properties of the complex structure come to the foreground as hyperbolicity, generality of type, negativity in the sense of Grauert etc. To a large extent algebraic-geometric methods are used. On the other hand, the influence of ideas from complex analysis is felt, especially the branch which gradually has become known as "hyperbolic analysis". In many situations a key rôle is played by the fact that the holomorphic maps in a compact complex space themselves form a complex space. This idea, actually going back to Riemann, has long been prominent in algebraic geometry. A clear instance of it is the Bochner-Montgomery theorem, which states that the automorphism group of a compact complex manifold is a complex Lie group (in general, with an infinite number of connected components). Finally, the idea was made full use of in complex analysis with the well-known thesis of Douady [75]. The results highlightened in Chap. 2 were to a large extent obtained in the last 10–15 years. Although we are here still far from the simplicity and completeness, as compared to the classical model, one can nevertheless speak of an interesting branch of the theory of complex spaces, which currently is in a stage of very active development.

A few words on terminology and notation are in order. By a *Riemann surface of finite type* we intend a compact Riemann surface with a finite number of "punctures" (synonym: irreducible smooth algebraic curve). All complex spaces are assumed to be reduced (and, as a rule, irreducible). If not stated otherwise, all algebraic varieties are taken over the field of complex numbers. We assume that the reader knows the basic notions of hyperbolic analysis (hyperbolicity, measure hyperbolicity, hyperbolic imbedding etc.; cf. Part III, [25], [26]). We denote by $\mathrm{Hol}(X, Y)$ (respectively $\mathrm{Mer}(X, Y)$ the set of all holomorphic (meromorphic) maps from X into Y. The set $\mathrm{Hol}(X, Y)$ (and all its subsets considered in what follows too) are equipped with the compact-open topology. We denote by Hol_* (respectively $\mathrm{Hol}_{\mathrm{dom}}$) the subset of Hol consisting of all nonconstant (dominant) maps; Mer_* and $\mathrm{Mer}_{\mathrm{dom}}$ have an analogous meaning. We denote by $\mathrm{Aut}\, X$ the group of automorphisms (i.e., biholomorphic transformations) of the space X. The notation D will be attached to the unit disk $\{z \in \mathbb{C}: |z| < 1\}$ and by $\mathbb{C}^*$ one intends, as usual, $\mathbb{C} \setminus \{0\}$ (if necessary $\mathbb{C}^*$ will be equipped with the usual structure of complex Lie group). The term *"regular map"* is always taken in its algebraic geometry meaning (for algebraic varieties over $\mathbb{C}$, rational maps without points of indeterminicy).

We would like to thank D.N. Akhieser, Sh.I. Kaliman and L.I. Potepun for several helpful discussions and precious advice. We also are obliged to the editor of the series V.V. Nikunin for a series of important remarks, which were taken account of in the final version of the text.

Chapter 1
Theorems on Nonexistence of Holomorphic Maps and Coincidence Theorems

§ 1. Liouville and Picard Spaces

1.1. Liouville Spaces. A complex space X is said to be a *Liouville space* (respectively a *Picard space*) if $\mathrm{Hol}_*(X, D)=\emptyset$ (respectively if $\mathrm{Hol}_*(X, \mathbb{C}\setminus\{0; 1\})=\emptyset$).

Classical examples of Liouville spaces are connected compact complex spaces and their connected Zariski open subsets (in particular, connected quasiprojective algebraic varieties). A finite sheeted connected unramified holomorphic covering X of a Liouville space Y is Liouville (every symmetric polynomial in the values of a holomorphic function $f: X\to D$ at all points x, lying over a point $y\in Y$, defines a bounded holomorphic function on Y and therefore must be a constant).

For infinitely sheeted coverings this is in general not true (the simplest example is the universal covering $D\to\mathbb{C}\setminus\{0; 1\}$). However, in appropriate restrictions on the base Y the Liouville property is preserved also for some classes of infinitely sheeted covers.

A connected unramified *cover* $p: X\to Y$ is termed *Abelian* if it is *regular* (i.e. $p_*(\pi_1(X))$ is a normal subgroup of the group $\pi_1(Y)$) and its group of covering transformations $G=\pi_1(Y)/p_*(\pi_1(X))$ is Abelian; in an analogous way, one defines other similar classes of coverings: nilpotent coverings, solvable ones etc. G.B. Shabat [51] proved the following theorem.

Theorem 1. *A holomorphic Abelian cover of an arbitrary Riemann surface R of finite type is a Liouville space.*

For compact R such a theorem was previously obtained by Mori [103] and in the special case $R=\mathbb{C}\setminus\{0; 1\}$ also by Wakabayashi [57] and (independently) by Demailly [73].

The methods employed in these papers, apparently, do not extend to the multidimensional case. However, there are other routes. Thus, it follows from recent results by Lyons and Sullivan [95], devoted to the analogous problem for harmonic functions on Riemannian manifolds, that nilpotent coverings of compact complex Kähler manifolds are Liouville [1]. The methods of this paper depend on the known connections between the theory of harmonic

---

[1] We are grateful to A.M. Vershik for pointing out this result to us and for informing us that related results have been obtained by his student V.A. Kaĭmanovich. Using an entropy approach due to A.M. Vershik, V.A. Kaĭmanovich [85] proved that on a polycyclic covering of a compact Riemannian manifold there exist no bounded harmonic functions. As on a Kähler manifold real parts of holomorphic functions are harmonic, it follows from this that polycyclic coverings of compact complex Kähler manifolds are Liouville.

functions and the theory of stochastic processes. We describe here a different approach based on an immediate use of an appropriate invariant mean on the group of covering transformations (assumed to be amenable) and which in the complex analytic case leads to stronger results. Let us point out that the method suggested here is also applicable to harmonic functions on Riemannian manifolds (or discrete groups with a nondegenerate probability measure).

Let us recall some definitions. A discrete *group* G is called *amenable* if on the Banach space $L_\infty(G)$ of all bounded functions $f\colon G\to C$ with the norm $\|f\|_\infty=\sup_{g\in G}|f(g)|$ there exists a (right) invariant mean, that is, a continuous linear functional m subject to the conditions:

a) $m(\bar{f})=\overline{m(f)}$ for all $f\in L_\infty(G)$ and $\inf_{g\in G} f(g)\leqq m(f)\leqq \sup_{g\in G} f(g)$ for all real $f\in L_\infty(G)$;

b) $m(f)$ is right invariant, that is, $m(f_h)=m(f)$ for all $f\in L_\infty(G)$ and all $h\in G$ (here $f_h(g)=f(gh)$ is the right shift of f).

Solvable (and thus, in particular, nilpotent) groups are amenable (cf. [78]). Also every *ω-nilpotent group* is amenable (i.e., a group which coincides with the union of all members of its upper central series; for finitely generated groups this condition is equivalent to being nilpotent). A *group G* is *almost ω-nilpotent* if it contains an ω-nilpotent subgroup of finite index; it is clear that these groups are amenable too.

We say that a discrete group G acts holomorphically on a complex space X from the left if there is given an antirepresentation $G\to\operatorname{Aut}X$, $G\ni g\mapsto(x\to gx)\in\operatorname{Aut}X$ of G in $\operatorname{Aut}X$. Such an *action* is said to be *properly discontinuous* if for each compact set $K\subset X$ the set $G_K=\{g\in G\colon gK\cap K\neq\emptyset\}$ is finite; by a theorem of H. Cartan's (cf. [71]) the quotient space $Y=X/G$ is then a complex space and the natural projection $X\to Y$ is holomorphic. (If $p\colon X\to Y$ is unramified regular holomorphic covering then the group of covering transformations $G=\pi_1(Y)/p_*(\pi_1(X))$ acts holomorphically and freely (all the more, properly discontinuously) on X, and $X/G=Y$.)

A complex space Y is termed an *ultra-Liouville space* if there exist no nonconstant bounded continuous plurisubharmonic functions [2] on Y. If $Z\to Y$ is a surjective holomorphic map of a Zariski open subset Z of a compact complex space onto a connected complex space Y, then Y is ultra-Liouville (cf. [8]). In particular, this comprises all connected compact complex spaces and their connected Zariski open subsets (and, therefore, also all connected quasiprojective varieties); likewise all pseudoconcave spaces are ultra-Liouville [8]. (As a matter of fact, all spaces of the above type enjoy a stronger property: they carry no nonconstant plurisubharmonic functions that are bounded from above, cf. [8].)

---

[2] A function $f\colon Y\to[-\infty,\infty)$ is called plurisubharmonic if it is continuous from above and if for each holomorphic map $\varphi\colon D\to Y$ the function $f\circ\varphi$ is subharmonic on D. We shall shortly give an example showing that ultra-Liouvilleness is a much stronger condition than just being Liouville.

A complex *space* X is termed *hyperbolic in the sense of Carathéodory* if for any two distinct points x_1, $x_2 \in X$ one can find a holomorphic function $f\colon X \to D$ such that $f(x_1) \neq f(x_2)$ (typical example: an analytic subset of a bounded domain in $\mathbb{C}^n$).

Let the discrete group G acts holomorphically on a connected complex space X. The following result is due to V.Ya. Lin.

Theorem 2. *Assume that either one of the following two conditions is fulfilled:* a) *there exists a nonconstant bounded holomorphic function on X and G is almost ω-nilpotent;* b) *X is hyperbolic in the sense of Carathéodory and G is amenable, acts effectively on X and contains a subgroup H of finite index with a nontrivial center $C(H) \neq \{e\}$. Then there exists a nonconstant bounded continuous G-invariant plurisubharmonic function on X.*

If the quotient space $Y = X/G$ is separable and admits a complex structure (compatible with the quotient topology) such that the natural projection $p\colon X \to Y$ is holomorphic then each continuous G-invariant plurisubharmonic function f on X descends to a continuous plurisubharmonic function $\tilde{f}(y) = f(p^{-1}(y))$ on Y [8], Prop. 1.3. Therefore we obtain from Theorem 2 the following result.

Theorem 3. *Let the action of G on X be properly discontinuous.* a) *If G is almost ω-nilpotent and the quotient space $Y = X/G$ is ultra-Liouville then X is Liouville. In particular, every connected unramified holomorphic almost ω-nilpotent cover of an ultra-Liouville space must be Liouville.* b) *If condition* b) *of Theorem 2 is fulfilled then $Y = X/G$ can not be ultra-Liouville. In particular, the quotient space U/G of a bounded domain $U \subset \mathbb{C}^n$ with respect to an effective holomorphic properly discontinuous action of an amenable group G with nontrivial center can not be a Zariski open subset of a compact complex space* (and, therefore, can not be an algebraic variety).

Theorem 2, in turn, is a simple corollary of Lemma 1 below, which lemma in our opinion has an independent interest.

Lemma 1. *Let the discrete amenable group G act holomorphically on a connected complex space X and assume that there are no nonconstant bounded continuous G-invariant plurisubharmonic functions on X. If a bounded holomorphic function f on X satisfies for some $s \in G$ the condition $f(g s x) = f(s g x)$ for all $g \in G$ and all $x \in X$ then $f(s x) = f(x)$ for all $x \in X$.*

Outline of proof. Let βG be the space of maximal ideals of the symmetric commutative Banach algebra $L_\infty(G) = C(\beta G)$. The group G is imbedded in βG as an open everywhere dense discrete subset; in addition, G acts continuously on βG from the right (and from the left). For each bounded holomorphic function f on X and arbitrary $g \in G$, $x \in X$ set $f_x(g) = f(g x) = f^g(x)$. Then $\{f^g\colon g \in G\}$ is a uniformly bounded and equicontinuous family of holomorphic functions on X (cf. [8]). It readily follows from this that for each continuous linear functional $m \in L_\infty(G)^*$ the function $m(f_x)$ is holomorphic on X. Let us denote by $\hat{f}_x \in C(\beta G)$ the Gel'fand transform of f_x (i.e. the continuous extension of f_x from G to βG).

a) Set $\|f\|_X = \sup_{x\in X}|f(x)|$ and $\varphi_f(x) = \sup_{g\in G}|f(gx)| = \sup_{\xi\in\beta G}|\hat{f}_x(\xi)| = \|\hat{f}_x\|$. Then $\varphi_f(x)$ is a continuous bounded G-invariant plurisubharmonic function on X; this entails that $\varphi_f = \text{const}$. It is easy to see that in fact $\varphi_f(x) \equiv \|f\|_X$. The function $\hat{f}_x$ is continuous on βG; therefore its peak set $M(\hat{f}_x) = \{\xi\in\beta G : |\hat{f}_x(\xi)| = \|f\|_X\}$ is a nonempty compact subset of βG. It turns out that the set $M(f) = M(\hat{f}_x)$ does not depend on $x\in X$ and is invariant with respect to the right action of G on βG. (In fact, if $x_0\in X$ and $\xi_0\in M(\hat{f}_{x_0})$ then the function $f^{\xi_0}(x) = \hat{f}_x(\xi_0)$ is holomorphic on X and its modulus takes on its maximum $\|f\|_X$ at the point $x = x_0$; therefore $\hat{f}_x(\xi_0) \equiv \hat{f}_{x_0}(\xi_0)$ so that $\xi_0\in M(\hat{f}_x)$ which again entails the assertion.) Let us apply this to the function $f^0(x) = f(x) - f(sx)$. It follows from the condition in the Lemma, viz. $f(gsx) \equiv f(sgx)$, that $\hat{f}^0_x = \hat{f}_x - \hat{f}_{sx}$. As the compact set $M(f^0)$ is right invariant and G amenable, there exists on βG a right invariant regular Borel probability measure μ with support $\operatorname{supp}\mu$ contained in $M(f^0)$ (G acts affinely from the right on the compact convex set of all probability measures supported by $M(f^0)$ and so by the amenability of G it follows that there must be a fixed point for this action; cf. the book by Greenleaf quoted above, [78], Theorem 3.3.1). By the definition of $M(f^0)$, the identity $|\hat{f}^0_x(\xi)| = \|f^0\|_X$ then holds for all $x\in X$ and all $\xi\in\operatorname{supp}\mu$.

b) Let us consider the space $L_2(\mu)$ where μ is the above probability measure. To each function $\psi\in C(\beta G)$ there corresponds an element $[\psi]\in L_2(\mu)$. It turns out that the element $[\hat{f}_x]\in L_2(\mu)$ does not depend on the point $x\in X$. For the proof we remark that the function

$$\Phi_f^2(x) = \|[\hat{f}_x]\|^2_{L_2(\mu)} = \int_{\beta G} |\hat{f}_x(\xi)|^2\, d\mu(\xi)$$

is bounded, continuous, G-invariant and plurisubharmonic on X; therefore $\Phi_f^2 = \text{const}$. This means that the image of X in $L_2(\mu)$ under the (holomorphic!) map $X \ni x \mapsto [\hat{f}_x] \in L_2(\mu)$ is entirely contained in the boundary of a ball of radius Φ_f; in view of the strict convexity of balls in Hilbert space it follows from this that indeed $[\hat{f}_x]$ does not depend on x. In particular $[\hat{f}_x - \hat{f}_{sx}] = 0$, so that $\hat{f}^0_x(\xi) = \hat{f}_x(\xi) - \hat{f}_{sx}(\xi) = 0$ for all $\xi\in\operatorname{supp}\mu$. If we combine this with the identity at the end of (a), we conclude that $\|f^0\|_X = 0$, i.e. $f(x) - f(sx) \equiv 0$. The proof is complete.

Let us turn to Theorem 2. Assume that there are no nonconstant bounded G-invariant plurisubharmonic functions on X. For each $s\in G$ denote by $[G, s]$ the subgroup of G generated by all commutators $[g, s] = gsg^{-1}s^{-1}$, $g\in G$. The condition $f(gsx) \equiv f(sgx)$ in Lemma 1 is equivalent to f being $[G, s]$-invariant and our statement is that (in the hypothesis that the other conditions are fulfilled) such a function must be s-invariant. Let $\{G^{(n)}\}_{n=0}^{\infty}$ be the upper central series of G, i.e. $G^{(0)} = \{e\}$ and $G^{(n+1)} = \{s\in G : [G, s] \subset G^{(n)}\}$ for $n \geq 0$. It follows from Lemma 1 that every $G^{(n)}$-invariant bounded holomorphic function on X must be $G^{(n+1)}$-invariant; this gives Theorem 2(a) for ω-nilpotent groups G (the general case of almost ω-nilpotent groups is easily brought back to this). For the proof of Theorem 2(b) it is sufficient to consider the case when G is an amenable group with nontrivial center. Let $s \neq e$ be a central element in G. Lemma 1 shows that $f(sx) \equiv f(x)$ for any bounded holomorphic function f on X. If X is hyperbolic in Carathéodory's sense, it follows from this that $sx = x$ for all $x \in X$ and this contradicts the assumption of the action being effective.

Let us remark that Theorem 2(a) and (b) are not true if we in them replace the condition of ω-nilpotence of the group G by the assumption of solvability (and even the stronger assumption of polycyclicity, i.e. solvability together with the assumption that all subgroups $H \subset G$ are finitely generated). The corresponding (counter)example, as observed by D.N. Akhieser, who communicated this to us, is provided by the surfaces of Inoue[3] [83]. Let A be an integervalued unimodular 3×3 matrix with one real eigenvalue $\alpha > 1$ and two complex conjugate eigenvalues β and $\bar{\beta}$ ($|\beta| < 1$). Let $a = (a_1, a_2, a_3)$ and $b = (b_1, b_2, b_3)$ be the real and complex eigenvectors of A corresponding to the eigenvalues α and β respectively. Set $X = \mathbb{P} \times \mathbb{C}$ ($\mathbb{P} = \{w \in \mathbb{C}: \operatorname{Im} w > 0\}$ being the upper halfplane) and consider the discrete group G of automorphisms of X generated by the following four transformations: $g_0(w, z) = (\alpha w, \beta z)$, $g_j(w, z) = (w + a_j, z + b_j)$, $1 \leqq j \leqq 3$, $(w, z) \in \mathbb{P} \times \mathbb{C} = X$. Consider the subgroup G_0 generated by the elements g_j ($1 \leqq j \leqq 3$). The latter is commutative (isomorphic to $\mathbb{Z}^3$) and, as is readily seen, a normal subgroup of G with the quotient group $G/G_0 \cong \mathbb{Z}$. Therefore G is solvable (and even polycyclic). It acts freely on X and the quotient space $Y = X/G$ is a smooth compact surface (which does not contain any analytic subsets of codimension 1 and, therefore, is nonalgebraic). This means that Y is ultra-Liouville, whereas X carries nonconstant bounded holomorphic functions (for example, the function $f(w, z) = w(w + \sqrt{-1})^{-1}$). Let us also consider the space $X_1 = X/G_0$. The group $G_1 = G/G_0 \cong \mathbb{Z}$ acts freely and holomorphically on X_1, and $X_1/G_1 \cong Y$; therefore Theorem 3(a) tells us that X_1 is Liouville. However, it is not ultra-Liouville, as otherwise (since G_0 is commutative) Theorem 3(a) would imply that X is Liouville. The same triple X_1, G_1, $Y = X_1/G_1$ shows that the property of being ultra-Liouville, in general, is not preserved if we pass from a space to an Abelian covering.

Example 1. Let $\{\Pi_n\}_{n=0}^{\infty}$ be the lower central series of the fundamental group $\Pi = \pi_1(Y)$ of an ultra-Liouville space Y (i.e. $\Pi_0 = \Pi$ and $\Pi_n = [\Pi, \Pi_{n-1}]$ for $n \geqq 1$) and let Y_n be the covering space of Y corresponding to the normal subgroup Π_n of Π. As, for every n, the quotient group $G_n = \Pi/\Pi_n$ is nilpotent

[3] From the results of V.M. Kaĭmanovich mentioned in footnote [1] it follows that the compact complex-analytic surfaces Y of Inoue described below are not only non-Kähler (this may have been known previously) but also in general can not be equipped with a Riemannian metric such that the real parts of the germs of all holomorphic functions become harmonic. In the Lyons-Sullivan paper it is observed that there exist "two step-solvable" non-Liouville covers of compact Riemann surfaces of genus $g > 1$.

and $Y_n/G_n \cong Y$, it follows from Theorem 3(a) that all the spaces Y_n are Liouville. In particular, the "maximal Abelian covering" $Y^{(ab)} = Y_1$ of any ultra-Liouville space Y is Liouville.

Here is a consequence of Theorem 3(a). Let $n \geqq 2$, H a hyperplane in $\mathbb{CP}^n$ and Z a hypersurface there such that outside some algebraic set $S \subset \mathbb{CP}^n \setminus H$ of codimension 3 all singularities of the hypersurface $V = H \cup Z$ are normal crossings [4].

Corollary 1. *The universal covering X of the domain $Y = \mathbb{CP}^n \setminus V \subset \mathbb{C}^n = \mathbb{CP}^n$ is Liouville.*

Indeed, by a theorem by Pham [111] $\pi_1(Y)$ must be Abelian so that $X \to Y$ is an Abelian covering.

Again, it follows from this corollary that the universal covering of the complement in $\mathbb{C}^n$ ($n \geqq 2$) of the union of any finite family of hypersurfaces in general position is Liouville (cf. [57]).

1.2. Picard Spaces. A Picard space must be Liouville. The converse, of course, is not true ($\mathbb{CP}^1 \setminus 3$ points).

The well-known proof of the little Picard theorem provides a simple tool with the aid of which for some Liouville spaces one can shown to be Picard. Here is the proof. If f: $\mathbb{C} \to \mathbb{C} \setminus \{0; 1\}$ is a holomorphic map then by the theorem on the covering of a map there exists a commutative triangle of holomorphic maps

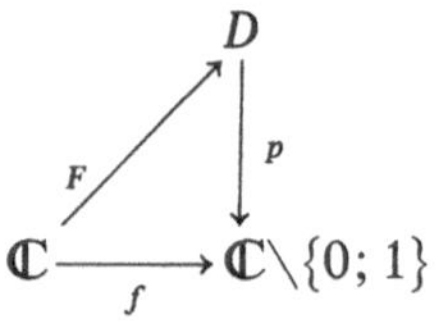

where p: $D \to \mathbb{C} \setminus \{0; 1\}$ is the universal covering; by Liouville's theorem F $=$const and so $f =$const. In exactly the same manner one shows that a simply connected Liouville space X is Picard. In fact, the assumption of X being simply connected can be relaxed considerably. Let us remark that each element of the fundamental group $\pi_1(X, x_0)$ of X can be interpreted as a homotopy class of continuous maps of the pair $(\mathbb{C}^*, 1)$ into the pair (X, x_0). Let us call an *element* $\alpha \in \pi_1(X, x_0)$ *holomorphic* if there exists a holomorphic map f: $\mathbb{C}^* \to X$ which is freely homotopic to some (and then any) map a: $(\mathbb{C}^*, 1) \to (X, x_0)$ of the class α. Denote by $\operatorname{hol} \pi_1(X, x_0)$ the set of all holomorphic elements in $\pi_1(X, x_0)$ and by $\operatorname{Hol} \pi_1(X, x_0)$ the subgroup of $\pi_1(X, x_0)$ generated by this set; it is easy to see that $\operatorname{Hol} \pi_1(X, x_0)$ is a normal subgroup of $\pi_1(X, x_0)$. If f: $(X, x_0) \to (Y, y_0)$ is a holomorphic map and f_*: $\pi_1(X, x_0)$

[4] One says that all singularities of a hypersurface V in a complex manifold M have *normal crossings* (or normal selfintersections) if for each point $x \in V$ there exists a chart $(U; z_1, \ldots, z_n) \ni x$ such that $V \cap U = \{u \in U : z_1(u) z_2(u) \ldots z_k(u) = 0\}$ for some $k \in \{1, 2, \ldots, n\}$.

$\to \pi_1(Y, y_0)$ is the homomorphism of fundamental groups induced by it, then $f_*(\mathrm{hol}\,\pi_1(X, x_0)) \subset \mathrm{hol}\,\pi_1(Y, y_0)$ and $f_*(\mathrm{Hol}\,\pi_1(X, x_0)) \subset \mathrm{Hol}\,\pi_1(Y, y_0)$.

A group *homomorphism* $\varphi: G \to H$ is termed *Abelian* if the image $\varphi(G)$ is an Abelian subgroup of H.

Theorem 4 (V.Ya. Lin [32], [35]). *If X is a Liouville space and the quotient group $G = \pi_1(X, x_0)/\mathrm{Hol}\,\pi_1(X, x_0)$ has no non-Abelian homomorphisms into the free group $\mathbb{F}_2$ of rank 2, then X is a Picard space.*

Proof. Let $f: X \to \mathbb{C}\setminus\{0; 1\}$ be a holomorphic map, $f(x_0) = z_0 \in \mathbb{C}\setminus\{0; 1\}$ and $f_*: \pi_1(X, x_0) \to \pi_1(\mathbb{C}\setminus\{0; 1\}, z_0)$ the induced homomorphism of the fundamental groups. Every holomorphic map from $\mathbb{C}^*$ into $\mathbb{C}\setminus\{0; 1\}$ must be constant; therefore $\mathrm{Hol}\,\pi_1(\mathbb{C}\setminus\{0; 1\}, z_0)) = \{1\}$ and so $f_*(\mathrm{Hol}\,\pi_1(X, x_0)) = \{1\}$. This means that there exists a commutative diagram of group homomorphisms

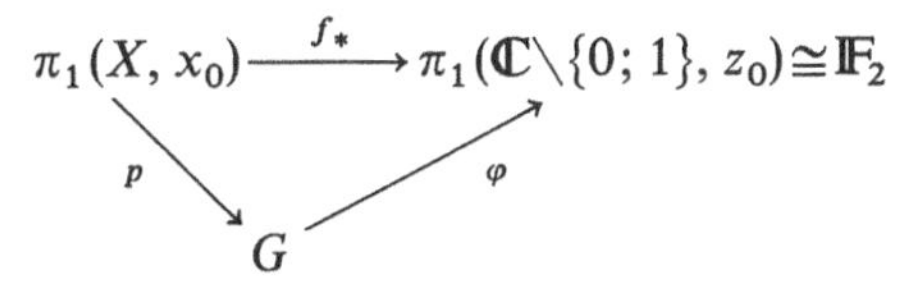

where p is the natural projection. By the assumption of the theorem, φ is an Abelian homomorphism, so that $H = f_*(\pi_1(X, x_0)) = \varphi(G)$ is an Abelian subgroup of $\pi_1(\mathbb{C}\setminus\{0; 1\}, z_0) \cong \mathbb{F}_2$. Therefore H is either isomorphic to $\mathbb{Z}$ or else trivial; in both cases by Riemann's theorem the covering Y_H of the manifold $\mathbb{C}\setminus\{0; 1\}$, corresponding to the subgroup H, must be isomorphic to a bounded domain. As X is Liouville, it follows from the commutative diagram of holomorphic maps

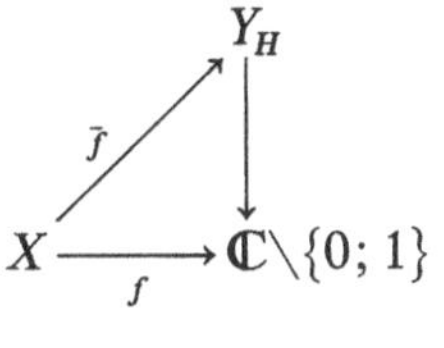

(which exists in view of the covering map theorem) that $f = \mathrm{const}$.

In exactly the same way we prove that if R is a Riemann surface with a noncommutative free fundamental group and X is a space satisfying the hypothesis of Theorem 4, then $\mathrm{Hol}_*(X, R) = \emptyset$.

The condition imposed in Theorem 4 on the fundamental group $\pi_1(X, x_0)$, wittingly, is fulfilled in each of the following cases:

1) X is a homogeneous space of a connected complex Lie group (in this case it follows from a theorem by Grauert-Ramspott [114] that $\mathrm{hol}\,\pi_1(X, x_0) = \pi_1(X, x_0)$ so that $G = \{1\}$.

2) The group $\pi_1(X, x_0)$ (or just the quotient group by the normal subgroup generated by all elements of finite order) is solvable.

3) The commutant $\pi_1'(X, x_0)$ of $\pi_1(X, x_0)$ is a *Cainian group*[4'] (i.e. it does not have nontrivial homomorphisms into Abelian groups).

4) $\operatorname{rank} H^1(X, \mathbb{Z}) \leqq 1$.

Example 2. Let $n \geqq 3$ and let E_n be the complement in $\mathbb{C}^n$, with coordinates $\lambda_1, \ldots, \lambda_n$, of the union of the hyperplanes $H_{ij} = \{\lambda \in \mathbb{C}^n \colon \lambda_i - \lambda_j = 0\}$ $(1 \leqq i < j \leqq n)$ and consider further the complement G_n in $\mathbb{C}^n$ (with coordinates $z_1, \ldots, z_n$) of the hypersurface $\{z \in \mathbb{C}^n \colon d_n(z) = 0\}$, where $d_n(z)$ is the discriminant (in t) of the polynomial $t^n + z_1 t^{n-1} + \ldots + z_n$. The Viète map $p \colon \mathbb{C}^n \to \mathbb{C}^n$,

$$z_k = p_k(\lambda) = (-1)^k \sum_{1 \leqq i_1 < \ldots < i_k \leqq n} \lambda_{i_1} \ldots \lambda_{i_k} \quad (1 \leqq k \leqq n),$$

induces a regular $n!$-sheeted unramified covering $p \colon E_n \to G_n$. The points z of the domain G_n identify in a natural way with *polynomials* $t^n + z_1 t^{n-1} + \ldots + z_n$ *without multiple roots*. It is easy to see that $G_n \cong \mathbb{C} \times G_n^0$, where $G_n^0 = G_n \cap \{z_1 = 0\}$. The restriction d_n^0 of the discriminant d_n to $G_n^0 \subset G_n$ defines a holomorphic locally trivial bundle $d_n^0 \colon G_n^0 \to \mathbb{C}^*$ whose standard fiber may be taken to be the smooth irreducible $(n-2)$-dimensional affine algebraic manifold $SG_n = \{z = (z_2, \ldots, z_n) \in \mathbb{C}^{n-1} \colon d_n^0(z) = 1\}$. The manifolds E_n, G_n, G_n^0 and SG_n have remarkable topological and analytic properties and play an important rôle in the study of algebraic functions of several complex variables. They are all Eilenberg-MacLane spaces of type $K(\pi, 1)$. Also $\pi_1(G_n) \cong \pi_1(G_n^0) \cong B(n)$ is the Artin braid group with n strings and $\pi_1(E_n) \cong I(n)$ is the pure braid group with n strings (i.e. the kernel of the standard epimorphism of the braid group $B(n)$ into the symmetric group $S(n)$); $\pi_1(SG_n) \cong B'(n)$ is the commutant of the braid group $B(n)$. Each homomorphism $B(n) \to \mathbb{F}_2$ is Abelian; therefore (Theorem 4) G_n and G_n^0 are Picard spaces. For $n > 4$ the group $B'(n)$ is Cainian (cf. [12], [35]); therefore SG_n is a Picard space if $n > 4$ (SG_3 and SG_4 are also Picard spaces, but this requires a separate proof, because there exist epimorphisms of $B'(3)$ and $B'(4)$ into $\mathbb{F}_2$). By Theorem 3 any Abelian covering X of G_n (G_n^0) is Liouville; as $B'(n) \subset \pi_1(X) \subset B(n)$ and $\operatorname{Hom}(B'(n), \mathbb{Z}) = 0$ for $n > 4$, then for $n > 4$ each homomorphism $\pi_1(X) \to \mathbb{F}_2$ is Abelian. Therefore it follows from Theorem 4 that for $n > 4$ each Abelian covering X of G_n (or G_n^0) is Picard[5]. The space E_n is Liouville but not Picard (cf. Sec. 4.3). For $n \geqq 3$ the manifold SG_n is C-hyperbolic (cf. Sec. 3.1). Moreover, Sh.I. Kaliman [86], [87] proved that its universal covering $SG_n^{\sim}$ can in a natural way be identified with the Teichmüller space $T(0, n+1)$ of the Riemann sphere with $n+1$ punctures and, therefore, it is homeomorphic to a ball and can be realized as a bounded and hyperbolically complete (in the sense of Carathéodory) Bergman domain in $\mathbb{C}^{n-2}$.

[4'] *Translator's note.* It took him quite a time to understand this pun(?).

[5] This is also true for $n = 3$ or 4. One can in fact show that for each $n \geqq 3$ the maximal Abelian covering $G_n^{0(ab)}$ of G_n^0 (i.e. the covering corresponding to the commutant $\pi_1'(G_n^0)$ of the group $\pi_1(G_n^0)$) is isomorphic to $\mathbb{C} \times SG_n$.

Let us further give a description of the holomorphic maps $\mathbb{C}^* \to G_n$. It will be convenient to identify the point $z=(z_1, \ldots, z_n) \in G_n$ with the polynomial $t^n + z_1 t^{n-1} + \ldots + z_n$ without multiple roots. Define on G_n holomorphic actions L and U of the groups $\mathbb{C}$ and $\mathbb{C}^*$ respectively, setting $L(v) z = (t+v)^n + z_1 (t+v)^{n-1} + \ldots + z_n$ ("shift") and $U(w) z = t^n + w z_1 t^{n-1} + \ldots + w^n z_n$ ("dilation"); here $v \in \mathbb{C}$, $w \in \mathbb{C}^*$, $z=(z_1, \ldots, z_n) \in G_n$. It turns out that each holomorphic map $f\colon \mathbb{C}^* \to G_n$ can be written in the form $f(\zeta) = L(v(\zeta))\, U(e^{\varphi(\zeta)} \zeta^{q/n(n-1)})\, c$, where $v, \varphi\colon \mathbb{C}^* \to \mathbb{C}$ are holomorphic functions, $q \in Z$, $c=(0, c_2, \ldots, c_n) \in SG_n$ where $c_m = 0$ for all m such that $mq \not\equiv 0 (mod\, n(n-1))$. It follows from this representation that each $f \in \mathrm{Hol}(\mathbb{C}^*, G_n)$ is homotopic (in the class of holomorphic maps) to either one of the maps $f_k\colon \mathbb{C}^* \to G_n$, $f_k = t^n + \zeta^k$ or else one of the maps $g_l\colon \mathbb{C}^* \to G_n$, $g_l(\zeta) = t^n + \zeta^l t$ $(k, l \in \mathbb{Z})$, with $f_k \sim g_l$ iff $k(n-1) = ln$. From this it is easy to get a description of the set $\mathrm{hol}\, \pi_1(G_n) \subset \pi_1(G_n) \cong B(n)$ of all holomorphic elements of $\pi_1(G_n)$; this set consists of all elements of the form $g a^k g^{-1}$, $h b^l h^{-1}$ where $k, l \in \mathbb{Z}$, $g, h \in B(n)$, $a = \sigma_1 \ldots \sigma_{n-1}$, $b = a \sigma_1$ ($\sigma_1, \ldots, \sigma_{n-1}$ is the standard system of generators of the braid group $B(n)$). It is well-known that the elements a and b generate $B(n)$; therefore $\mathrm{Hol}\, \pi_1(G_n) = \pi_1(G_n)$ and $\pi_1(G_n)/\mathrm{Hol}\, \pi_1(G_n) = 0$, which (in view of Theorem 4) provides yet another proof of the fact that G_n is Picard. One has also an "invariant" description of the set $\mathrm{hol}\, \pi_1(G_n)$ (V.Ya. Lin, Murasugi, cf. [35], § 7, Sec. 2 and [104]): it consists of all elements $g \in \pi_1(G_n)$, the images of which in the quotient group of $\pi_1(G_n)$ by its center are of finite order.

§ 2. Hurwitz's Inequality and its Generalizations

2.1. The Classical Hurwitz's Inequality. Let R_1, R_2 be Riemann surfaces of finite type and assume further that the Euler characteristic $\chi(R_2)$ is <0 (i.e. that R_2 is hyperbolic). Let $f\colon R_1 \to R_2$ be a nonconstant holomorphic map. The classical Hurwitz theorem states that the degree $\deg f$ of the map f (the number of preimages of a generic point $\zeta \in R_2$) satisfies the inequality

$$\deg f \leqq \chi(R_1)/\chi(R_2). \tag{1}$$

Indeed, let $k = \deg f$; as R_2 is hyperbolic, f must be regular and finitely sheeted and there exist finite subsets $N \subset R_2$, $\# N = n$ and $M = f^{-1}(N) \subset R_1$, $\# M = m$, such that the map

$$f|(R_1 \setminus M)\colon R_1 \setminus M \to R_2 \setminus N$$

is k-fold unramified covering. Therefore

$$\chi(R_1 \setminus M) = k\, \chi(R_2 \setminus N). \tag{2}$$

As $\chi(R_1 \setminus M) = \chi(R_1) - m$, $\chi(R_2 \setminus N) = \chi(R_2) - n$ and $\# f^{-1}(\zeta) \leqq k$ for all $\zeta \in R_2$, it follows that $m \leqq kn$ and

$$\chi(R_1 \setminus M) \geqq \chi(R_1) - kn. \tag{3}$$

From (2) and (3) we get $k[\chi(R_2\setminus N)+n]\geqq\chi(R_1)$, that is, $k\chi(R_2)\geqq\chi(R_1)$, whence, in view of the assumption $\chi(R_2)<0$, (1) follows.

For compact surfaces of genera g_1, g_2 ($g_2>1$) the inequality (1) takes the form $\deg f\leqq(g_1-1)/(g_2-1)$ and for open surfaces of finite types (g_1, n_1) and (g_2, n_2) in the hypothesis $\operatorname{rank} H^1(R_2, \mathbb{Z})=2g_2+n_2-1>1$ (which is equivalent to the hyperbolicity of R_2) it takes the form

$$\deg f\leqq\frac{\operatorname{rank} H^1(R_1, \mathbb{Z})-1}{\operatorname{rank} H^1(R_2, \mathbb{Z})-1}. \tag{4}$$

It follows from (4) that if $\operatorname{rank} H^1(R_2, \mathbb{Z})>\max\{1, \operatorname{rank} H^1(R_1, \mathbb{Z})\}$ then $\operatorname{Hol}_*(R_1, R_2)=\emptyset$ and that if $\operatorname{rank} H^1(R_1, \mathbb{Z})=\operatorname{rank} H^1(R_2, \mathbb{Z})>1$ and $\operatorname{Hol}_*(R_1, R_2)\neq\emptyset$ then R_1 and R_2 are isomorphic and each map $f\in\operatorname{Hol}_*(R_1, R_2)$ is an isomorphism.

2.2. Generalizations of Hurwitz's Inequality. The inequalities (1) and (4) have multidimensional analogues; some of them will be reproduced in this Section.

Let X and Y be smooth irreducible quasiprojective algebraic varieties and assume further that Y is hyperbolically imbedded into a smooth projective variety $\bar{Y}$. For each continuous map $f\colon X\to Y$ we denote by $\kappa(f)$ the index of the subgroup $f_*(\pi_1(X))$ in $\pi_1(Y)$. If $y\in Y$ we denote by $k_y(f)$ the cardinality of the set of all connected components of the f-fiber $f^{-1}(y)$. Recall (for more details about this see Chap. 2, Sec. 2.1) that a *holomorphic map* $f\colon X\to Y$ is *dominant* if the image $f(X)$ contains a nonempty Zariski open subset of Y; the set of all dominant holomorphic maps from X into Y will be denoted $\operatorname{Hol}_{\text{dom}}(X, Y)$.

Theorem 1 (V.Ya. Lin). *Let* $f\in\operatorname{Hol}_{\text{dom}}(X, Y)$. a) *There exists a proper algebraic subset* $Z_f\subset Y$ *such that for all* $y\in Y\setminus Z_f$ *the quantity* $k_y(f)$ *takes one and the same value* $k(f)$ *and*

$$\kappa(f)\leqq k(f)<\infty, \tag{5}$$

so that the homomorphism $f^*\colon H^1(Y, \mathbb{Z})\to H^1(X, \mathbb{Z})$ *must be injective.* b) *If* $\pi_1(Y)$ *is a free group of finite rank* >1 *then*

$$\kappa(f)\leqq\frac{\operatorname{rank} H^1(X, Z)-1}{\operatorname{rank} H^1(Y, Z)-1}<\infty. \tag{6}$$

In particular, if Y is an open hyperbolic Riemann surface of finite type, then (5) *and* (6) *hold for any* $f\in\operatorname{Hol}_*(X, Y)$.

Proof. It follows from a theorem by Kiernan and Kobayashi [25], [26] and Hironaka's theorem on the resolution of singularities that f extends to a holomorphic map of projective varieties and therefore must be regular. This means that $k_y(f)<\infty$ for any point $y\in Y$ and, moreover, there exists a proper algebraic subset $Z_f \subset Y$ such that the map

$$f' = f \mid (X \setminus f^{-1}(Z_f)):\ X \setminus f^{-1}(Z_f) \to Y \setminus Z_f$$

is a smooth locally trivial fibration (this follows from a well-known theorem by A.N. Varchenko (cf. [119])). Therefore the fibers $f^{-1}(y)$ at all points $y \in Z \setminus Z_f$ must be homeomorphic to each other and so $k(f) = k_y(f)$ does not depend on the point $y \in Y \setminus Z_f$. As $f^{-1}(Z_f)$ and Z_f are proper algebraic subsets of smooth varieties, the imbeddings $X \setminus f^{-1}(Z_f) \to X$ and $Y \setminus Z_f \to Y$ induce epimorphisms of the corresponding fundamental groups. From this together with the exact homotopy sequence of the fibration f' we obtain the commutative diagram

$$\begin{array}{ccc}
\pi_1(X \setminus f^{-1}(Z_f)) & \xrightarrow[f'_*]{} & \pi_1(Y \setminus Z_f) \to \pi_0(f^{-1}(y)) \to 1 \\
\downarrow & & \downarrow \\
\pi_1(X) & \xrightarrow[f_*]{} & \pi_1(Y) \\
\downarrow & & \downarrow \\
1 & & 1
\end{array}$$

As $\# \pi_0(f^{-1}(y)) = k(f)$ for all $y \in Y \setminus Z_f$, this diagram leads to the inequality (5).

If G is a free group of finite rank $r > 1$ and H a subgroup of finite index κ (free by the Nielsen-Schreier theorem) then $m = \operatorname{rank} H = \kappa(r-1)+1$ so that $\kappa = (m-1)/(r-1)$. Applying this to the subgroup $H = f_*(\pi_1(X))$ of $G = \pi_1(Y)$ in the case when $\pi_1(Y) \cong \mathbb{F}_r$, $1 < r < \infty$, and taking into account that $f_*(\pi_1(X))$ is free and

$$\begin{aligned} m &= \operatorname{rank} f_*(\pi_1(X)) = \operatorname{rank} \operatorname{Hom}(f_*(\pi_1(X)), \mathbb{Z}) \leqq \\ &\leqq \operatorname{rank} \operatorname{Hom}(\pi_1(X), \mathbb{Z}) = \operatorname{rank} H^1(X, \mathbb{Z}) < \infty, \end{aligned}$$

we obtain inequality (6).

Corollary 1. *If X and Y are as in Theorem 1(a) and* $\operatorname{rank} H^1(X, \mathbb{Z}) < \operatorname{rank} H^1(Y, \mathbb{Z})$ *then* $\operatorname{Hol}_{\mathrm{dom}}(X, Y) = \emptyset$.

Example 1. Let X be an irreducible smooth quasiprojective algebraic variety, $r = \operatorname{rank} H^1(X, \mathbb{Z})$ and $m = \max\{r, 1\}$. A holomorphic function f on X, which omits $m+1$ different values $a_1, \ldots, a_{m+1}$, gives a holomorphic map $f: X \to Y_m = \mathbb{C} \setminus \{a_1, \ldots, a_{m+1}\}$. As $m+1 > 1$, Y_m must be hyperbolic and hyperbolically imbedded in $\mathbb{CP}^1$, with $\operatorname{rank} H^1(Y_m, \mathbb{Z}) = m+1 > \operatorname{rank} H^1(X, \mathbb{Z})$. Therefore $f = \text{const}$. Sh.I. Kaliman proved that if X is a Liouville manifold and $r = \operatorname{rank} H^1(X, \mathbb{Z}) < \infty$ then $\kappa(f) < \infty$ for every nonconstant holomorphic map $f: X \to Y$ into a hyperbolic Riemann surface Y of finite type, so that the homomorphism $f^*: H^1(Y, \mathbb{Z}) \to H^1(X, \mathbb{Z})$ is injective (if Y is open, then

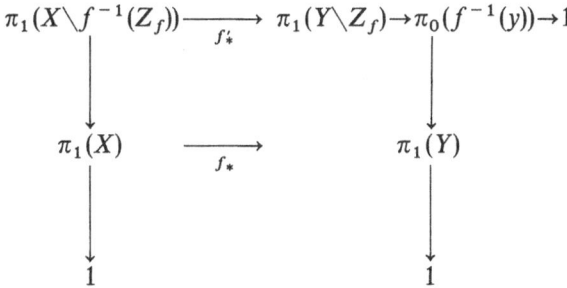

$\kappa(f)<\infty$ also entails the validity of inequality (6)). The proof is based on a factorization of X, reminding of the Stein factorization: two points x', $x''\in X$ are said to be R_f-equivalent if for each commutative triangle of holomorphic maps

$$\begin{array}{ccc} X & \xrightarrow{\ f\ } & Y \\ & {\scriptstyle p}\searrow \quad \nearrow{\scriptstyle q} & \\ & Z & \end{array},$$

where Z is a Riemann surface, p a surjective map and q everywhere nondegenerate on Z, one has $p(x')=p(x'')$ and there exist a path $\varphi\colon [0, 1]\to X$, $\varphi(0)=x'$, $\varphi(1)=x''$, such that the loop $p\circ\varphi\colon [0, 1]\to Z$ is contractible on Z. The quotient space $X'=X/R_f$ turns out to be a Riemann surface of finite type (here one uses the fact that X is Liouville and the condition $r<\infty$) and $f_*\colon \pi_1(X)\to\pi_1(X')$ is an epimorphism. The map f has thus the composition $f=q\circ p$: $X\xrightarrow{p}X'\xrightarrow{q}Y$ and it remains to invoke the fact that q is finite sheeted. From Sh.I. Kaliman's theorem it now follows that on a Liouville manifold X with $r=\operatorname{rank} H^1(X,\mathbb{Z})<\infty$ a nonconstant holomorphic function cannot omit more than $m=\max\{r, 1\}$ different values.

By the Gauss-Bonnet formula, if R is a hyperbolic Riemann surface then $\chi(R)=-(2\pi)^{-1}v(R)$ where $v(R)$ is the hyperbolic area of R. Therefore the classical Hurwitz's inequality (1) can be written in the form $\deg f\leqq v(R_1)/v(R_2)$. Pelles (formerly Eisenman) in 1975 and Yau in the same year (cf. [26]) obtained multidimensional analogues of this inequality. Let v_X be the Kobayashi-Eisenman volume on X (cf. Part III).

Theorem 2. *Let X, Y be n-dimensional complex manifolds, let $f\in\operatorname{Hol}(X, Y)$, and assume that one of the following conditions is fulfilled*: a) *X and Y are compact, X measure hyperbolic*[6] *(Pelles)*; b) $0<v_Y(Y)<\infty$ *and f proper (Yau).*

Then[7] $\deg f\leqq v_X(X)/v_Y(Y)$. *In particular, if $v_X(X)<v_Y(Y)$ then* $\operatorname{Hol}_{\mathrm{dom}}(X, Y)$ *does not contain any proper maps* (so that in case (a) $\operatorname{Hol}_{\mathrm{dom}}(X, Y)=\emptyset$). *If $v_X(X)<2\,v_Y(Y)$ then each nondegenerate proper holomorphic map from X into Y is biholomorphic* (so that in case (a) $\operatorname{Hol}_{\mathrm{dom}}(X, Y)=\operatorname{Bihol}(X, Y)$).

§ 3. Coincidence Theorems

3.1. The Borel-Narasimhan Theorem. A *complex space* Y is called *C-hyperbolic* if there exists an unramified holomorphic cover $p\colon Y'\to Y$ where Y'

[6] Cf. Part III.

[7] In the compact situation one can define $\deg f$ to be the number of f-images of a generic point $y\in Y$; equivalently: $\deg f=f^*(b)/a$ where a, b are the generators of the groups $H^{2n}(X,\mathbb{Z})$ and $H^{2n}(Y,\mathbb{Z})$, corresponding to the "complex" orientations. The last definition also works for proper maps; one has then to consider cohomology with compact supports.

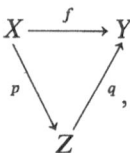

is hyperbolic in Carathéodory's sense (cf. Sec. 1.1). Borel and Narasimhan [8] proved the following theorem.

Theorem 1. *Let X be an ultra-Liouville complex space* (cf. Sec. 1.1) *and let Y be C-hyperbolic. If two holomorphic maps $f^{(1)}, f^{(2)}: X \to Y$ coincide at the point $x_0 \in X$ and induce the same homomorphism of fundamental groups $f_*^{(1)} = f_*^{(2)}: \pi_1(X, x_0) \to \pi_1(Y, f^{(1)}(x_0))$, then $f^{(1)} = f^{(2)}$.*

Outline of proof. Let $\tilde{X}$ be the universal cover of X. By the theorem on lifting of maps there exist commutative diagrams of holomorphic maps

$$\begin{array}{ccc} \tilde{X} & \xrightarrow{\tilde{f}^{(i)}} & Y' \\ {\scriptstyle q}\downarrow & & \downarrow{\scriptstyle p} \\ X & \xrightarrow{f^{(i)}} & Y \end{array}$$

($i=1, 2$). As $f^{(1)}(x_0) = f^{(2)}(x_0)$ and $f_*^{(1)} = f_*^{(2)}$, one can choose $\tilde{f}^{(1)}, \tilde{f}^{(2)}$ in such a way that they coincide on the fiber $q^{-1}(x_0) \subset \tilde{X}$. Let $h: Y' \to C$ be a bounded holomorphic function. Then

$$\varphi_h(x) = \sup_{\tilde{x} \in q^{-1}(x)} |h(\tilde{f}^{(1)}(\tilde{x})) - h(\tilde{f}^{(2)}(\tilde{x}))|$$

is a bounded continuous plurisubharmonic function on X with $\varphi_h(x_0) = 0$; consequently $\varphi_h = 0$. As bounded holomorphic functions separate points on Y', it follows from this that $\tilde{f}^{(1)} = \tilde{f}^{(2)}$ and so $f^{(1)} = f^{(2)}$. The proof is complete.

In [8] it is shown that in Theorem 1 one can take as X any connected complex space which can be represented as the image of a surjective holomorphic map $f: X' \to X$ from a Zariski open subset X' of a compact complex space (for example, every connected algebraic variety will do). Another class of admissible spaces X are connected pseudoconcave spaces (cf. [8]). As Y may serve a quotient space $Y = U/\Gamma$ of any bounded domain U in $\mathbb{C}^n$ under the free action on U of a discrete subgroup $\Gamma \subset \operatorname{Aut} U$ (for example, any hyperbolic Riemann surface R). In the case when the domain U is strictly pseudoconvex and X quasiprojective, Imayoshi [18] proved that any two homotopic nonconstant holomorphic maps $f, g: X \to Y = U/\Gamma$ must coincide.[8] In particular, if Y is compact then the set $\operatorname{Hol}_*(X, Y)$ is finite (cf. Theorem 1 in Sec. 1.1, Chap. 2).

On the other hand, V.V. Rabotin [44] observed that in Theorem 1 one cannot replace the hypothesis of C-hyperbolicity; it suffices to take for $X = Y$ a simply connected projective hyperbolic manifold M and for $f^{(1)}, f^{(2)}$ the identity map and a constant map. An example of such an M was earlier exhibited by Brody and Green (1977): the surface of degree $2m$ in

[8] In contrast to Theorem 1, here it is not required *a priori* that f and g coincide at some point $x_0 \in X$; moreover, even if it is known that $f(x_0) = g(x_0)$, it follows in general from the free homotopy of f and g only that the homomorphisms $f_*, g_*: \pi_1(X, x_0) \to \pi_1(Y, f(x_0))$ are conjugate, not that they coincide.

$\mathbb{CP}^3$ defined by the equation $z_0^{2m}+z_1^{2m}+z_2^{2m}+z_3^{2m}+\varepsilon(z_0 z_1)^m+\varepsilon(z_0 z_2)^m=0$ where m is sufficiently large ($m \geqq 25$) and $0<\varepsilon \ll 1$.

In [44] likewise the following generalization of Theorem 1 is obtained. Let $Z \subset Y$; let us say that Y is C-hyperbolic modulo Z if there exists an unramified holomorphic covering p: $Y' \to Y$ and an analytic subset $Z' \subset Y'$ such that $p(Z')=Z$ and such that the bounded holomorphic functions on Y' separate pairs of points y', $y'' \in Y'$, provided one of them is not in Z'. It turns out that if f, $g \in \mathrm{Hol}((X, x_0), (Y, y_0))$ and $f_*=g_*$ then either $f=g$ or else $f(X) \cup g(X) \subset Z$.

Let us mention a useful corollary to Theorem 1.

A smooth irreducible algebraic *curve* Γ lying in an algebraic variety X, is called *π_1-determative* if the natural imbedding i: $\Gamma \to X$ induces an epimorphism of fundamental groups $i_*: \pi_1(\Gamma) \to \pi_1(X)$.

Lemma 1. *On every smooth irreducible quasiprojective algebraic variety X of dimension $\geqq 1$ there exists a π_1-determative curve Γ.*

For smooth projective X the existence of such a curve follows, for instance, from Lefschetz's theorem on hyperplane sections. If X is affine then by the normalization theorem there exists a finite regular map p: $X \to \mathbb{C}^n$ ($n=\dim X$). Let Z be a closed algebraic subset of codimension 1 in $\mathbb{C}^n$ such that the restriction $\tilde{p}$ of the projection p to $X \setminus p^{-1}(Z)$ is a finite unramified covering over $\mathbb{C}^n \setminus Z$. By a well-known theorem of Zariski's (cf. [119]) there exists in $\mathbb{C}^n$ a complex line l, which intersects Z transversally only at nonsingular points, such that the imbedding j: $l \setminus Z \to \mathbb{C}^n \setminus Z$ induces an epimorphism j_*: $\pi_1(l \setminus Z) \to \pi_1(\mathbb{C}^n \setminus Z)$. It is easy to see that the curve $\Gamma=p^{-1}(l) \subset X$ has all the required properties. The general case of an irreducible smooth quasiprojective X can easily be brought back to the cases just worked out.

Corollary 1. *On each irreducible quasiprojective algebraic variety X there exists a smooth irreducible curve Γ such that for any C-hyperbolic space Y and arbitrary maps f, $g \in \mathrm{Hol}(X, Y)$ it follows from $f|\Gamma=g|\Gamma$ that $f=g$.*

In fact, let Γ be a π_1-determative curve on the manifold X_{reg} of all nonsingular points of X. It follows from Theorem 1 that $f|\Gamma=g|\Gamma$ implies $f|X_{\mathrm{reg}} = g|X_{\mathrm{reg}}$; but then $f=g$.

Let us further remark that with the aid of Lemma 1 one can establish the Liouvilleness of Abelian coverings of algebraic varieties without using the techniques of amenable groups (cf. Theorem 3(a), Sec. 1.1). Indeed, this lemma and the method used in the proof of the Borel-Narasimhan theorem allow us to carry out a reduction to the case of Abelian coverings of nonsingular algebraic curves, as considered by G.B. Shabat (Theorem 1 in Sec. 1.1).

In conclusion, let us state a theorem by Kwack [30]. Let f: $X \to X$ be a meromorphic map of a compact complex manifold X with negative first Chern class $c_1(X)$ (cf. Chap. 2, Sec. 2.1), which is holomorphic in the neighborhood of a fixed point $x_0 \in X$, and let $J_f(x_0)$ be the matrix giving the differential $d_{x_0} f$: $T_{x_0} X \to T_{x_0} X$ of f at x_0 in some local chart.

Theorem 2. (a) $|\det J_f(x_0)| \leqq 1$, *where equality holds iff* f *is an automorphism;* (b) *if* $d_{x_0} f = \mathrm{id}$ *then* $f = \mathrm{id}$.

For holomorphic maps of hyperbolic manifolds an analogous theorem was earlier proved by Wu (1967) and Kaup (1967); cf. [24]–[26]. Such a theorem is also true for holomorphic maps of a non-simply connected hyperbolic Riemann surface into itself (Minda [40]); in the case $|\det J_x(f)| = 1$ here f is an automorphism of finite order.

3.2. Coincidence Theorems in Dimension 1. A rational function of degree k, not identically zero, on a compact Riemann surface can vanish at most k distinct points. Here is an analogue of this result in a somewhat more general context.

Theorem 3. *Let* f_i: $R_g \to R_{g'}$ *be regular maps of compact Riemann surfaces,* $k_i = \deg f_i$ $(i = 1, 2)$. *If* $f_1 | A = f_2 | A$ *for a finite set* $A \subset R_g$ *of cardinality* $\# A > (k_1 + k_2)(g' + 1)$ *then* $f_1 = f_2$.

In fact, let φ: $R_{g'} \to \mathbb{CP}^1$ be a rational function of degree $\leqq g' + 1$. Then $h_\varphi = \varphi \circ f_1 - \varphi \circ f_2$ is a rational function on R_g, vanishing on A, with $\deg h_\varphi \leqq (k_1 + k_2)(g' + 1)$; therefore $h_\varphi = 0$. But by the Riemann-Roch theorem (or by Riemann's inequality) it follows that the rational functions of degree $\leqq g' + 1$ separate points on $R_{g'}$; therefore $f_1 = f_2$.

Corollary 2 (Hurwitz 1893; cf. [82]). *If two automorphisms of a compact Riemann surface of genus* g *coincide at* $2g + 3$ *points then they are equal.*

The canonical involution of a hyperelliptic curve of genus g has $2g + 2$ fixed points; therefore the number $2g + 3$ in the formulation of the corollary cannot be replaced by a smaller one. If an automorphism of a compact Riemann surface R_g of genus $g \geqq 2$ has exactly $2g + 2$ fixed points, then R_g must be hyperelliptic and f is its canonical involution.

If R is a hyperbolic Riemann surface and the map $f \in \mathrm{Hol}_*(R, R)$ has at least two fixed points then f is an automorphism of finite order (the hyperbolicity of R is essential) [40].

If f is an automorphism of an open Riemann surface R of finite genus g, there exists a compact Riemann surface R^* of genus g and a holomorphic imbedding $R \to R^*$ such that f is the restriction of some automorphism f^* of R^* (Oikawa [107], Maskit [97]; cf. also [40]). Therefore Corollary 2 holds also for open surfaces of finite genus g (Minda [40]). In [40] it is further shown that if R is a compact Riemann surface with nonempty boundary ∂R, different from the disk $\bar{D}$, then any two automorphisms, coinciding at some boundary point $x_0 \in \partial R$, must coincide identically.

A classical result on the coincidence of automorphisms is Hurwitz's theorem (1893).

Theorem 4. *If two automorphisms f_1, f_2 of a compact Riemann surface R_g of genus $g \geqq 2$ induce the same automorphisms of the homology $H_1(R_g, \mathbb{Z}) = \mathbb{Z}^{2g}$ then they coincide.*

As the group $H_1(R_g, \mathbb{Z})$ does not have torsion and as each automorphism $f\colon R_g \to R_g$ induces the identity automorphism on $H_0(R_g, \mathbb{Q})$ and $H_2(R_g, \mathbb{Q})$, the theorem follows from Theorem 5 below.

Let X be a complex manifold and let $f \in \mathrm{Hol}(X, X)$. Denote by $\mathrm{Fix}(f)$ the fixed point set of f. Let $f_{*k}\colon H_k(X, \mathbb{Q}) \to H_k(X, \mathbb{Q})$ be the homology homomorphisms induced by f and let $L(f) = \sum_{k \geqq 0} (-1)^k \operatorname{tr} f_{*k}$ be the *Lefschetz number* of f (we assume that $\dim_{\mathbb{Q}} H_k(X, \mathbb{Z}) < \infty$ for all $k \geqq 0$). It is well-known that the *Lefschetz index* of any isolated fixed point $x \in \mathrm{Fix}(f)$ is positive: $i_f(x) \geqq 1$ (for a simple proof, cf. the Appendix of Milnor's book [100]).

Remark 1. Recall that a *fixed point* $x_0 \in M$ of a smooth map f of a smooth (real) manifold M into itself is called *nondegenerate* if the spectrum $\sigma(f, x_0)$ of the differential of f at x_0 does not contain the number 1. In this case the Lefschetz index equals $i_f(x_0) = \operatorname{sign}(\det(I - J_f(x_0))$, where I is the unit matrix and $J_f(x_0)$ the Jacobi matrix of f at x_0 (in some chart). In the case when $f\colon X \to X$ is a holomorphic map of a complex manifold, the determinant figuring in the above formula for $i_f(x_0)$ coincides with the square of the modulus of the analogous complex determinant; therefore, if x_0 is nondegenerate, then $i_f(x_0) = +1$. If $f \in \mathrm{Aut}\, X$ and the set of iterates $\mathrm{It}(f) = \{f^{(m)}\colon m \in \mathbb{Z}\}$ is relatively compact in $\mathrm{Aut}\, X$, then any isolated point $x \in \mathrm{Fix}(f)$ is nondegenerate and $\sigma(f, x)$ is contained in the unit circumference $S_1 = \{\lambda \in \mathbb{C}\colon |\lambda| = 1\}$. (In fact, from the compactness of $\overline{\mathrm{It}(f)}$ it follows that there exists an f-invariant metric ρ on X; a small ball $B_\varepsilon(x) = \{y \in X\colon \rho(y, x) < \varepsilon\}$ must then be isomorphic to a bounded domain in $\mathbb{C}^n$ and is f-invariant. By H. Cartan's theorem [70] the automorphism f then linearizes in suitable coordinates in $B_\varepsilon(x)$; that is, if $1 \in \sigma(f, x)$ then there passes through x a curve consisting of fixed points of f. Finally, the inclusion $\sigma(f, x) \subset S_1$ follows from the compactness of $\overline{B_\varepsilon(x)}$.) If X is hyperbolic then the stabilizer $\mathrm{Aut}_x X$ at any point $x \in X$ is compact; therefore, for each $f \in \mathrm{Aut}\, X$ all isolated fixed points $x \in \mathrm{Fix}(f)$ are nondegenerate and $\sigma(f, x) \subset S_1$. (In the case, when f is an automorphism, other than the identity, of a compact Riemann surface of genus $g \geqq 1$, the nondegeneracy of all fixed points $x \in \mathrm{Fix}(f)$ is almost selfevident: it suffices to lift f to the universal covering; an automorphism $f \in \mathrm{Aut}\, \mathbb{CP}^1$ can have degenerate fixed points: $f(z) = z(z+1)^{-1}$.) This remark will be useful in the application of the holomorphic Lefschetz formula.

If X is a compact complex manifold, $f\colon X \to X$ a holomorphic map and $\operatorname{int} \mathrm{Fix}(f) = \emptyset$, one can apply to f Lefschetz's formula: $L(f) = \sum_{x \in \mathrm{Fix}(f)} i_f(x)$. This leads to the following result.

Theorem 5 (V.Ya. Lin). *Let X be a compact complex manifold, $\beta(X) = \sum_{k \geqq 0} \operatorname{rank} H_k(X, Z)$ and $f \in \mathrm{Hol}(X, X)$. Assume that one of the following conditions is fulfilled:* a) $L(f) < 0$; b) $\chi(X) < 0$ *and* $f_{*k} = \mathrm{id}$ *for all* $k \in \mathbb{N}$; c)

$\#\mathrm{Fix}(f)>\beta(X)$ and for each $k\in\mathbb{N}$ there exist numbers m_k, $n_k\in N$, $m_k<n_k$, such that

$$(f_{*k})^{m_k}=(f_{*k})^{n_k}.$$

Then[9] $\dim_{\mathbb{C}}\mathrm{Fix}(f)\geqq 1$.

Remark 2. If $\#\mathrm{Fix}(f)>\beta(X)$ the condition (b) is fulfilled in each of the following cases: 1) there exist $m, n\in\mathbb{N}$, $m<n$, such that the iterates $f^{(m)}$, $f^{(n)}$ are homotopic (within the class of continuous maps $X\to X$); 2) there exists a sequence $m_i\to+\infty$ such that the set of iterates $\{f^{(m_i)}\}$ is relatively compact in $C(X,X)$; 3) the map f is contained in a subsemigroup $H\subset \mathrm{Hol}(X,X)$, having only finitely many connected components; 4) X is hyperbolic.

Proof of Theorem 5. Let us assume that $\dim_{\mathbb{C}}\mathrm{Fix}(f)=0$. Then the Lefschetz formula is applicable. As $i_f(x)>0$ for all $x\in\mathrm{Fix}(f)$ it follows that $L(f)\geqq 0$ (and even >0, provided only $\mathrm{Fix}(f)\neq\emptyset$). If $f_{*k}=\mathrm{id}$ for all $k\geqq 0$ then $\chi(X)=L(f)\geqq 0$. In case (c) we consider the endomorphisms A_k induced by f on the finite dimensional vector spaces $H_k(X,\mathbb{C})$; it is clear that $A_k^{m_k}=A_k^{n_k}$. Let $v\neq 0$ be an eigenvector of A_k with eigenvalue λ; then $\lambda^{m_k}v=A_k^{m_k}v=A_k^{n_k}v=\lambda^{n_k}v$ so that either $\lambda=0$ or $|\lambda|=1$. Consequently $|\mathrm{tr}\, f_{*k}|=|\mathrm{tr}\, A_k|\leqq\dim_{\mathbb{C}}H_k(X,\mathbb{C})=\mathrm{rank}\, H_k(X,\mathbb{Z})$; therefore $\#\mathrm{Fix}(f)\leqq|L(f)|\leqq\beta(X)$. The proof is complete.

Notice that Theorem 5(c) gives another proof of Corollary 2, not based on the Riemann-Roch theorem.

Theorem 4 admits various strengthenings and generalizations; one of these can easily be obtained with the aid of the *holomorphic Lefschetz formula*[10] (cf., for example, [15], Chap. 3, § 4, p. 426). Let us first remark that if $g>1$, $f\in\mathrm{Aut}\, R_g$ and $f\neq\mathrm{id}$, then (cf. Remark 1) each point $x\in\mathrm{Fix}(f)$ is nondegenerate and, in a suitable chart (U_x, z_x), f takes the form $f(z_x)=\lambda_x z_x$ where $\lambda_x\in\mathbb{C}$, with $|\lambda_x|=1$ and $\lambda_x\neq 1$. Therefore, the holomorphic Lefschetz formula applied to f takes the form:

$$1-\mathrm{tr}(f^*|H_{\bar\partial}^{0,1}(R_g))=\sum_{x\in\mathrm{Fix}(f)}(1-\lambda_x)^{-1}.$$

Using the Dolbeault isomorphism and the Kodaira-Serre duality

$$H_{\bar\partial}^{0,1}(R_g)\cong H^1(R_g,\mathcal{O})\cong H^0(R_g,\Omega^1)^*$$

(where $\mathcal{O}$ is the structure sheaf of R_g and Ω^1 the sheaf of germs of holomorphic 1-forms on R_g) and taking into account that $\mathrm{Re}(1-\lambda_x)^{-1}=1/2$ we can write the formula as $\mathrm{Re}[1-\mathrm{tr}(f^*|H^0(R_g,\Omega^1))]=(\#\mathrm{Fix}(f))/2$. The trace of f^* in $H^0(R_g,\Omega^1)$ here is evaluated as follows. Let $\{a_i, b_i\}$ $(1\leqq i\leqq g)$ be a canonical basis for $H_1(R_g,\mathbb{Z})$ (i.e. the intersection indices satisfy $a_i\cdot a_j=b_i\cdot b_j=0$, $a_i\cdot b_j=\delta_{ij}$) and let $\{\omega_i\}$ $(1\leqq i\leqq g)$ be a basis for $H^0(R_g,\Omega^1)$ such that $\langle\omega_i, a_j\rangle$

---

[9] We have in mind the maximum of the (complex) dimensions of the irreducible components of the analytic set Fix f.

[10] The holomorphic Lefschetz formula is due to Atiyah and Bott (and for Riemann surfaces to Eichler and Shimura).

$=\int_{a_j}\omega_i=\delta_{ij}$. Then

$$\mathrm{tr}(f^*|H^0(R_g,\Omega^1))=\sum_i\langle f^*(\omega_i),a_i\rangle=\sum_i\langle\omega_i,f_*(a_i)\rangle;$$

in particular, if $f_*(a_i)=a_i$ for all $i=1,\ldots,g$ the trace equals g. This gives the following result. [11]

Theorem 6. *If* $g>1$, $f_1, f_2\in\mathrm{Aut}\,R_g$ *and* $f_{1*}(a_i)=f_{2*}(a_i)$ *for all* $i=1,\ldots,g$ *then* $f_1=f_2$.

Indeed, if we assume that $f_1^{-1}f_2=f\neq\mathrm{id}$, then, by the above, the holomorphic Lefschetz formula would lead to the identity $1-g=(\#\mathrm{Fix}(f))/2$, which is impossible, as $1-g<0$.

Accola [63] has proved that if $g>1$ and $c_i\in H_1(R_g,\mathbb{Z})$ $(1\leqq i\leqq 4)$ are independent elements with intersection indices $c_1\cdot c_3=c_2\cdot c_4=1$, $c_i\cdot c_j=0$ for $i+j\equiv 1 \pmod 2$, then two automorphisms $f', f''\in\mathrm{Aut}\,R_g$, satisfying the requirement $f'_*(c_i)=f''_*(c_i)$, $1\leqq i\leqq 4$, must coincide. Kato [22] proved that if R is an open Riemann surface of genus $g\geqq 1$, $c_1, c_2\in H_1(R,\mathbb{Z})$, $c_1\cdot c_2=1$, and $f\in\mathrm{Hol}(R,R)$ a map satisfying $f_*(c_i)=c_i$, $i=1,2$, then $f=\mathrm{id}$ (one first shows that f is an automorphism of finite order; this allows one to assume that f is an automorphism of a compact Riemann surface of genus $g\geqq 1$ with boundary, so one can apply Accola's theorem to the extension of f to the double of the surface).

Grothendieck and Serre [79] showed that in Theorem 4 for the two automorphisms f_1, f_2 to coincide it is sufficient that for some $m>2$ the induced automorphisms $f_{1*}^{(m)}$, $f_{2*}^{(m)}$ of the homology $H_1(R_g,\mathbb{Z}/m\mathbb{Z})$ coincide. The same is likewise true for noncompact Riemann surfaces with a non-Abelian fundamental group (Kato [18]). The restriction $m>2$ in the Grothendieck-Serre theorem is essential (the canonical involution of a hyperelliptic curve R induces the identity automorphism on the homology $H_1(R,\mathbb{Z}/2\mathbb{Z})$). However, if R_g is a compact Riemann surface of genus $g>1$, which is not hyperelliptic, and $f_{1*}^{(2)}=f_{2*}^{(2)}$ then $f_1=f_2$ ([23]).

It follows from Theorem 4 that for $g>1$ $\mathrm{Aut}\,R_g$ is a discrete group, and, as (in view of the hyperbolicity of R_g) it is also compact (indeed, automorphisms preserve the hyperbolic metric on R_g so that $\mathrm{Aut}\,R_g$ is equicontinuous) we must have $\#\mathrm{Aut}\,R_g<\infty$ (theorem of Schwarz, 1879). Hurwitz's famous theorem (1893) gives an estimate for the order of this group:

Theorem 7. *If* $g>1$ *then* $\#\mathrm{Aut}\,R_g\leqq 84(g-1)$.

Proof. Put $G=\mathrm{Aut}\,R_g$, $n=\#G$. Consider the natural projection $\pi\colon R_g\to Y=R_g/G$. Its branchpoints on R_g are the points $x\in R_g$ with nontrivial stabilizer G_x and on Y the orbits of these points. The orbit of a point $x\in R_g$ with

[11] We do not make an effort to trace the authors of this well-known theorem. I.I. Pyatetskiĭ-Shapiro and I.R. Shafarevich [113] applied the Lefschetz formula in the proof of the analogue of Theorem 4 for algebraic $K3$-surfaces, while Peters [109], using both variants of the Lefschetz formula, obtained analogous results for a much larger class of compact Kähler surfaces.

$\# G_x = r$ points consists of n/r points. Therefore the Riemann-Hurwitz formula gives

$$2g-2=n\left[(2g_Y-2)+\sum_{i=1}^{s}(1-1/r_i)\right],$$

where $g_Y \geqq 0$ is the genus of Y, $s \geqq 0$ the number of singular orbits y_i and $r_i \geqq 2$ the orders of the stabilizers of the points lying on the orbit y_i. As $2g-2>0$ then $2g_Y-2+\sum_{i=1}^{s}(1-1/r_i)>0$. Therefore, the theorem follows from the statement: if $m \geqq 0$, $s \geqq 0$ and $r_i \geqq 2$ $(1 \leqq i \leqq s)$ are integers such that

$$\alpha = 2m-2+\sum_{i=1}^{s}(1-1/r_i)>0,$$

then $\alpha > 1/42$ (for the proof of this elementary number theoretic result cf. Kra's book [91], Chap. 2, § 3, p. 77–78).[12]

3.3. Conditions on the Preimages of Divisors. For two nonconstant polynomials $p, q \in \mathbb{C}[z]$ to be equal it is sufficient that the preimages of two points to coincide, for instance, that $p^{-1}(0)=q^{-1}(0)$ and $p^{-1}(1)=q^{-1}(1)$. (Here is a well-known question whose answer is unknown: let $\deg p = \deg q > 1$; is it true that $p^{-1}(\{0; 1\})=q^{-1}(\{0, 1\})$ implies that either $p=q$ or $p=1-q$?) Polynomials are rational functions on $\mathbb{CP}^1$ with poles only at the distinguished point $\infty \in \mathbb{CP}^1$, so that for two nonconstant polynomials p, q always $p^{-1}(\infty)=q^{-1}(\infty)$. Is it true that two rational functions on $\mathbb{CP}^1$ must coincide if the preimages of three points coincide? The answer is no: for the functions $f_1=-4z^3(z-1)^{-3}(z+1)^{-1}$ and $f_2=-4z(z-1)^{-1}(z+1)^{-3}$ the complete preimages of the points 0, 1, ∞ coincide (Pizer [110]; it seems that this is the only known example of this kind[13]). But four points suffice (Pólya [112]); we prove this in a somewhat more general form.

Theorem 8. *Let $f_1, f_2\colon R_g \to R_{g'}$ be nonconstant holomorphic maps of compact Riemann surfaces with genera g and g′ such that $f_1^{-1}(\zeta_i)=f_2^{-1}(\zeta_i)$ for m disjoint points $\zeta_i \in R_{g'}$. If $m>3$ and $(m-2)^2>4(gg'+g-g')$ then $f_1=f_2$.*

Proof. Put $B=\{\zeta_1, \dots, \zeta_m\}$, $A=f_1^{-1}(B)=f_2^{-1}(B)$ and let $l=\#A$. Then f_1, f_2 map $R_g \setminus A$ into $R_{g'} \setminus B$; by Hurwitz's inequality $\deg f_1 + \deg f_2 \leqq 2(2g+l-2)/(2g'+m-2)$. Hence (taking into account the trivial inequality $l \geqq m$) by the inequality imposed on m it follows that $l>(\deg f_1+\deg f_2)(g'+1)$. As $f_1|A = f_2|A$ therefore $f_1=f_2$ (Theorem 3).

Notice that if $g'=0$ then the conditions in Theorem 8 reduce to $m>3$, $(m-2)^2>4g$ and if $g=g'=0$ only the condition $m>3$ remains. Until now we have spoken only of maps of compact surfaces and then every holomorphic map is regular. In the noncompact case the following result is due to Nevanlinna [105].

[12] Much useful information about Riemann surfaces and their mappings can be found in the book by S.L. Krushkal', B.N. Apanasov and N.A. Gusevskiĭ [92].

[13] We are obliged to A.E. Eremenko for showing this example to us.

Theorem 9. *Let $f_1, f_2 \in \mathrm{Hol}_*(\mathbb{C}, \mathbb{CP}^1)$ and assume that for four distinct points $\zeta_i \in \mathbb{CP}^1$ holds $f_1^{-1}(\zeta_i) = f_2^{-1}(\zeta_i)$* (taking into account multiplicities). *Then either* (after a suitable relabeling) *$f_1^{-1}(\zeta_1) = f_1^{-1}(\zeta_2) = \emptyset$ and $f_2 = L \circ f_1$, where $L: \mathbb{CP}^1 \to \mathbb{CP}^1$ is a fractional linear transformation, or else $f_1 = f_2$.*

Another theorem of Nevanlinna's states that if $f_1, f_2 \colon \mathbb{C} \to \mathbb{CP}^1$ are nonconstant meromorphic functions and $f_1^{-1}(\zeta_i) = f_2^{-1}(\zeta_i)$ for five disjoint points $\zeta_i \in \mathbb{CP}^1$ then $f_1 = f_2$. The proof depends on value distribution theory (cf. [80]).[14] Analogous results hold for holomorphic maps of affine algebraic varieties into elliptic curves (Schmid [115], Drouilhet [10]). Fujimoto has obtained multidimensional analogues of the preceding theorems by Nevanlinna; let us state one of them (cf. [11] and also [76] and Part II).

Theorem 10. *Let $f_1, f_2 \colon \mathbb{C}^n \to \mathbb{CP}^N$ be meromorphic maps and assume that f_1 is algebraically nondegenerate* (i.e. $f_1(\mathbb{C}^n)$ is not contained in any projective hypersurface $V \subset \mathbb{CP}^N$). *If for $2n+3$ distinct hyperplanes $H_i \subset \mathbb{CP}^N$ the divisors $f_1^{-1}(H_i)$ and $f_2^{-1}(H_i)$ coincide, then $f_1 = f_2$.*

Further generalizations of the Nevanlinna theorems were gotten by Drouilhet [10]; in view of the akwardness of the formulations we state only one beautiful corollary in [10].

Theorem 11. *Let $f, g \colon \mathbb{C} \to \mathbb{CP}^1$ be a nonconstant meromorphic function and $p(x, y)$ a polynomial of degree k in x and of degree l in y. Assume that there exist finite sets $A, B \subset \mathbb{CP}^1$, each consisting of $m \geqq k+l+3$ distinct points, such that $f^{-1}(A) = g^{-1}(B) = E$ and $p(f(z), g(z)) | E = 0$. Then $p(f(z), g(z)) \equiv 0$.*

Example 1. Let $f = \cos z$, $g = \sin z$, $p(x, y) = x^2 + y^2 - 1$ $(k = l = 2)$, $A = B = \{0; \pm 1, \pm 1/2, \pm\sqrt{3}/2\}$ $(m = 7 = k + l + 3)$. It is easy to see that the hypothesis of Theorem 11 is fulfilled; this implies the fundamental trigonometric identity $\cos^2 z + \sin^2 z = 1$.

The above results are close to the ideas of Sakai [47].

§ 4. Diverse Results on the Nonexistence of Maps

In this Section we list some results on nonexistence of holomorphic maps which primarily pertain to some concrete manifold.

4.1. Maps of Annuli. In Sec. 1.2 we defind holomorphic elements of the group $\pi_1(X)$ as elements realized by maps $f \in \mathrm{Hol}(\mathbb{C}^*, X)$. If we use instead of $\mathbb{C}^*$ the annulus $K_r = \{r^{-1} < |z| < r\}$ the result will in essential way depend on the conformal modulus r^2. This is witnessed, for instance, at the hand of the well-known "annulus theorem".

[14] Generalizations of these theorems by Nevanlinna can be found in papers by Gross and Yang; cf. Gross, F., Yang, Ch.-ch.: Meromorphic functions covering certain finite sets at the same points. Ill. Math. J. 26, 432–441 (1982).

Theorem 1. *Let* $f \in \mathrm{Hol}_*(K_r, K_R)$. *Then*

$$n = |\deg f| \leqq \frac{\log R}{\log r};$$

equality holds only if $f(z) = e^{i\varphi} z^{\pm n}$.

The idea of the proof goes essentially back to Carathéodory [9]. One replaces the given function $f(z) = z^{\pm n} e^{g(z)}$ by $f_1(z) = z \exp(1/n \cdot g(z^{\pm 1}))$; it is clear that $f_1 \in \mathrm{Hol}_*(K_r, K_{R_1})$, where $R_1 = \sqrt[n]{R}$ and $\deg f_1 = 1$. The unit circle S_1 is the unique geodesic in the hyperbolic metric of K_r in its free homotopy class; its hyperbolic length l_r decreases monotonically as r increases (this is easily verified by passing to the universal covering). As the hyperbolic metric does not increase under holomorphic maps, we have

$$l_r = l_{K_r}(S_1) \geqq l_{K_{R_1}}(f_1(S_1)) \geqq l_{K_{R_1}}(S_1) = l_{R_1},$$

whence $R_1 \geqq r$ or $\log R/\log r \geqq n$. The verification of the second statement is easy.

Let us remark that there exists a homotopy $f_t(z) = z^{\pm n} e^{tg(z)}$, $t \in [0, 1]$, connecting the maps f and $f_0 = z^{\pm n}$ in $\mathrm{Hol}(K_r, K_R)$.

Example 1. Bedford [6], using a seminorm introduced by him on real homologies in middle dimensions of complex manifolds (analogous to extremal length), which has the property that it does not increase under injective holomorphic maps, proved the following result on the imbedding of annuli into tori.

Set $T_\tau = \mathbb{C}/(\mathbb{Z} + \tau\mathbb{Z})$, where $\operatorname{Im}\tau > 0$, and let γ_1, γ_2 be the natural generators of $H_1(T_\tau, \mathbb{Z})$, α being the one in $H_1(K_r, \mathbb{Z})$. If $f\colon K_r \to T_\tau$ is an injective holomorphic map and $f_*(\alpha) = m\gamma_1 + n\gamma_2$ then $|m + n\tau|^2 \leqq \pi \cdot \operatorname{Im}\tau/\log r$.

Thus, only a finite number of classes in $H_1(T_\tau, \mathbb{Z})$ (depending on τ and r) can be realized by holomorphic imbeddings of an annulus of "given width"; moreover, the width of the annulus realizing the class $m\gamma_1 + n\gamma_2$ is bounded from above: $\log r \leqq \pi \cdot \operatorname{Im}\tau/|m + n\tau|^2$.

Here the injectivity is essential: the map $g\colon \mathbb{C}^* \to T_\tau$, $g(z) = ((m + n\tau)/2\pi i) \cdot \log z$, satisfies $g_*(\alpha) = m\gamma_1 + n\gamma_2$. Notice that if $(m, n) = 1$ it is injective on the unit circumference S_1. Therefore the classes $m\gamma_1 + n\gamma_2$ with relatively prime m and n are realized by holomorphically imbedded annuli which are sufficiently narrow.

Example 2. In Example 2 in Sec. 1.2 we described the set $\operatorname{hol}\pi_1(G_n)$ of holomorphic elements of the group $\pi_1(G_n) \cong B(n)$, that is, group elements realized by maps $f \in \mathrm{Hol}(\mathbb{C}^*, G_n)$: it is the set of elements of finite order modulo the center. Every element $g \in \pi_1(G)$ may be realized by a map $f \in \mathrm{Hol}(K_r, G_n)$ with sufficiently small $r = r(g) > 1$; however, for large r and fixed n there are few such elements.

An element $g \in B(n)$ is called irreducible if its image under the natural epimorphism $B(n) \to S(n)$ is an n-cycle. Let us denote by $\operatorname{irred}_n \pi_1(G_n)$ the set of all irreducible elements $g \in \pi_1(G_n)$ such that $\deg(d_n \circ g)$ (d_n: $G_n \to \mathbb{C}^*$ is the discriminant map) is divisible by all prime factors of n. It turns out that for a suitable constant $c(n) \gg 1$ no element $g \in \mathrm{irred}_n \pi_1(G_n)$ can be realized (up to free homotopy) by a map $f \in \mathrm{Hol}(K_r, G_n)$ for $r \geqq c(n)$ (E.A. Gorin, V.Ya. Lin [13]). In particular, no such element is realized by a map $f \in \mathrm{Hol}(D^*, G_n)$. It follows from this that each polynomial $p(t, z) = t^n + a_2(z)\, t^{n-2} + \ldots + a_n(z)$, with coefficients holomorphic in K_r ($r \geqq c(n)$) and discriminant $d(z) \equiv 1$, is reducible in the ring $H(K_r)$ of holomorphic functions on K_r.

4.2. On the Rigidity of Bounded Domains. Carathéodory (1932) proved the following theorem on holomorphic maps of multiply connected plane domains.

Theorem 2 ([9], Theorem 14). *Let G be a bounded connected and multiply connected domain and $z_0 \in G$. Then there exists a constant $c < 1$ such that each map $f \in \mathrm{Hol}(G, G)$ with $f(z_0) = z_0$ and $|f'(z_0)| \geqq c$ is an automorphism of G.*

Minda [101] generalized this theorem to the case of hyperbolic Riemann surfaces.

In [9], § 27 Carathéodory writes: "Theorem 14 shows that the rigidity of multiply connected planar domains with respect to conformal maps is much greater than one usually thinks. One expects that similar properties hold also in multidimensional spaces, not only, which is quite obvious, for multiply connected domains [15] but also (in suitable circumstances) for domains having the topological connectivity of the hyperball. In fact, all our previous experience in working with analytic maps of multidimensional domains points to that already rather simple circular bodies resemble by their properties more multiply connected planar domains than simply connected ones."

In 1971 Hirschovitz [81] showed that the polynomial polyhedron $P_H = \{|(x^2-1)(y^2-1)| < 5\}$ in $\mathbb{C}^2$ is topologically contractible but not holomorphically contractible (i.e. there exists no continuous family of holomorphic maps f_t: $P_H \to P_H$, $0 \leqq t \leqq 1$, such that $f_0 = \mathrm{id}_{P_H}$ and $f_1 = \mathrm{const}$). The rigidity of P_H arises from the configuration of the four lines $x = \pm 1$, $y = \pm 1$, which is invariant for all $f \in \mathrm{Hol}_*(P_H, P_H)$ (and, apparently, is not topologically contractible in itself to a point). However, this polyhedron is not bounded. In our paper [61] the following results were obtained.

Theorem 3. *Let $p(x, y)$ be a polynomial such that all irreducible components of the curves $\Gamma_\zeta = \{(x, y) \in \mathbb{C}^2 \colon p(x, y) = \zeta\}$, $\zeta \in \mathbb{C}$, are hyperbolic. Then for each polynomial $q(x, y)$ and sufficiently large numbers $c_1 \ll c_2$ the polynomial polyhedron $P = \{|p(x, y)| \leqq c_1, |q(x, y)| \leqq c_2\}$ is topologically contractible, but not holomorphically.*

[15] This is acknowledged, for instance, in the papers Bedford [64] and Mok [102].

As a concrete example we may cite the bounded polyhedron $P_0=\{|x^3+y^3-3x|^2<c_1, |x|<c_2\}$. Analogous properties are enjoyed by the bounded strictly pseudoconvex domain $Q_0=\{|x^3+y^3-3x|+\varepsilon(|x|^2+|y|^2)<r\}$ (with real analytic boundary), where $r>2$ and $0<\varepsilon\ll 1$, obtained from P_0 by perturbation. The identity map on Q_0 (or P_0) has the following rigidity property: there exists a $\delta>0$ such that if $f\in\mathrm{Hol}(Q_0, Q_0)$ and $\overline{f(Q_0)}\subset Q_0$ or $f\in\mathrm{Hol}(V, Q_0)$, where V is an arbitrary domain containing $\bar{Q}_0$, then $\sup_{z\in Q_0}|f(z)-z|\geqq\delta$. This shows, in particular, that (in contrast to the one dimensional case) for $n\geqq 2$ the well-known theorem on approximation of functions holomorphic in a strictly pseudoconvex domain $Q\subset\mathbb{C}^n$ and continuous on $\bar{Q}$ by functions holomorphic in neighborhood of $\bar{Q}$ can not be obtained by approximating id_Q by holomorphic maps $\bar{Q}\subset V\to Q$ (because such approximations may not exist). Such an "obstruction" remains in force also for polynomial polyhedra.

For polynomial polyhedra in $\mathbb{C}^2$ defined by one single polynomial $p(x, y)$ one has the following criterion for holomorphic contractibility (M.G. Zaĭdenberg, V.Ya. Lin [122]): a polyhedron $P_c=\{|p(x, y)|<c\}$ is holomorphically contractible iff p can be written in the form $p=\varphi(q\circ\alpha)+a$ where $\varphi\in\mathbb{C}[z]$, $q=q(x, y)$ is a quasihomogeneous polynomial[16], α a polynomial automorphism of $\mathbb{C}^2$ and $a\in\mathbb{C}$, $|a|<1$. If p is a primitive polynomial (i.e. its generic level curves Γ_ζ are connected) then this representation simplifies: $p=q\circ\alpha+a$ where q, α and a are as before. The *family of curves* $\Gamma_\zeta=\{p=\zeta\}$, defined by such a polynomial, is *isotrivial* (i.e. all curves Γ_ζ with the exception of at most a finite number are isomorphic to each other). It turns out that the polynomials $p(x, y)$ for which the family $\{\Gamma_\zeta\}$ is isotrivial admits a comparatively simple description; it was obtained with subsequent efforts by a series of authors: Gutwirth (1961), Nagata (1971), Saito (1972 and 1977), Miyanishi and Sugie (1980), M.G. Zaĭdenberg and V.Ya. Lin (1984) and, finally, the last step was taken by Sh.I. Kaliman (1984) (see Kaliman, Sh.I.: Polynomials on $\mathbb{C}^2$ with isomorphic generic fibers. Dokl. Akad. Nauk SSSR *288*, 39–42 (1986) [Russian]; for a list of the corresponding "normal forms" see [60]). Let us remark that each such polynomial plays the rôle of a quasi-invariant for some $\mathbb{C}^*$-action, which is regular on the complement of a suitable algebraic curve in $\mathbb{C}^2$; these actions themselves can also be written down (M.G. Zaĭdenberg [60]).

Theorem 3 shows that many bounded polynomial polyhedra in $\mathbb{C}^2$ are topologically contractible, but not holomorphically; however, it gives not explicit example of such a polyhedron, because the constants c_1, c_2 entering there are not effective. Recently V.V. Petunin (to appear) found an explicit example: the polyhedron obtained from Hirschowicz's polyhedron P_H by "cutting" with a ball $|x|^2+|y|^2<r^2$ where $r\sim 3\cdot 10^6$.

Let us mention yet another result on the rigidity of bounded domains with respect to holomorphic maps. We say that a *domain* $U\subset\mathbb{C}^n$ is *rigid*

[16] I.e., $q(t^k x, t^l y)=t^n q(x, y)$ for some $k, l\in\mathbb{Z}_+$, $n\in\mathbb{N}$ and all $t\in\mathbb{C}$, $(x, y)\in\mathbb{C}^2$ (this definition is somewhat wider than the usual one, as one the powers k, l is allowed to be zero).

in Carathéodory's sense at the point $z_0 \in U$ if there exists a constant $c<1$ such that each holomorphic map $f\colon U \to U$, such that $f(z_0)=z_0$ and $|\det J_f(z_0)|>c$, is an automorphism of U. It is well-known that for each nonconstant polynomial $p(x, y)$ the family of curves $\Gamma_\zeta = p^{-1}(\zeta)$, $\zeta \in \mathbb{C}$, defined by it has only a finite number of degenerate fibers, i.e. there exists a finite set $S \subset \mathbb{C}$ such that $p\colon \mathbb{C}^2 \setminus p^{-1}(S) \to \mathbb{C} \setminus S$ is a topologically locally trivial fibration; we may always assume that S is the minimal set possessing this property. Let $r>0$, $P=\{|p(x, y)|<r\}$ and let $\{P_n\}$ be an exhaustion of P by bounded domains, e.g. $P_n \subset P_{n+1}$ and $P=\bigcup P_n$.

Theorem 4 (M.G. Zaĭdenberg). *Fix a compact set* $K \subset P \setminus p^{-1}(S)$. *Assume that one of the following two conditions is fulfilled:* a) *p is a polynomial of general type, i.e. the curves* Γ_ζ $(\zeta \in \mathbb{C} \setminus S)$ *are hyperbolic, and that* $S \cap D_r \neq \emptyset$ (here $D_r=\{\zeta \in \mathbb{C} : |\zeta|<r\}$); b) *p is not isotrivial. Then one can find a number* $n_0 = n_0(K)$ *such that for* $n \geqq n_0$ *each of the domains* P_n *is rigid in Carathéodory's sense at each point* $z \in K$.

It is clear that if $S \subset D_r$ then one can pick the exhaustion $\{P_n\}$ in such a manner that all P_n are homeomorphic to a ball and are either polynomial polyhedra of the form $\{|p|<a_n, |q|<b_n\}$ or else strictly pseudoconvex domains with real analytic boundaries. Thus, for instance, in view of Theorem 4 the starshaped "quasihomogeneous" circular domains $P_n=\{|x^3-y^3|<1, |x|<n\}$ are rigid in the sense of Carathéodory at points of a fixed compact set $K \subset \{0<|x^3-y^3|<1\}$ if n is sufficiently large (of course, these domains, as well as arbitrary quasihomogeneous polyhedra, are not rigid at the origin). Let us remark that if the polynomial p is isotrivial and $S \cap D_r = \emptyset$ then there exists an exhaustion $\{P_n\}$ of P by bounded domains each of which is not rigid in the sense of Carathéodory at any point.

Let us give a *sketch of the proof* for Theorem 4 in the assumption that condition (b) is fulfilled, i.e. that p is not isotrivial, and that K consists of just one point z_0. Let us assume that P_n is not rigid at z_0 for any n; then one can find a holomorphic map $f_n\colon P_n \to P_n$ such that $f_n(z_0)=z_0$, $J_n = |\det J_{f_n}(z_0)| \geqq 2/3$ but f_n is not an automorphism of P_n. By a well-known theorem of H. Cartan's on automorphisms of bounded domains (cf. [65], Chap. 3, § 3, p. 76) $J_n<1$. Upon replacing each f_n by a suitable iterate (and passing, if necessary, to a subsequence), we may assume that $1/2 \leqq J_n \leqq 3/4$ for all n and that $\{p \circ f_n\}$ converges to a bounded holomorphic function φ. Let $\zeta_0 = p(z_0)$ and $|\zeta_0 - s|>\varepsilon>0$ for all $s \in S$. For $\delta \leqq \varepsilon$ set $T_\delta = p^{-1}(D_\delta(\zeta_0))$ and $T_{\delta,n} = T_\delta \cap P_n$. As $f_n(z_0)=z_0$ for all n, then $\varphi | \Gamma_{\zeta_0} \equiv \varphi(z_0) = \zeta_0$. Therefore one can find $\delta\,(0<\delta \ll \varepsilon)$ such that $f_m(T_{\delta,n}) \subset T_\varepsilon$ for $m \gg n$. The polyhedron T_ε, fibered by the hyperbolic curves Γ_ζ $(|\zeta-\zeta_0|<\varepsilon)$, is hyperbolically complete in Kobayashi's sense (cf. Zaĭdenberg [121]); therefore the space $\mathrm{Hol}((T_{\delta,n}, z_0), (T_\varepsilon, z_0))$ is compact ([25]). Taking into account that $f_m(z_0)=z_0$ for all m, we may assume that $f_m \to f_0 \in \mathrm{Hol}(T_\delta, T_\varepsilon)$. As p is not isotrivial and $f | \Gamma_{\zeta_0} \neq \mathrm{const}$, then for $|\zeta - \zeta_0| < \delta_1 \ll \delta$ all curves Γ_ζ can not be isomorphic to each other and $f_0 | \Gamma_\zeta \neq \mathrm{const}$. On the other hand, $f_0(\Gamma_\zeta) \subset \Gamma_{\zeta'}$ (because $(p \circ f_0) | \Gamma_\zeta = \mathrm{const} = \zeta'$)

and as the curves Γ_ζ, $\Gamma_{\zeta'} \subset T_\varepsilon$, are homeomorphic and hyperbolic, then for $|\zeta-\zeta_0|<\delta_1$ the map $f_0|\Gamma_\zeta\colon \Gamma_\zeta \to \Gamma_{\zeta'}$ must be an isomorphism (cf. Sec. 2.1); this means that $\zeta'=\zeta$ and $f_0|\Gamma_\zeta \in \operatorname{Aut}\Gamma_\zeta$ for $|\zeta-\zeta_0|<\delta_2 \ll \delta_1$. It follows that $f_0|T_{\delta_2} \in \operatorname{Aut} T_{\delta_2}$ and therefore (once more by H. Cartan's theorem, which holds also in the case of automorphisms of hyperbolic manifolds; cf. [25], Theorem 3.3, Chap. 5) $|\det J_{f_0}(z_0)|=1$; this contradicts the inequality $J_n \leqq 3/4$.

4.3. Maps of Spaces Connected with Polynomials Without Multiple Roots. We list now some results on maps of the spaces G_n, G_n^0 and SG_n (cf. Sec. 1.2, Example 2).

Let k, $n \geqq 3$. A continuous *map* $f\colon G_n \to G_n$ is called *splittable* if there exists a continuous map $g\colon \mathbb{C}^* \to G_k$ such that f is homotopic to the compositions $g \circ d_n$ where $d_n\colon G_n \to \mathbb{C}^*$ is the standard map defined by the discriminant polynomial d_n. A necessary and sufficient condition for f to be splittable is that the homomorphism $f_*\colon \pi_1(G_n) \to \pi_1(G_k)$ is Abelian (the necessity is obvious and the sufficiency follows from the fact that G_k is a $K(\pi, 1)$-space and $B(n)/B'(n) \cong \mathbb{Z}$). Using the description given in Sec. 2.1 (Example 2) of holomorphic maps $\mathbb{C}^* \to G_k$, one can write down all splittable holomorphic maps $G_n \to G_k$. Recall that the $\mathbb{C}$-action L and the $\mathbb{C}^*$-action U on the space of polynomials without multiple roots $G_k = \{w = t^n + w_1 t^{n-1} + \ldots + w_k\colon d_k(w) \neq 0\}$ are defined by the formulae

$$L(v)\, w = (t+v)^k + w_1 (t+v)^{k-1} + \ldots + w_k$$

and

$$U(\zeta)\, w = t^k + \zeta\, w_1\, t^{k-1} + \ldots + \zeta^k\, w_k$$

respectively ($v \in \mathbb{C}$, $\zeta \in \mathbb{C}^*$).

Theorem 5 ([33], [35]). *Every splittable holomorphic map $f\colon G_n \to G_k$ can be written in the form*

$$f(z) = L(v(z))\, U(e^{\varphi(z)}\, d_n^{q/n(n-1)}(z))\, w^0, \tag{1}$$

where v, $\varphi\colon G_n \to \mathbb{C}$ are holomorphic functions, $q \in \mathbb{Z}$, $w^0 = (0, w_2^0, \ldots, w_k^0) \in SG_k \subset G_k$ and $z \in G_n$, also $w_m^0 = 0$ for all m with $m q \not\equiv 0 (\operatorname{mod} k(k-1))$.

For $n>k$ and $n>4$ every homomorphism $B(n) \to B(k)$ is Abelian ([94], [35]) so that in this case all holomorphic maps $G_n \to G_k$ are splittable and admit the representation (1).

The nonsplittable holomorphic maps $G_n \to G_k$ and $G_n^0 \to G_k^0$ with $4 < n \leqq k$ are most interesting. Let $\operatorname{Hol}^U(G_n^0)$ be the additive group of all holomorphic U-invariant (i.e. quasihomogeneous of degree 0 with exponents 2, …, n) functions $G_n^0 \to \mathbb{C}$.

Theorem 6 ([33], [35]). a) *For $n>4$ each nonsplittable holomorphic map $f\colon G_n \to G_n$ (respectively $f^0\colon G_n^0 \to G_n^0$) can be represented in the form*

$$f(z) = f_{q,\varphi,v}(z) = L(v(z))\, U(e^{\varphi(z)}\, d_n^q(z))(z)$$

(respectively

$$f^0(z)=f^0_{q,\varphi}(z)=U(e^{\varphi(z)}\,d_n^q(z))(z)),$$

where $z\in G_n$ *(or* $z\in G_n^0$*),* $q\in\mathbb{Z}$ *and* φ *and* v *are holomorphic functions on* G_n *(on* G_n^0*).*

b) *The fibers of each map* $f^0_{q,\varphi}\colon G_n^0\to G_n^0$ *are discrete; this map is proper iff* $\varphi\in\mathrm{Hol}^U(G_n^0)$.

c) *Each proper holomorphic map* $f^0\colon G_n^0\to G_n^0$ *is everywhere nondegenerate and constitutes an unramified regular finitely sheeted cover with the number of sheets* $N\equiv 1\,(\mathrm{mod}\,n(n-1))$*; any two such coverings* f_1*,* f_2 *with the same* N *are equivalent to each other (i.e.* $f_2=f_1\circ h$ *for a suitable biholomorphic* h*:* $G_n^0\to G_n^0$*).*

d) *The map* $f^0_{\varphi,q}$ *is biholomorphic iff* $q=0$ *and* $\varphi\in\mathrm{Hol}^U(G_n^0)$*, so that the group* $\mathrm{Aut}\,G_n^0$ *is Abelian and naturally isomorphic to* $\mathrm{Hol}^U(G_n^0)/2\pi i\mathbb{Z}$*. The orbits of* $\mathrm{Aut}\,G_n^0$ *in* G_n^0 *are smooth algebraic curves isomorphic to* $\mathbb{C}^*$ *and coincide with the orbits of the* $\mathbb{C}^*$*-action* U *on* G_n^0*.*

The central step in the proof is to obtain the representation $f=f_{q,\varphi,v}$. For this one establishes first that for $n\neq 4$ the subgroup of pure braids $I(n)\subset B(n)$ is mapped into itself under any non-Abelian endomorphism $\psi\colon B(n)\to B(n)$, that is, $\psi(I(n))\subset I(n)$ (for automorphisms of $B(n)$ an analogous statement was proved already by Artin (1947)). This purely algebraic result allows one to lift any nonsplittable map $f\colon G_n\to G_n$ $(n\neq 4)$ to a map $F\colon E_n\to E_n$ of the cover E_n corresponding to the subgroup $I(n)$. If f is holomorphic then F is holomorphic too; moreover, F is equivariant under the natural action of the symmetric group $S(n)$ on E_n (G_n is obtained from E_n upon factoring out this action). All such maps can be written down in the following way. One shows that each function $h\in\mathrm{Hol}_*(E_n,\mathbb{C}\setminus\{0;1\})$ is either a simple ratio $(\lambda_p-\lambda_r)/(\lambda_q-\lambda_r)$ or else a double ratio $(\lambda_p-\lambda_r)(\lambda_q-\lambda_s)/(\lambda_p-\lambda_s)(\lambda_q-\lambda_r)$ (here $\lambda_1,\ldots,\lambda_n$ are the coordinates in E_n and p, q, r, s pairwise distinct indices). Let $F_i(\lambda)$ be the coordinate functions of the map F. For $i\geqq 3$ the functions $h_i(\lambda)=(F_i(\lambda)-F_1(\lambda))/(F_2(\lambda)-F_1(\lambda))$ belong to $\mathrm{Hol}_*(E_n,\mathbb{C}\setminus\{0;1\})$ (because F is equivariant they are not constant); therefore each of them must be either a simple or a double ratio of coordinate differences. It turns out that the second case is impossible, i.e. all h_i are simple ratios, and renumbering them suitably we may assume that $h_i=(\lambda_i-\lambda_1)/(\lambda_2-\lambda_1)$ for all $i=3,\ldots,n$. From this follows readily that (in a suitable renumeration of the coordinates)

$$F_i(\lambda)=V(\lambda)+\lambda_i\,D_n^q(\lambda)\,e^{\Phi(\lambda)},\quad 1\leqq i\leqq n,$$

where $V(\lambda)$, $\Phi(\lambda)$ are symmetric holomorphic functions on E_n, $q\in\mathbb{Z}$ and $D_n(\lambda)=\prod_{i\neq j}(\lambda_i-\lambda_j)$. Transplantation to G_n gives the desired representation of the map f.

Theorem 6 can be applied to the study of superposition of algebraic functions of several variables [34].

Concerning maps $G_n\to G_k$ for $4<n<k$ the following is known. Let $\Pi(n)$ be the union of four increasing arithmetic progressions with the same differ-

ence $d=n(n-1)$ and starting, respectively, from the numbers $a_0^{(1)}=n(n-1)$, $a_0^{(2)}=n$, $a_0^{(3)}=n(n-1)+1$ and $a_0^{(4)}=(n-1)^2$.

Theorem 7 ([33], [35]). *If $n>4$ and $k\notin\Pi(n)$ then each holomorphic map $G_n\to G_k$ splits.*

The proof is based on the fact that a homomorphism of fundamental groups $f_*\colon \pi_1(G_n)\to\pi_1(G_k)$, induced by a holomorphic map $f\colon G_n\to G_k$, must satisfy $f_*(\operatorname{hol}\pi_1(G_n))\subset\operatorname{hol}\pi_1(G_k)$. The description of the set $\operatorname{hol}\pi_1(G_n)$ given in Sec. 1.2 (Example 2) allows one to prove that if $n>4$ and $k\notin\Pi(n)$ then each homomorphism of braid groups $f_*\colon B(n)\to B(k)$, satisfying the above property, must be Abelian and this guarantees that f is splittable.

It would be interesting to find out whether there exist nonsplittable holomorphic maps $G_n\to G_k$ for $n>4$ and $k\in\Pi(n)$ (cf. [31]).

Let us briefly turn to maps of the manifolds SG_n. Let $n>4$ and $n>k$; then the group $\pi_1(SG_n)\cong B'(n)$ admits no nontrivial homomorphisms into the group $\pi_1(G_k)\cong B(k)$ and *a fortiori* not into $\pi_1(SG_k)\cong B'(k)\subset B(k)$ (V.Ya. Lin 1972; cf. [35]). Therefore each map $f\in\operatorname{Hol}(SG_n, SG_k)$ lifts to a holomorphic map

$$F\colon SG_n\to SE_k^+=\left\{\lambda=(\lambda_1,\dots,\lambda_k)\colon \sum_{j=1}^{k}\lambda_j=0, \prod_{p<q}(\lambda_p-\lambda_q)=i^{k(k-1)/2}\right\}.$$

and is therefore constant (the functions $(\lambda_p-\lambda_1)/(\lambda_2-\lambda_1)$ belong to $\operatorname{Hol}_*(SE_k^+, \mathbb{C}\setminus\{0;1\})$ and "almost separate" the points of SE_k^+, while SG_n is Picard). Thus $\operatorname{Hol}_*(SG_n, SG_k)=\emptyset$ if $n>4$ and $n>k$. Moreover one can show that for $n>4$ the normal subgroup $J(n)=I(n)\cap B'(n)$ is a completely characteristic subgroup of $B'(n)$, i.e. $\varphi(J(n))\subset J(n)$ for each endomorphism $\varphi\colon B'(n)\to B'(n)$ (V.Ya. Lin [93], [35]). Therefore for $n>4$ each continuous (holomorphic) map $f\colon SG_n\to SG_n$ lifts to an equivariant (with respect to the action on SE_n^+ of the alternating group $A(n)$) continuous (holomorphic) map $F\colon SE_n^+\to SE_n^+$ of the cover SE_n^+ of SG_n corresponding to the subgroup $J(n)\subset B'(n)$. The equivariant holomorphic maps $F\colon SE_n^+\to SE_n^+$ can be described by a method close to the one employed in the description of the analogous maps $E_n\to E_n$; after this a transition to SG_n leads to the following result.

Theorem 8 (Sh.I. Kaliman [20], cf. further [35]). *For $n>4$ each nonconstant holomorphic map $f\colon SG_n\to SG_n$ is a biregular automorphism and can be written in the form $f(0, z_2, \dots, z_n)=(0, \varepsilon^2 z_2, \dots, \varepsilon^n z_n)$ where $\varepsilon^{n(n-1)}=1$; in particular,* $\operatorname{Aut} SG_n\cong\mathbb{Z}/n(n-1)\mathbb{Z}$.

The same is true also for $n=3$ [17], but not for $n=4$; there exist nonconstant but everywhere degenerate holomorphic maps $f\colon SG_4\to SG_4$ (cf. [35], § 8, Sec. 2.2).

Concerning the set $\operatorname{Hol}_*(SG_n, SG_k)$ for $4<n<k$ one knows only that it is finite (E.A. Gorin, V.Ya. Lin, cf. [31]) [18]; it is plausible that it is empty.

[17] SG_3 is isomorphic to the curve $\{x^3+y^3=1\}$ (a torus with one puncture) so the holomorphic maps $SG_3\to SG_3$ can be studied directly.

[18] In fact, $\operatorname{Hol}_*(X, SG_k)$ is finite for any algebraic variety X.

In SG_n we have the smooth algebraic curve $\Gamma_n' = SG_n \cap \{z_2 = \ldots = z_{n-2} = 0\}$ isomorphic to the curve $\Gamma_n = \{x^n + y^{n-1} = 1\} \subset \mathbb{C}^2$. The curve Γ_n' is π_1-determative in SG_n (cf. Sec. 3.1), and as SG_k is C-hyperbolic, then Γ_n' must be a uniqueness set for holomorphic maps $SG_n \to SG_k$. Therefore for the verification of the conjecture $\mathrm{Hol}_*(SG_n, SG_k) = \emptyset$ $(4 < n < k)$ it would be sufficient to show that (for such n and k) $\mathrm{Hol}_*(\Gamma_n, SG_k) = \emptyset$ (let us remark that for appropriate $n > k$ the set $\mathrm{Hol}_*(\Gamma_n, SG_k)$ is nonempty). In connection with this it is of interest to find out whether SG_k $(k > 4)$ contains any smooth algebraic curve of genus $g < (k-1)(k-2)/2$.

The space G_n^0 is a "complex Weyl chamber" for the group $S(n)$, i.e. the group A_{n-1} in Coxeter's classification of finite groups generated by reflexions. Analogous spaces are connected with other finite Coxeter groups (Brieskorn 1971, cf. [68], [69]). Complex-analytic properties of such spaces for the groups of the series B and D (and, especially, their holomorphic maps) have been studied by V.M. Zinde, Sh.I. Kaliman, V.Ya. Lin, O.V. Lyashko (cf. the references in [35]).

Chapter 2
Finiteness Theorems

§ 1. Nonconstant Maps

1.1. Conditions of Hyperbolic Kind. The proofs of many finiteness theorems are based on the fact that the space of maps under consideration turns out to be compact and discrete. Here is an example how this mechanism works (Kaup [24]).

Theorem 1. *Let X, Y be connected compact complex spaces and assume that Y is C-hyperbolic. Then for any two points $x_0 \in X$, $y_0 \in Y$ the set $F_{x_0, y_0} = \mathrm{Hol}((X, x_0), (Y, y_0))$ is finite.*

Indeed, by the Borel-Narisimhan theorem (Chap. 1, Sec. 3.1) F_{x_0, y_0} must be discrete. As X and Y are compact and Y is C-hyperbolic, $\mathrm{Hol}(X, Y)$ is compact ([25], [26], Part III); therefore F_{x_0, y_0} must be compact too.

Example 1. If R_g is a compact Riemann surface of genus $g > 1$, then the manifold $R_g \times R_g$ is C-hyperbolic but $\mathrm{Hol}(R_g, R_g \times R_g)$ is not discrete. Therefore it is necessary in Theorem 1 to fix the points x_0, y_0.

Urata [55] proved, in the same hypothesis on X and Y, the finiteness of the set $F_{A,B} = \{f \in \mathrm{Hol}(X, Y) : f(A) = B\}$, where A and B are arbitrary connected analytic subsets of X and Y respectively. This can be derived from Theorem 1 with the aid of another result of his, according to which

$\mathrm{Hol}_{\mathrm{dom}}(A, B)$ is finite (cf. § 2). In [54] the finiteness of F_{x_0, y_0} is proved in the assumption that X is a connected compact complex space and Y is taut [19] (cf. [25], [26]; let us remark that for a compact Y tautness is equivalent to hyperbolicity). The set $F_{A,B}$ is finite if X, A and B are compact, A and B connected and Y a complete Hermitean manifold such that all holomorphic sectional curvatures are bounded from above by a negative constant [55]; in the same assumptions on X and Y the space $\mathrm{Hol}_*(X, Y)$ will be discrete (and finite if Y is compact); cf. [21] and further § 2 and § 3.

1.2. Negativity in the Sense of Grauert. A holomorphic vector *bundle* $p: E \to F$ is called *negative in the sense of Grauert* (or weakly negative), if there exists a continuous function $\varphi: E \to \mathbb{R}_+$, twice smooth and strictly plurisubharmonic outside the zero section $Z \subset E$, with $\varphi^{-1}(0) = Z$. For complete algebraic varieties over an arbitrary field k one has a different description equivalent to the given one when $k = \mathbb{C}$: the bundle $p: E \to Y$ is negative in the sense of Grauert if there exists a birational morphism $\psi: E \to S$ onto an affine algebraic variety S under which the zero section collapses to a point $s_0 \in S$ and $\psi | (E \setminus Z): E \setminus Z \to S \setminus \{s_0\}$ is a biregular isomorphism ([77]).

The negativity of the holomorphic sectional curvatures implies negativity in the sense of Grauert of the tangent bundle TY for a Hermitean manifold Y (cf., for instance [21], Lemma 1). For a compact complex manifold Y negativity in the sense of Grauert of TY implies the hyperbolicity of Y (Kobayashi; cf., for example, [55], Prop. 5) [20]. Therefore, the statements formulated at the end of the last paragraph concerning the discreteness or finiteness of the space $\mathrm{Hol}_*(X, Y)$ follows from the following theorem (Kalka, Shiffman, B. Wong [21]; cf. also Urata [54], Noguchi [41]).

Theorem 2. *If X is a connected compact complex space and Y is a complex manifold, with the tangent bundle TY negative in the sense of Grauert, then* $\mathrm{Hol}_*(X, Y)$ *is discrete (and finite, if Y is compact).*

The proof is based on the following three lemmata (essentially, the first is due to Grauert [77] and the second to Douady [75]).

Lemma 1. *If TY is negative in the sense of Grauert, then each connected compact analytic subset $A \subset TY$ of positive dimension is contained in the zero section Z.* [21]

---

[19] In Part III such spaces are called Montel spaces.

[20] The converse is in general not true (B. Wong [120]); a counterexample is provided by the Brody-Green surfaces, which are smooth hyperbolic surfaces of degree $\geqq 25$ in $\mathbb{CP}^3$ (cf. Chap. 1, Sec. 3.1). This depends on the fact that for the Chern classes $c_i = c_i(E)$ $(i = 1, 2)$ of any negative in the sense of Grauert bundle E of rank 2 over a surface Y one has Kleiman's inequality $c_1^2 - c_2 > 0$. Now, if $E = TY$, Y a smooth surface of degree m in $\mathbb{CP}^3$, we have: $c_1^2 - c_2 = (10 - 4m) h^2$, where h is the first Chern class of the hyperplane section bundle, so that $c_1^2 - c_2 < 0$ for $m \geqq 3$ and, thus, TY can not be Grauert negative.

[21] It follows from this lemma that the tangent bundle TY of the direct product $Y = Y_1 \times Y_2$ of two compact manifolds of positive dimension can not be Grauert negative.

Indeed, as φ is (strictly) plurisubharmonic and A compact, φ must be constant on A. If $A \not\subset Z$ one can find $f \in \text{Hol}_*(D, A)$ such that $f(0) \notin Z$ and the tangent vector $v = df/dt|_{t=0}$ to TY at the point $f(0)$ is different from 0. But the number $\partial^2(\varphi \circ f)/\partial t \, \partial \bar{t}|_{t=0}$ coincides with the value of the Levi form of φ on the vector v and, therefore, must be positive, which again contradicts the fact that $\varphi \circ f$ is constant.

Lemma 2. *Let X, Y be complex spaces and X compact. There exist a complex structure on $\text{Hol}(X, Y)$ such that the natural map $X \times \text{Hol}(X, Y) \to Y$, $(x, f) \mapsto f(x)$ is holomorphic.*

Lemma 3. *Let X be a connected compact complex space, Y a complex manifold and F a nonempty open subset of $\text{Hol}_*(X, Y)$. Let $p: TY \to Y$ be the natural projection. If F is not discrete then there exists a commutative diagram of holomorphic maps*

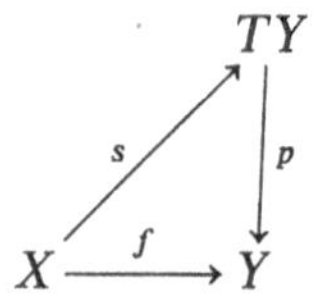

$$
\begin{array}{ccc}
 & TY & \\
{\scriptstyle s}\nearrow & & \downarrow{\scriptstyle p} \\
X & \xrightarrow{\ f\ } & Y
\end{array}
\tag{1}
$$

where $f \in F$ and $s \not\equiv 0$, such that $s(X) \subset TY$ is a connected compact analytic subset of positive dimension not contained in the zero section of the bundle TY.

Proof. As F is open in $\text{Hol}_*(X, Y)$ and not discrete, by Lemma 2 there exist a nonconstant map $\Phi: D \to F$ such that the map

$$
F: X \times D \xrightarrow{\ (\text{id}, \Phi)\ } X \times F \xrightarrow{\ h|(XF)\ } Y
$$

is holomorphic. As $\Phi \neq \text{const}$, one can find a point $x_0 \in \infty$ such that the map $F(x_0, \cdot): D \to Y$ is nonconstant; therefore for some $t_0 \in D$ the tangent vector $w_0 = \partial F(x_0, t)/\partial t|_{t=t_0}$ to Y does not vanish. Set $f(x) = F(x, t_0)$ and $s(x) = \partial F(x, t)/\partial t|_{t=t_0}$ (the value of the differential of the map $F(x, \cdot): D \to Y$ at the point t_0 on the vector d/dt). As $s(x_0) = w_0 \neq 0$, we see that the maps s and f satisfy the desired requirements.

The validity of Theorem 2 follows from Lemmata 1 and 3.

Theorem 3 (Noguchi and Sunada [42]). *Let X be a Zariski open subset of a compact complex space and Y a smooth compact algebraic variety with negative in the sense of Grauert tangent bundle TY. Then the set $\text{Mer}_*(X, Y)$ of all nonconstant meromorphic maps from X into Y is finite.*

An analogous statement is true for separable rational maps of algebraic varieties over an arbitrary algebraically closed field (cf. [42] and also [41] for the case of fields of characteristic 0).

It is easy to derive this theorem from Theorem 2 (also without assuming Y to be algebraic). Indeed, let $\bar{X} \supset X$ be the compactification of X, $\pi: \bar{X}' \to \bar{X}$

a resolution of singularities of $\bar{X}$ by Hironaka and $X' = \pi^{-1}(X) \subset \bar{X}'$. Then one has a natural imbedding $\mathrm{Mer}_*(X, Y) \subset \mathrm{Mer}_*(X', Y)$. As Y is compact, the negativity of TY forces Y to be hyperbolic. But each meromorphic map of a smooth complex manifold X' into a hyperbolic complex space Y is holomorphic (cf. Kodama [28]; the smoothness of X' is essential). Thus $\mathrm{Mer}_*(X', Y) = \mathrm{Hol}_*(X', Y)$. By a theorem of Kwack (cf. [25]) each holomorphic map $f: X' \to Y$ extends to a holomorphic map $\bar{f}: \bar{X}' \to Y$. In view of Theorem 2 and the compactness of Y then $\mathrm{Hol}_*(\bar{X}', Y)$ is finite; consequently, also $\mathrm{Mer}_*(X, Y)$ must be finite.

Example 2. Let Γ be a discrete subgroup of the automorphism group of the ball $B^n \subset \mathbb{C}^n$, acting freely on B^n, and assume that $Y = B^n/\Gamma$ is compact. Pulling down to Y the Bergman metric on B^n we obtain a Hermitean metric with negative holomorphic sectional curvatures. Therefore the tangent bundle TY must be negative in the sense of Grauert. This leads to the following strengthening of a theorem by Imayoshi [18]: for each Zariski open subset X of a compact complex space, $\mathrm{Hol}_*(X, B^n/\Gamma)$ is finite (in [18] X is assumed to be quasiprojective). An analogous statement (in the case X quasiprojective) is proved in [18] for compact quotients of the polydisc $Y = D^n/\Gamma$ where $\Gamma \subset (\mathrm{Aut}\, D)^n \subset \mathrm{Aut}\, D^n$ is a freely acting discrete irreducible subgroup (cf. [42], Prop. 4.2); the hypothesis that Γ is irreducible is essential (Example 1 in Sec. 1.1).

1.3. Maps of Rank $\geqq k$. Let $\mathrm{Hol}_k(X, Y)$ (respectively $\mathrm{Mer}_k(X, Y)$) be the set of all holomorphic (meromorphic) maps $f: X \to Y$ having at some point $x_0 \in X$ a rank $\geqq k$ (the points x_0 and $f(x_0)$ are assumed to be nonsingular and x_0 not a point of indeterminacy for f).

In Noguchi and Sunada [42] it is proved that if in the hypotheses of Theorem 3, instead of negativity in the sense of Grauert of the tangent bundle TY, one assumes that for some $k \geqq 1$ the bundle $\wedge^k TY$ is negative in the sense of Grauert (if $k > 1$, this somewhat weaker) then the set $\mathrm{Mer}_k(X, Y)$ is finite. In particular, Y can contain only a finite number of irreducible analytic subsets of dimension $n \geqq k$, which are biholomorphic to each other.

A holomorphic vector *bundle* $p: E \to Y$ over a connected complex manifold Y is called *G_k-negative* if there exists on E a continuous non-negative function φ, doubly smooth outside the zero section $Z \subset E$, such that $\varphi^{-1}(0) = Z$ and, outside Z, the Levi form of φ has not more than k negative eigenvalues (taken with their multiplicities). In [21] it is shown that if Y is a complex manifold with G_{k-1}-negative tangent bundle TY, then for each compact complex space X the set $\mathrm{Hol}_k(X, Y)$ is discrete. In particular, if $k = \dim Y$ then we have the discreteness of the space $\mathrm{Hol}_{\mathrm{dom}}(X, Y)$ of all dominant holomorphic maps (cf. the following Section; let us notice that if $\mathrm{Hol}_{\mathrm{dom}}(X, Y) \neq \emptyset$ then the compactness of X implies the compactness of Y). If Y is also hyperbolic then $\mathrm{Hol}_{\mathrm{dom}}(X, Y)$ is finite.

Example 3 (Shiffman [117]; cf. also [42]). Let Y_m be a smooth Fermat surface in $\mathbb{CP}^3$, in homogeneous coordinates defined by the equation $z_0^m - z_1^m + z_2^m - z_3^m = 0$. This surface is not hyperbolic, as it contains a projective line $\mathbb{CP}^1 = L = \{z_0 = z_1,\ z_2 = z_3\}$. One can however show that for $m \geqq 5$ the bundle $\wedge^2 TY_m$ is negative in Grauert's sense (cf. Sec. 2.1, Example 1). In $\mathbb{CP}^2$ with homogeneous coordinates $[u:v:w]$ consider the curve $\Gamma_m = \{u^m + v^m - w^m = 0\}$ and set $X_m = \Gamma_m \times \Gamma_m$. Then X_m is a smooth projective surface. The map

$$X_m \ni ([u_1 : v_1 : w_1], [u_2 : v_2 : w_2]) \to [u_1 w_2 : u_1 v_2 : v_1 u_2 : w_1 u_2] \in Y_m$$

is rational and has rank 2. It is not holomorphic at points of the form $([0:1:\omega_1], [0:1:\omega_2])$ where $\omega_j^m = 1\ (j = 1, 2)$. Therefore $\mathrm{Hol}_2(X_m, Y_m) \neq \mathrm{Mer}_2(X_m, Y_m)$, although for $m \geqq 5$ both sets are finite. Contrariwise, the set $\mathrm{Hol}_*(X_m, Y_m) = \mathrm{Hol}_1(X_m, Y_m)$ is infinite (and furthermore nondiscrete), as for each nonconstant holomorphic map $X_m \to \mathbb{CP}^1$ the composition $X_m \to \mathbb{CP}^1 = L \hookrightarrow Y_m$ is nonconstant.

Let Y be a compact quotient of a bounded symmetric domain U by the free action of a discrete subgroup $\Gamma \subset \mathrm{Aut}\, U$. In many cases one can compute explicitly the least k for which the bundle $\wedge^k TY$ is negative in Grauert's sense (it is 1 + the maximal dimension of the proper boundary components of U, [42], Theorem 3.4). This provides us with a large class of examples where the above theorem on the finiteness of $\mathrm{Mer}_k(X, Y)$ is applicable.

§ 2. Dominant Maps

2.1. Preliminary Material. A *meromorphic map* $f\colon X \to Y$ of irreducible complex spaces is called *dominant* if $f(X \setminus N_f)$ contains a Zariski open subset of Y (here N_f is the set of points of indeterminacy of f). If $f\colon X \to Y$ is a dominant map then $\dim X \geqq n = \dim Y$ and $\operatorname{rank} f = n$ on a Zariski open subset of X. The converse is in general not true, as is shown by the Fatou-Bieberbach example of a biholomorphic imbedding $f\colon \mathbb{C}^2 \to \mathbb{C}^2$ for which $f(\mathbb{C}^2)$ does not contain an entire ball (cf. for example, Part II). However, if X is compact (or X and Y both algebraic and f rational) then the two conditions are equivalent. Therefore in such cases the spaces of dominant maps $\mathrm{Hol}_{\mathrm{dom}}(X, Y)$ and $\mathrm{Mer}_{\mathrm{dom}}(X, Y)$ coincide with the spaces $\mathrm{Hol}_n(X, Y)$ respectively $\mathrm{Mer}_n(X, Y)$ defined in Sec. 1.3 (here $n = \dim Y$). Moreover, if X is compact and the map $f\colon X \to Y$ is holomorphic and dominant, then Y must be compact too and f is surjective.

We will make no distinction between a holomorphic vector bundle E over a complex space X and the (locally free) sheaf $\mathcal{O}(E)$ of germs of its holomorphic sections. In particular, $H^k(X, E)$ (or just $H^k(E)$) will stand for the k-th cohomology group (a complex vector space) of the sheaf $\mathcal{O}(E)$ and we write $h^k(X, E) = \dim_{\mathbb{C}} H^k(X, E)$. We denote by mE the m-th tensorial power of the bundle E; analogously, for line bundles (or, likewise, invertible sheaves) L', L'', we write $L'+L''$ for $L'\otimes L''$.

Let L be a holomorphic line bundle on a compact complex manifold X. If $h^0(X, m_0 L)>0$ for some $m_0\in\mathbb{N}$, then there exists α, $\beta>0$ and a (unique) integer k $(0\leqq k\leqq n=\dim_{\mathbb{C}} X)$ such that for all sufficiently large $m\in\mathbb{N}$ holds

$$\alpha m^k \leqq h^0(X, m m_0 L)\leqq \beta m^k$$

(this result as well as the definitions below are due to Iitaka; cf., for example, [17], [48] and further [47]). The number k is termed the *L-dimension* of X and is written $k(L, X)$; if $H^0(X, mL)=0$ for all $m\in N$ one puts $k(L, X)=-\infty$. For the canonical bundle $K_X=\wedge^n T^* X=\Omega^n_X$ one writes $k(X)$ instead of $k(K_X, X)$; the number $k(X)$ (or $-\infty$) is called the *Kodaira dimension* of X.

Let X be a complex manifold which admits a smooth compactification $\bar{X}$ such that $V=\bar{X}\setminus X$ is a hypersurface in $\bar{X}$ with simple normal crossings (i.e. all irreducible components of V are smooth and all singularities are normal crossings; cf. Chap. 1, § 1). The *logarithmic Kodaira dimension* $\bar{k}(X)$ is defined to be the $(K_{\bar{X}}+[V])$-dimension of $\bar{X}$ where $[V]$ is the invertible sheaf corresponding to the divisor V. This is a consistent definition, i.e. $\bar{k}(X)$ does not depend on the compactification $\bar{X}$, subject to the above properties; if X is compact then $\bar{k}(X)=k(X)$. We say that X is a *manifold of hyperbolic type* if $\bar{k}(X)=\dim X$ (for compact X one often says *"of general type"*).

Warning: if V is a hypersurface in $\bar{X}$ with "bad" singularities, it is not possible to conclude from $k(K_{\bar{X}}+[V], \bar{X})=\dim\bar{X}$ that $X=\bar{X}\setminus V$ is a manifold of hyperbolic type, i.e. there may not exist a compactification $\bar{X}'$ of X such that $V'=\bar{X}'\setminus X$ is a hypersurface with simple normal intersections and $k(K_{\bar{X}}+[V'], \bar{X}')=\dim X$ (although there exist such an $\bar{X}'$, if $k(\bar{X})>0$ (Iitaka)). Here is a counterexample (Sakai [47]). Let $\bar{X}=\mathbb{CP}^2$, $m\geqq 4$, $V=\{z_0^{m-1} z_1+z_2^m=0\}$ and $X=\mathbb{CP}^2\setminus V$; in this case $k(K_{\mathbb{CP}^2}+[V])=2$. The holomorphic map f: $\mathbb{C}^2\to X$, $f(x, y)=[1:(e^x-y^m):y]$ is everywhere nondegenerate so that X cannot be measure hyperbolic (cf. [26]; Part III). On the other hand [48], a manifold of hyperbolic type by necessity must be measure hyperbolic (but need not be hyperbolic in Kobayashi's sense, even if it is compact; cf. Example 3 in Sec. 1.3).

A compact complex space X is termed a *space of general type* if it has a smooth compact model[22] which is a manifold of general type (this definition is consistent, i.e. it does not depend on the choice of $\tilde{X}$).

Let L be a holomorphic line bundle over a compact complex manifold X, $h^0(X, L)=N+1\geqq 1$, and write $\mathscr{L}=H^0(X, L)$. For each point $x\in X$ fix an isomorphism φ_x: $L_x\xrightarrow{\approx}\mathbb{C}$ defined by some local holomorphic trivialization of L. Then one can associate with each $x\in X$ an element of the dual

[22] I.e., a compact complex manifold bimeromorphically equivalent to X. Hironaka's theorem on the resolution of singularities guarantees the existence of a compact complex manifold $\tilde{X}$ and a holomorphic map $\tilde{X}\to X$ which is bimeromorphic isomorphism.

space $\mathscr{L}^* \cong \mathbb{C}^{N+1}$, i.e. a linear functional Φ_x on L, defined by the condition

$$\Phi_x(s) = \varphi_x(s(x)), \quad s \in H^0(X, L) = \mathscr{L}.$$

If we pick another trivialization, the functional Φ_x will be multiplied by a number $\lambda \in \mathbb{C}^*$. Therefore the correspondance $\Phi_L: x \to \Phi_x$ defines in a consistent manner a meromorphic map Φ_L of the manifold X into the projective space $\mathbb{P}_L = \mathbb{P}(\mathscr{L}^*) = (\mathscr{L}^* \setminus \{0\})/\mathbb{C}^*$. This map is holomorphic on the Zariski open set $X_L \subset X$ of those points x for which there exists at least one section $s \in L$ which does not vanish at x. (Each basis $s_0, \ldots, s_N$ for L gives homogeneous coordinates in $\mathbb{P}(\mathscr{L}^*)$; in these coordinates $\Phi_L(x) = [s_0(x): \ldots; s_N(x)]$).

If there exists an integer $m_0 \in N$ such that $H^0(X, m_0 L) \neq 0$, then the map Φ_{mL} is defined for each $m \geqq m_0$. It turns out that in this case $k(L, X)$ coincides with the maximal dimension of the images $\Phi_{mL}(X_{mL}) \subset \mathbb{P}_{mL}$ for $m \geqq m_0$. It is known that a compact complex space X is of general type iff for some $m_0 \in \mathbb{N}$ the map $\Phi_{m_0 K_X}$ defines a bimeromorphic imbedding of X into $\mathbb{CP}^N$.

A holomorphic line bundle (invertible sheaf) L over a compact complex space X is called *very ample* if the map Φ_L defines a biholomorphic imbedding of X into the projective space $\mathbb{P}_L$. If $L = [V]$ is the bundle of a divisor V in X, then V is the Φ_L-preimage of a hyperplane section; in this case the *divisor* V is likewise called *very ample*. A *bundle* (invertible sheaf) L, or the corresponding divisor V, is called *ample* if some power $mL (m \in \mathbb{N})$ is very ample. The *ampleness of a line bundle* L is equivalent to its *positivity*, i.e. that the first Chern class $c_1(L)$ can be represented (in the de Rham cohomology $H^2_{DR}(X, \mathbb{R})$) by a closed positive definite (1, 1)-form. The ampleness (positivity) of L is also equivalent to the negativity in the sense of Grauert of the dual bundle $L^* = -L$.

Thus, if the canonical line bundle K_X of a compact complex manifold X is positive (and this is equivalent to the *negativity of the first Chern class* $c_1(X) = c_1(TX)$, because $c_1(TX) = c_1(\wedge^n TX) = -c_1(K_X)$), then X is of general type. The converse is not true; if X is of general type then for large m the map $\Phi_{mK_X}: X \to \mathbb{P}_{mK_X}$ is a bimeromorphic imbedding, but it need not be biholomorphic for any single value of m.

Example 1. Let $n \geqq 2$ and $Y = \mathbb{CP}^n \setminus V$, where V is a hypersurface of degree m with simple normal crossings. Then for the first Chern class we have $c_1(K_{\mathbb{CP}^n} + [V]) = (m - n - 1)h$, where $h = c_1(\mathbb{CP}^n)$. Therefore if $m \geqq n + 2$ the bundle $K_{\mathbb{CP}^n} + [V]$ must be ample, so that $\bar{k}(K_{\mathbb{CP}^n} + [V], \mathbb{CP}^n) = n$; i.e. Y is a manifold of hyperbolic type. If V is smooth and $m \geqq n + 2$ then K_V is positive, so that V is of general type.

A compact irreducible complex space X is called a *Moĭshezon space*, if the degree of transcendency of its field of meromorphic functions equals $\dim_C X$. Each Moĭshezon space is bimeromorphically equivalent to a projective algebraic variety; each projective algebraic manifold is a Moĭshezon space. According to a theorem of Siegel's, the degree of transcendency of the field of meromorphic functions on an irreducible compact complex space

X does not exceed its dimension $\dim X$. Thus, the Moĭshezon spaces are in a sense close to algebraic varieties (cf. [116], Chap. 8, § 2.3, Theorem 3, p. 478).

2.2. Maps in Spaces of General Type. One of the first results concerning the finiteness of the set of meromorphic maps is due to Kobayashi and Ochiai [27].

Theorem 1. *Let X be a Zariski open subset of a compact complex Moĭshezon space $\bar{X}$ and Y an n-dimensional compact complex space of general type. Then the set* $\mathrm{Mer}_n(X, Y)$ *of meromorphic maps $X \to Y$ of rank n is finite.*

Plan of proof. a) The assumption that $\bar{X}$ is Moĭshezon allows one to reduce the situation to the case when X and Y are smooth projective varieties of general type and the same dimension $\dim X = \dim Y = n$ (cf. Remark 1 below). In the sequel we consider this case only.

b) Let Y be a smooth projective variety lying in $\mathbb{CP}^q$ and let H be one of its hyperplane sections, denoting the corresponding holomorphic line bundle over Y by $[H]$. It is clear that $[H]$ is very ample (the restriction to Y of the homogeneous coordinates $z_0, \ldots, z_q$ in $\mathbb{CP}^q$ gives the identical biregular imbedding $Y \to \mathbb{CP}^q$ and, on the other hand, these may also be considered as holomorphic sections of the bundle $[H] \cong \mathcal{O}_{\mathbb{CP}^q}(1)|Y$). As $[-H] \cong J_H$ (the sheaf of ideals of H), then, upon multiplying tensorially the exact sequence of sheafs $0 \to [-H] \to \mathcal{O}_Y \to \mathcal{O}_H \to 0$ by $l K_Y$ ($l \in \mathbb{N}$) and passing to cohomology, we get the exact sequence of vector spaces

$$0 \to H^0(Y, l K_Y - [H]) \to H^0(Y, l K_Y) \to H^0(H, l K_Y | H).$$

If Y is a manifold of general type then for suitable $\alpha > 0$, $m_0 \in \mathbb{N}$ and all sufficiently large $m_1 \in \mathbb{N}$ we have $h^0(Y, m_1 m_0 K_Y) \geqq \alpha m_1^n$ (cf. Sec. 2.1). As $\dim H = n-1$, then $h^0(H, m_1 m_0 K_Y | H) \leqq \beta m_1^{n-1}$. Therefore, it follows from the exactness of the above sequence that $\dim H^0(Y, m_1 m_0 K_Y - [H]) > 0$ for some m_0 and all sufficiently large m_1 (we have reproduced here the proof of a lemma by Kodaira; cf. [47], Lemma 2, and further [89]). Let us fix such m_0 and m_1 and put $L = m K_Y - [H]$, where $m = m_1 m_0$. Then $H^0(L) = H^0(Y, L) \neq 0$ so that the bundle $m K_Y - L = [H]$ is very ample. As X too is a manifold of general type, we may also assume that the canonical map

$$i = \Phi_{m K_X}\colon X \to \mathbb{P}_X = \mathbb{P}(H^0(m K_X)^*) \cong \mathbb{CP}^M$$

is a bimeromorphic imbedding.

c) Let $0 \neq a \in H^0(L)$; then the correspondence

$$H^0(m K_Y - L) \ni s \mapsto a s \in H^0(m K_Y)$$

defines a linear bijection between $H^0(m K_Y - L)$ and a suitable subspace $W \subset H^0(m K_Y)$. Let $t_0, \ldots, t_N$ be a basis in W; the correspondence $y \to [t_0(y) : \ldots : t_N(y)]$ extends to a biregular projective imbedding $j\colon Y \hookrightarrow \mathbb{P}_Y = \mathbb{P}(W^*) \cong \mathbb{CP}^N$. Each $f \in \mathrm{Mer}_{\mathrm{sur}}(X, Y)$ induces an injective linear map f^*:

$W \to H^0(mK_X)$; the surjective dual map f_* of f^* defines a meromorphic map $\tilde{f}: \mathbb{P}_X \to \mathbb{P}_Y$. Furthermore, the diagram

$$\begin{array}{ccc} X & \xrightarrow{f} & Y \\ {\scriptstyle i}\downarrow & & \downarrow{\scriptstyle j} \\ \mathbb{P}_X & \xrightarrow{\tilde{f}} & \mathbb{P}_Y \end{array} \tag{1}$$

is commutative.

d) Let $\mathscr{H} = \mathrm{Hom}(H^0(mK_X)^*, W^*)$. Each element φ of the projective space $\mathbb{P}(\mathscr{H})$ induces a meromorphic map $\tilde{\varphi}: \mathbb{P}_X \to \mathbb{P}_Y$; the set $S \subset \mathbb{P}(\mathscr{H})$ of all $\varphi \in \mathbb{P}(\mathscr{H})$ such that $\tilde{\varphi}$ maps $i(X)$ onto the whole of $j(Y)$ is an algebraic variety. From the commutativity of the diagram (1) we infer that $S = \{\tilde{f}: f \in \mathrm{Mer}_{\mathrm{sur}}(X, Y)\}$ and that the correspondence $f \to \tilde{f}$ between $\mathrm{Mer}_{\mathrm{sur}}(X, Y)$ and S is a bijective one.

e) S is compact (this is the central step in the proof). In order to establish this, we first remark that S is homeomorphic to the space $K = \{f^*: f \in \mathrm{Mer}_{\mathrm{sur}}(X, Y)\} \subset \mathrm{Hom}(W, H^0(mK_X)) = \mathscr{E}$ (cf. (c)). It is, therefore, sufficient to prove that K is compact. To this end one introduces on $\mathscr{E}$ a "norm" $\|\cdot\|_{\mathscr{E}}$ (a nondegenerate positive homogeneous (but not convex) bounded functional satisfying the triangle inequality with some multiplicative constant)[22'] such that $1 \leqq \|f^*\|_{\mathscr{E}} \leqq c$ for all $f^* \in K$ and some $c > 0$. Here the rightmost inequality is an analogue of the Schwarz-Ahlfors lemma for meromorphic maps; this depends on the existence of a volume form on Y with negative definite Ricci form (cf. [25], [26], [27], [47]). The leftmost inequality expresses, essentially, the fact that for surjective meromorphic maps of projective varieties of the same dimension it is possible to define the degree, a natural number. The compactness of K readily follows from these inequalities.

f) S has dimension 0. In fact, let τ be the restriction to S of the standard $\mathbb{C}^*$-bundle $\mathscr{H} \setminus \{0\} \to \mathbb{P}(\mathscr{H})$. The bundle τ has an obvious section $S \ni f \mapsto f_* \in \mathscr{H} \setminus \{0\}$ and is therefore trivial; then also the line bundle θ_S associated with τ is trivial, so that $c_1(\theta_S) = 0$. But θ_S is the restriction to S of the tautological line bundle θ on $\mathbb{P}(\mathscr{H})$, so that $c_1(\theta | S) = 0$. If Γ is a projective curve contained in S then

$$0 = c_1(\theta | \Gamma) = \int_\Gamma c_1(\theta) = -\int_\Gamma c_1(\mathbb{P}(H)) = -\deg \Gamma,$$

which is impossible.

Thus, S is compact and zero dimensional and, therefore, finite. By d) then $\mathrm{Mer}_{\mathrm{sur}}(X, Y)$ too is finite.

Remark 1. If $\dim X = \dim Y$ then it is not necessary to assume in Theorem 1 that $\bar{X}$ is Moĭshezon; it suffices to take $\bar{X}$ to be any compact complex space. The same holds true also for $\dim X \geqq \dim Y$, provided Y is smooth and its canonical bundle is positive (in this case one may, in view of Kodaira's theo-

[22'] *Translator's note.* Thus a quasi-norm!

rem on projective imbeddings, assume that Y is projective); indeed, it suffices to take $k=n$ in the Noguchi-Sunada theorem cited in Sec. 1.3. An analogue of Theorem 1 for separable dominant rational maps of smooth projective varieties over a field of any characteristic was obtained in Martin-Deschamps and Lewin-Menegaux [39] and Kurke [29].

Tsushima [53] proved the following finiteness theorem for maps of non-compact manifolds.

Theorem 2. *Let X and Y be Zariski open subsets of compact complex manifolds $\bar{X}$ and $\bar{Y}$ respectively and assume that $\bar{X}$ is Moĭshezon and Y of hyperbolic type. Then the set $\overline{\mathrm{Mer}}_{\mathrm{dom}}(X, Y)$ of all dominant meromorphic maps $f: X \to Y$ admitting a meromorphic extension $\bar{f}: \bar{X} \to \bar{Y}$ is finite.*

Let us underscore that $\bar{Y}$ need not be a manifold of general type. The proof of this theorem follows to a large extent the plan outlined above. One establishes that the image S' of the space $\overline{\mathrm{Mer}}_{\mathrm{dom}}(X, Y)$ in H is contained in an algebraic variety F; the points of its completion $\bar{F}$ can likewise be interpreted as elements of $\overline{\mathrm{Mer}}_{\mathrm{dom}}(X, Y)$ (this is the central step in the proof); thus, the affine variety F turns out to be complete, and therefore finite. An analogous finiteness theorem is obtained in [53] for strictly rational dominant maps $f: X \to Y$ of smooth algebraic varieties over an algebraically closed field of characteristic 0 in the hypothesis that Y is a manifold of hyperbolic type (a rational map is said to be strictly rational if there exists a proper regular birational map $\varphi: X' \to X$ such that the composition $X' \xrightarrow{\varphi} X \xrightarrow{f} Y$ is regular).

Remark 2. Let X be Zariski open subset of a compact complex space $\bar{X}$ (for $\dim X > n$ we assume in addition that $\bar{X}$ is Moĭshezon) and Y a smooth quasiprojective n-dimensional variety of hyperbolic type. Then $\mathrm{Hol}_n(X, Y)$ is finite. Indeed, Lemma 10 in [27] allows one to restrict oneself to the case $\dim \bar{X} = n$. Resolving the singularities, we may assume that $\bar{X}$ is smooth and $\bar{X} \setminus X$ a hypersurface with normal crossings. Then by Sakai [48] each map $f \in \mathrm{Hol}_n(X, Y)$ extends to a map $\bar{f} \in \mathrm{Mer}_n(\bar{X}, \bar{Y})$, so it remains to invoke Theorem 2. (This statement contains, in particular, Proposition 3.5 in [45]; cf. Example 1 in Sec. 2.1.)

2.3. Restrictions on Chern Classes. Let X be a connected compact complex space and Y a compact n-dimensional Kähler manifold. Kalka, Shiffman and B. Wong [21] proved the following theorem, generalizing an analogous result by Lichnerowicz on automorphisms groups.

Theorem 3. *The space* $\mathrm{Hol}_n(X, Y)$ *is discrete in each of the following cases:*

a) $c_1(Y) \leq 0$ (*for example* $c_1(Y) = 0$) *and* $\chi(Y) \neq 0$;

b) $c_1(Y) \leq 0$ *and* $c_1^n(Y) \neq 0$;

c) *Y is a Kähler $K3$-surface.*[23]

[23] Recently Siu [118] proved that all $K3$-surfaces are Kähler.

The condition $c_1(Y) \leqq 0$ means that the class $c_1(Y) = c_1(TY)$ in the de Rham cohomology $H^2_{DR}(Y, \mathbb{R})$ can be represented by a negative semidefinite closed (1, 1)-form ω; condition (b) is equivalent to requiring that ω can be chosen in such a way that it is negative definite at least at one point $y \in Y$. If $c_1(Y) < 0$, the supplementary conditions on Y are not needed: by the theorem by Noguchi and Sunada $\mathrm{Hol}_n(X, Y)$ is finite (cf. Sec. 1.3 and Remark 1 in Sec. 2.2).

Here is a sketch of the proof in the case (a). If $\mathrm{Hol}_n(X, Y)$ is not discrete then by Lemma 3 in Sec. 1.2 there exists a commutative triangle of holomorphic maps (1), where f is dominant and $s \not\equiv 0$. By Stein's factorization theorem there exits a commutative triangle

$$\begin{array}{ccc} & X' & \\ {}^{g}\nearrow & & \searrow^{f'} \\ X & \longrightarrow & Y, \end{array} \tag{2}$$

where X' is a connected compact complex space and g and f' are holomorphic maps, such that all fibers of g are connected and the fibers of f' are finite. As X is compact, s is then constant on the fibers of g, and therefore can be written in the form $s = s' \circ g$, where $s': X' \to TY$ is a holomorphic map, $s' \equiv 0$ and $p \circ s' = f'$. The essential thing in the proof is that $s'(x') \neq 0$ for all $x' \in X'$. In fact, we shall prove that $\|s'(x')\| = \mathrm{const} > 0$, where the norm $\|\cdot\|$ corresponds to the Kähler metric on Y whose Ricci form coincides with the given form $\omega \in c_1(Y)$ (the existence of such a metric results from a famous theorem by Yau [123]). Let $x'_0 \in X'$ be an arbitrary maximum point for the function $\|s'(x')\|$ and let $X'_0 \subset X'$ and $Y_0 \subset Y$ be neighborhoods of the points x'_0 and $y_0 = f'(x'_0)$ respectively such that $f'(X'_0) = Y_0$, the map $f'_0 = f'|X'_0$: $X'_0 \to Y_0$ being a branched finite sheeted covering and x'_0 the unique f'-preimage of y_0 lying in X'_0. It is clear that

$$\sigma(y) = \sum_{x' \in (f'_0)^{-1}(y)} s'_0(x') \tag{3}$$

(the sum is with multiplicities taken into account) is a holomorphic vector field on Y_0. Moreover, $\sigma(y_0) = k s'(x'_0)$ (k = the multiplicity of f'_0) and the function $\|\sigma(y)\|$ takes its maximal value on Y_0 at the point y_0, the maximal value being $k\|s'(x'_0)\|$. It turns out that

$$\Delta \|\sigma\|^2 \geqq -2\,\mathrm{Ric}(\sigma, \bar{\sigma}) = -2\omega(\sigma, \bar{\sigma}) \tag{4}$$

(Δ = Laplacean). Therefore, as ω is negative, $\|\sigma\|^2$ is subharmonic. By the maximum principle it is the constant on Y_0. Therefore, $\|s'(x')\|$ is constant on X'_0. This means that the set of maximum points of $\|s'\|$ in X' is open, as it obviously also is closed, we must have $\|s'\| = \mathrm{const}$. By resolution of singularities, we may assume that X' is smooth. The map s' defines a nonvanishing section of the induced bundle $E = (f')^* \, TY$ over X' of rank n; therefore $c_n(E) = 0$. On the other hand, a computation of the value of $c_n(E)$ on the fundamental class of the manifold X' gives

$$\langle c_n(E), X'\rangle = \langle (f')^* c_n(TY), X'\rangle = m\langle c_n(TY), Y\rangle = m\chi(Y) \neq 0$$

(m = the multiplicity of f' at generic points) and we have a contradiction.

The case (b) is treated in an analogous manner: in a neighborhood of y_0, where the form ω is negative definite, inequality (4) $\|\sigma\|$ is nonconstant, which leads to a contradiction of the maximality of $\|\sigma(y_0)\|^2$. Finally, case (c) reduces to (a), as for Kähler $K3$-surfaces holds $c_1=0$ and $\chi=24$ (cf. [15], Chap. 4, § 5, p. 583, 590).

Example 2. Let $n \geqq 2$, $Y \to_i \mathbb{CP}^{n+1}$ a smooth hypersurface of degree m, N_Y its normal bundle, H a hyperplane, $h = c_1(\mathbb{CP}^{n+1})$ and, finally, $\alpha = i^*(h)$. Topologically, $T(\mathbb{CP}^{n+1})\,|\,Y = TY \oplus N_Y$ and, as the total Chern class is given by $c(T(\mathbb{CP}^{n+1})) = (1+h)^{n+2}$, it follows that $(1+\alpha)^{n+2} = c(TY)\cdot c(N_Y)$. By the adjunction formula $N_Y = [Y]\,|\,Y = m[H]\,|\,Y$ so that $c(N_Y) = 1 + m\alpha$ and $c(TY) = (1+\alpha)^{n+2}(1+m\alpha)^{-1}$; in particular, $c_1(TY) = (n+2-m)\,\alpha$ and

$$c_n(TY) = \frac{m(n+2)+(1-m)^{n+2}-1}{m^2}\,\alpha^n.$$

Let us remark that $\alpha^n \neq 0$, as $\langle\alpha^n, Y\rangle = \langle h^n, mH\rangle = m$. Therefore for $m = n+2$ we get $c_1(Y)=0$ and $c_n(Y)\neq 0$; i.e., condition (a) of Theorem 3 is fulfilled and the set $\mathrm{Hol}_n(X, Y)$ is discrete for any compact X. If $m > n+2$ then $c_1(Y) < 0$ and so by the Noguchi-Sunada result $\mathrm{Mer}_n(X, Y)$ is finite (cf. Example 3 in § 1).

2.4. Conditions of Hyperbolic Kind. The following theorem is due to Urata [54], [55].

Theorem 4. *Let X, Y be compact complex spaces.*

(a) *If Y is C-hyperbolic then* $\mathrm{Hol}_{\mathrm{dom}}(X, Y)$ *is finite.*

(b) *If Y is hyperbolic then the set S of all dominant holomorphic maps $X \to Y$ with connected fibers is finite.*

Statement (a) of Theorem 4 follows from Lemmata 2 and 3 of § 1 and the following analogue of Lemma 1 in the same section. Recall that in the category of reduced complex spaces the notion of tangent bundle TY is defined; its fiber at a point $y \in Y$ coincides with the Zariski tangent space, i.e. with the vector space of all derivations of the local ring $\mathcal{O}_y(Y)$. This notion is functorial; the differential $df\colon TX \to TY$, where $f \in \mathrm{Hol}(X, Y)$, is holomorphic. Let us remark that Lemma 3 in § 1 is applicable in this more general situation.

Lemma 1. *Let Y be a compact C-hyperbolic complex space, $p\colon TY \to Y$ being the natural projection, and let $A \subset TY$ be an irreducible compact analytic subset such that $p_A = p\,|\,A\colon A \to Y$ is a surjective map. Then $A \subset Z$, where $Z \subset TY$ is the zero section.*

In the proof one uses the continuity and pseudoconvexity (plurisubharmonicity) of the infinitesimal pseudometric of Carathéodory on TY (a function φ on a complex space X is pseudoconvex if each function $\varphi \circ f$, where $f \in \mathrm{Hol}(D, X)$, is subharmonic).

In (a) we may take for Y a compact quotient space of a bounded domain by a discrete torsion free subgroup of its automorphism group.

Simha [52] gave another proof of Theorem 4(b) where the discreteness of S is established with the aid of the following lemma.

Lemma 2 ([52], Theorem 2.1). *For each map $f_0 \in S$ there exists a neighborhood $U \subset S$ such that each $f \in U$ can be written in the form $f = g \circ f_0$, where* $g \in \mathrm{Aut}^* Y$ (here $\mathrm{Aut}^* Y$ is the set of all homeomorphisms of Y which lift to automorphisms of the normalization of Y).

The neighborhood U can be chosen such that each $f \in U$ is constant on the fibers of f_0 and conversely; this gives the representations $f = g \circ f_0$ and $f_0 = g' \circ f$, where $g' = g^{-1}$; it is easy to see that $g \in \mathrm{Aut}^* Y$.

In view of this lemma the finiteness of S follows from the finiteness of the group of automorphisms of a compact hyperbolic complex space (cf. § 3, Sec. 3.1).

§ 3. Automorphism Groups

3.1. Compact Spaces. We do not set forth here results on the discreteness or finiteness of automorphism groups which follow from theorems in §§ 1, 2.

Theorem 1 (B. Wong [58], Yau, cf. [20]). *The group* Aut X *of all automorphisms of a compact complex measure hyperbolic manifold X is discrete.*

Remark 1. In view of Theorem 1 in § 2 (and Remark 1 to it) the group of bimeromorphic transformations of a compact complex space of general type is finite[24]. As we already noted in Sec. 2.1, compact complex manifolds of general type (and also quasiprojective varieties of hyperbolic type) are measure hyperbolic (cf. [48]); whether the converse is true is not known.

On the proof of Theorem 1. By the Bochner-Montgomery theorem [66], [67], Aut X is a complex Lie group of transformations of X. If it is not discrete, one can find a complex one parameter subgroup $\mathbb{C} \to \mathrm{Aut}\, X$, which allows us to construct a not everywhere degenerate holomorphic map $D^{n-1} \times \mathbb{C} \to X$. But this contradicts the measure hyperbolicity of X, because the Kobayashi-Eisenman volume form vanishes identically on $D^{n-1} \times \mathbb{C}$.

In an analogous manner one shows that the group Aut X of automorphisms of a compact Kobayashi hyperbolic complex space X is finite [25] (and coincides with the group Bim X of bimeromorphic transformations, pro-

[24] The finiteness of the group of birational automorphisms of a smooth projective variety of general type was earlier established by Matsumura [98].

vided X is smooth [28]; the smoothness is essential). Moreover, in this case for any $f \in \mathrm{Hol}(X, X)$ one can find a natural number m such that $f^{2m} = f^m$ and the map $f^m: X \to f^m(X)$ is a fibration, whose base $f^m(X)$ is smooth, provided X is smooth (Kaup [24]).

Theorem 2. *Let X be a compact n-dimensional Kähler manifold with a least one nonvanishing Chern number* (for example, this is so if $\chi(X) \neq 0$ or $\chi(\mathcal{O}_X) \neq 0$). *Assume that one of the following two conditions is fulfilled:* (a) $c_1(X) \leqq 0$; (b) $H^{n,0}(X) \neq 0$ (i.e. there exists a nonzero holomorphic n-form on X). *Then* Aut X *is discrete.*

Proof. If the group is not discrete, then it contains a complex one parameter subgroup $\mathbb{C} \to \mathrm{Aut}\, X$, which generates on X a holomorphic vector field s, not everywhere 0. The zero set $Z_s = \{x: s(x) = 0\}$ of this field is nonempty (because by a theorem of Bott[25] all Chern numbers of a compact complex manifold on which there exists a holomorphic vector field without zeros must vanish). In the case (a) (analyzed in [21]) this contradicts the fact that the function $\|s(x)\|^2$, taken in the Yau metric, is subharmonic and, consequently, constant (cf. the proof of Theorem 3 in Sec. 2.3). In case (b) (V.Ya. Lin) we apply to s a theorem by Carrel and Liebermann (cf. [15], Chap. 5, § 4, p. 708), which says that $H^{p,q}(X) = 0$ for $|p-q| > \dim_{\mathbb{C}} Z_s$. As $s \neq 0$, we have $\dim_{\mathbb{C}} Z_s \leqq n-1$ and therefore $H^{n,0}(X) = 0$, contradicting (b).

Let us turn to automorphisms of compact surfaces.

a) For a smooth compact surface X of general type the groups Aut X and Bim X are finite (cf. Remark 1) and for a surface with negative first Chern class $c_1(X)$ these groups coincide[26] (in particular, this is the case for smooth projective surfaces of degree $m \geqq 5$ in $\mathbb{CP}^3$). One has estimates for $\#\mathrm{Aut}\, X$ in terms of numerical invariants of the surface (Andreotti 1950; T.M. Bandman [5], cf. Remark 2 in Sec. 4.1).

b) For a compact Kähler[27] $K3$-surface X (in particular, for a smooth projective surface of degree 4 in $\mathbb{CP}^3$) Aut X is discrete (§ 2, Theorem 3(c); this follows also from the above Theorem 2(b), because for such a surface $\chi(X) \neq 0$ and $H^{2,0}(X) \neq 0$). For a $K3$-surface X, the study of Aut X can to a large extent be reduced to the study of the automorphisms of the sublattice S_X of algebraic cycles in the lattice $H_2(X, \mathbb{Z})$. One can prove (I.I. Pyatetskiĭ-Shapiro and I.R. Shafarevich [113][28]) that modulo a finite group Aut X coincides with the quotient group $G = (\mathrm{Aut}\, S_X)/W(S_X)$, where $W(S_X)$ is the subgroup generated by reflexions of S_X in cycles with the selfintersection

[25] This theorem is a simple consequence of the Bott residue formula; cf., for example, [15], Chap. 3, § 4, p. 427.

[26] For more general results of this nature, see Sec. 3.2.

[27] Cf. footnote [23].

[28] In that paper it is shown, among other things, the following analogue of Hurwitz's theorem (Chap. 1, Sec. 3.2, Theorem 4) for $K3$-surfaces: an automorphism of a $K3$-surface which induces the identity on the cohomology $H^2(X, \mathbb{Z})$ is the identity (for the proof one likewise uses the Lefschetz formula).

index -2 (there exists surfaces for which this quotient group is infinite). The study of the structure of G is carried out by methods of the theory of groups generated by reflexions in Lobachevskiĭ[28'] space, connected with the lattice S_X. This reduction (together with the global Torelli theorem proved in the same paper by I.I. Pyatetskiĭ-Shapiro and I.R. Shafarevich) led finally to a complete solution of the question of finiteness of the groups Aut X and Bir X for algebraic $K3$-surfaces and for Enriques surfaces, that is, to an "enumeration" of the surfaces for which these groups are finite (I.I. Pyatetskiĭ-Shapiro and I.R. Shafarevich, Shioda and Inose, V.V. Nikulin, E.B. Vinberg, and others; cf. Nikulin [106], [107] and further Kondo [90]).

c) For a rational surface X the group Bir X is isomorphic to the Cremona group Bir $(\mathbb{CP}^2)$ and therefore infinite dimensional. But Aut X may still be discrete (finite as well as infinite). A rational surface X with Aut X discrete can be obtained by blowing up $k \geqq 4$ points $p_1, \ldots, p_k$ in general position in $\mathbb{CP}^2$ (in this way the general smooth cubic in $\mathbb{CP}^3$ arises, as it can be gotten by blowing up six points in general position in $\mathbb{CP}^2$; cf. [15], Chap. 4, § 1. p. 480). In fact, let $\pi: X \to \mathbb{CP}^2$ be the natural morphism and $L_i = \pi^{-1}(p_i)$ the exceptional curves of first kind, pasted in the blow up $(1 \leqq i \leqq k)$. If g is an automorphism of X, sufficiently close to the identity map, then the set $\pi \circ g(L_i)$ is contained in an affine neighborhood of p_i; consequently, $\pi \circ g \mid L_i = \text{const}$. As $\pi \mid (X \setminus \bigcup_i L_i)$ is an isomorphism, it is clear that $g(L_i) = L_i$ for all i. Consequently, we can find $g' \in PGL(3, \mathbb{C})$ such that $\pi \circ g = g' \circ \pi$. But then $g'(p_i) = p_i$ for all i and as $k \geqq 4$ then $g' = \text{id} \mid \mathbb{CP}^2$, whence $g = \text{id}_X$. This means that Aut X is discrete.

Example 1 (a rational surface with an infinite automorphism group; cf. [1], Chap. 7, § 1). We apply the above construction in the case $k=9$, the $\{p_i\}_{i=1}^9$ being the points of intersection of two cubics $F_1 = 0$ and $F_2 = 0$ in $\mathbb{CP}^2$. Then X is a rational surface with discrete automorphism group Aut X and a pencil of elliptic curves $\varrho: X \to \mathbb{CP}^1$ $(\varrho = F_1/F_2)$, this family being also relatively minimal (i.e. its fibers contain no exceptional curves of the first kind). The family ϱ has 9 trivial sections $L_i = \pi^{-1}(p_i)$; on its general fibers (which are elliptic curves Γ) one can introduce a group structure taking as 0 the point of intersection between Γ and L_1. The shifts τ_i by the elements $\Gamma \cap L_i$ $(2 \leqq i \leqq 9)$ generate a free Abelian subgroup of rank 7 in Bir X (the single relation $\Sigma_{i=2}^9 \tau_i = id$ is given by Abel's theorem). In fact, the maps $\tau_i: X \to X$ are biregular. Indeed, as τ_i is a birational morphism of the family ϱ, the points of indeterminicy of τ_i would blow up to exceptional curves of the first kind lying in the fibers of ϱ, but there are no such curves. This means that τ_i and τ_i^{-1} have no points of indeterminicy and therefore must be biregular. Thus, Aut $X \supset \mathbb{Z}^7$ is an infinite discrete group.

Each automorphism of a smooth projective hypersurface V in $\mathbb{CP}^n$ $(n \geqq 3)$ of degree $m \neq n+1$ extends to a projective transformation in $\mathbb{CP}^n$ ([15], Chap. 1, § 4, p. 178 and also Chap. 2, § 6, p. 326).

[28'] *Translator's Note.* Non-Euclidean!

In a well-known paper by V.A. Iskovskikh and Yu.I. Manin [84] it is proved that for smooth projective hypersurfaces of degree 4 in $\mathbb{CP}^4$ the groups Bir and Aut coincide and are finite (in particular, such hypersurfaces are not rational, as Bir $\mathbb{CP}^3$ is infinite). As there exist smooth unirational quartics in $\mathbb{CP}^4$ (Segre), they constitute counterexamples to the Lüroth problem in dimension 3.

3.2. Noncompact Spaces. Let X be a complex space. Denote by SBim X the subgroup of Bim X generated by all strictly meromorphic $f \in \operatorname{Bim} X$ (recall that a map $f \in \operatorname{Bim} X$ is strictly meromorphic if there exists a proper bimeromorphic holomorphic morphism $p: \tilde{X} \to X$ such that the map $f \circ p$ is holomorphic). The group SBir X, where X is an algebraic variety, is defined in an analogous manner.

The Iitaka-Sakai theorem [17], [48] states that (a) the groups Aut X and SBim X of a smooth complex manifold X of hyperbolic type are finite (this follows also from Theorem 3 of § 2); (b) if $X = \bar{X} \setminus V$, where $\bar{X}$ is a smooth projective variety and V a hypersurface in $\bar{X}$ with simple normal crossings, assuming also that the line bundle $K_{\bar{X}} + [V]$ is ample (in particular: X of hyperbolic type), then each strictly rational dominant map $f: X \to X$ is the restriction to X of some $\bar{f} \in \operatorname{Aut} \bar{X}$; thus the groups SBir X, Aut X and Aut $X \cap \operatorname{Aut} \bar{X}$ coincide and are finite.

Example 2 ([17], [48], cf. also [45]). If $V \subset \mathbb{CP}^n$ is a hypersurface of degree $m \geqq n+2$ with simple normal crossings then each bimeromorphic map of the manifold $X = \mathbb{CP}^n \setminus V$ extends to a projective transformation and the group of all bimeromorphic transformation of X is finite. The assumption $m \geqq n+2$ is essential; in fact, the group Aut $(\mathbb{CP}^n \setminus V)$, where V is the union of $n+1$ hyperplanes in general position in $\mathbb{CP}^n$, has a connected component isomorphic to $(\mathbb{C}^*)^n$.

Theorem 3 (Iitaka [17]). *Let X be a nonsingular algebraic variety of hyperbolic type. Then every proper dominant morphism $f: X \to X$ is an automorphism. If further X is affine then* SBir $X = \operatorname{Aut} X$.

Example 3. Let $X = \mathbb{CP}^n \setminus V$ be as in Example 2; then each nondegenerate proper holomorphic map $f: X \to X$ is biholomorphic and such maps are in finite number ([45], Cor. 3.4). In particular, one may take for V the union of $n+2$ hyperplanes in general position.

Remark 2. For some applications of finiteness (discreteness) theorems for automorphism groups to the study of the structure of holomorphic bundles and to the cancellation theorems, see [58], [56].

§ 4. Estimates for the Number of Maps

In this Section we will reproduce some results concerning estimates of the number of holomorphic maps by suitable characteristics of the target space, which generalize the theorem of de Franchis-Severi; it is question of estimates for $\#\mathrm{Hol}_{\mathrm{dom}}(X, Y)$ for fixed X and Y. A modern proof of this part of the de Franchis-Severi theorem can be found, for example, in [38] and in its full extent and with effective estimates in [16] (cf. also Sec. 5.1).

4.1. Maps of Quasiprojective Varieties. *The Hilbert polynomial* $P_{X,L}(m)$ of an invertible sheaf L over a compact complex space $\bar{X}$ is by definition the Euler characteristic of the sheaf mL:

$$P_{\bar{X},L}(m)=\chi(mL)=\sum_{i=0}^{\dim X}(-1)^i \dim H^i(\bar{X}, mL).$$

$P_{X,L}$ is a polynomial of degree $\leqq \dim X$ in m with rational coefficients. For a compact manifold $\bar{X}$, the Hilbert polynomial $P_{\bar{X}}(m)$ is defined to be the Hilbert polynomial of its canonical sheaf.

A theorem by Matsusaka [99] states: for each polynomial $P=P(m)$ there exists $m_0=m_0(P)$ such that if L is an ample invertible sheaf over a projective variety $\bar{X}$ with $P_{\bar{X},L}=P$ then kL is very ample for $k\geqq m_0$.

Let us fix a polynomial $P=P(m)$ of degree n and denote by $\mathscr{H}(P)$ the class of all n-dimensional manifolds X which can be represented in the form $X=\bar{X}\setminus V_X$ ($\bar{X}$ = a nonsingular projective variety, V_X a hypersurface in $\bar{X}$) such that the sheaf $K_X+[V_X]=L$ is ample and has the Hilbert polynomial $P_{\bar{X},L}=P$. Further, denote by $\mathscr{B}_n$ the class of all n-dimensional manifolds Y which can be represented in the form $Y=\bar{Y}\setminus V_Y$ (where V_Y is a divisor with normal crossings in a nonsingular projective variety $\bar{Y}$) such that the sheaf $K_Y+[V_Y]$ is ample. Let $\mathrm{Hol}_{\mathrm{sur}}(X, Y)$ (respectively $\mathrm{Mer}_{\mathrm{sur}}(X, Y)$) be the set of all proper surjective holomorphic (meromorphic) maps $X\to Y$. By Theorem 2 in § 2, $\mathrm{Hol}_{\mathrm{sur}}(X, Y)$ is finite for all $X\in\mathscr{H}(P)$, $Y\in\mathscr{B}_n$. From this and a theorem by T.M. Bandman [4] we get the following theorem.

Theorem 1. *There exists a constant* $\alpha=\alpha(P)$ *such that* $\#\mathrm{Hol}_{\mathrm{sur}}(X, Y)\leqq\alpha$ *for arbitrary* $X\in\mathscr{H}(P)$, $Y\in B_n$.

Remarks. 1. If (in the hypothesis of Theorem 1) X and Y are compact (i.e. $V_X=\emptyset$, $V_Y=\emptyset$) then $\mathrm{Mer}_{\mathrm{sur}}(X, Y)=\mathrm{Hol}_{\mathrm{sur}}(X, Y)$ (T.M. Bandman [31]). Moreover, in this case it follows from the Riemann-Roch-Hirzebruch formula that the coefficients of the Hilbert polynomial $P_X(m)$ can be expressed linearly in the Chern numbers of X. Therefore $\#\mathrm{Mer}_{\mathrm{sur}}(X, Y)$ can be estimated by topological characteristics of X, namely by its Chern numbers. In [3] there is given also an estimate which involves only n, $[c_1(X)]^n$ and the number $m_0(P)$ in Matsusaka's theorem.

2. It is not known whether Theorem 1 is true for $\dim X>\dim Y$ (in Sec. 5.1 we show that this is the case if X and Y are compact). Nevertheless, if Y is a smooth hyperbolic curve and $X\in\mathscr{H}(P)$ then even $\#\mathrm{Hol}_*(X, Y)$ can be estimated in terms of P (T.M. Bandman [5]). If Y is a smooth compact curve of genus $g\geqq 2$ and X is a smooth projective surface then $\#\mathrm{Hol}_*(X, Y)$

can be estimated by the Betti numbers $b_1(X)$, $b_2(X)$ of X ([3]). For smooth projective surfaces X, Y of general type, $\#\mathrm{Hol}_{\mathrm{dom}}(X, Y)$ can be estimated by $b_2(X)$ ([5]).

3. For a compact manifold X with universal cover $\tilde{X} \cong D^n$ one can estimate $\#\mathrm{Hol}_{\mathrm{dom}}(X, Y)$ in terms of n and the Euler characteristic $\chi(X)$ for any smooth projective variety Y with ample canonical sheaf [5].

4. If $X \subset \mathbb{CP}^N$ is an *n-dimensional complete intersection*[29] of degree $m \geqq N+2$ then $\#\mathrm{Hol}_{\mathrm{dom}}(X, Y)$ can be estimated by n, m and N for an arbitrary smooth projective variety Y with ample canonical bundle [5].

Let us outline the plan of the proof of Theorem 1. Let $\mathscr{B}_n(X)$ be the set of all $Y \in B_n$ such that $\mathrm{Hol}_{\mathrm{sur}}(X, Y) \neq \emptyset$. First one shows that there exists a finite collection of polynomials $\mathscr{P}(P) = \{P_1, \ldots, P_s\}$, defined only by the Hilbert polynomial $P = P_X$, such that the Hilbert polynomial P_Y of any $Y \in \mathscr{B}_n(X)$ coincides with one of the polynomials $P_i \in \mathscr{P}(P)$. Using this and the above mentioned theorem of Mazusaka's, one sees that one can imbed all manifolds $\bar{X}$ and $\bar{Y}$ with $X \in \mathscr{H}(P)$, $Y \in \mathscr{B}_n(X)$ into a fixed projective space $\mathbb{CP}^N$, where N and the degrees of the images of $\bar{X}$, $\bar{Y}$, V_X, V_Y under this imbedding can be estimated from above by a quantity $\alpha_1(P)$ depending only on P. Moreover, there exists an $\alpha_2(P)$ estimating from above the degree of the graph $\Gamma_f \subset XY \subset \mathbb{CP}^N \times \mathbb{CP}^N$ of arbitrary maps $f \in \mathrm{Hol}_{\mathrm{sur}}(X, Y)$ for manifolds $X \in \mathscr{H}(P)$, $Y \in \mathscr{B}_n(X)$. Finally, the proof is completed by the following lemma.

Lemma 1. *Let $\bar{X}, \bar{Y} \subset \mathbb{CP}^N$, $X = \bar{X} \backslash V_X$, $Y = \bar{Y} \backslash V_Y$, $X \in \mathscr{H}(P)$, $Y \in \mathscr{B}_n(X)$ and let the degrees d_1, d_2, d_3, d_4 of $\bar{X}$, V_X, $\bar{Y}$, V_Y be fixed. Let $F_k(X, Y)$ denote the set of all $f \in \mathrm{Hol}_{\mathrm{sur}}(X, Y)$ such that $\deg f = k$. Then $\# F_k(X, Y)$ can be estimated from above by a number α_3 depending only on N, n, d_i and k.*

For the proof of the lemma one constructs universal (i.e., depending only on N, n, d_i, k) quasiprojective varieties H and G possessing the following properties: (a) H parametrizes in a natural way all pairs (X, Y) (satisfying the hypothesis of the lemma) such that $F_k(X, Y) \neq \emptyset$; (b) G parametrizes all triples (X, Y, f) such that $(X, Y) \in H$ and $f \in F_k(X, Y)$. The "forgetful" morphism $\pi: G \to H$, $\pi(X, Y, f) = (X, Y)$ is surjective. It follows from Theorem 2, § 2 that each fiber $\pi^{-1}(X, Y) \subset \mathrm{Hol}_{\mathrm{dom}}(X, Y)$ is finite. As the triple (G, π, H) depends only on N, n, d_i, k, we may choose for α_3 the degree of the morphism π.

4.2. Functions Without Two Values. Let X be a smooth irreducible affine algebraic variety. Is it possible to estimate the number $\gamma(X) = \#\mathrm{Hol}_*(X; \mathbb{C} \backslash \{0; 1\})$ only by topological characteristics of X? This question arose in connection with some problems in the theory of algebraic functions of several variables (cf. [31]). Let us first observe that $\gamma(X) < \infty$ (of course, this follows from Theorem 2, § 2, but there are also simple direct proofs).

[29] That is, the transversal intersection of smooth hypersurfaces in $\mathbb{CP}^N$; the degree of X equals the product of the degrees of the hypersurfaces. The condition $m \geqq N+2$ is equivalent to K_X being ample ([26], Example 13).

Set $r_i = \operatorname{rank} H^i(X, \mathbb{Z})$. If $\dim X = 1$ then Hurwitz's inequality (Chap. 1, Sec. 2.1) shows that $\deg f \leqq r_1 - 1$ for each $f \in \operatorname{Hol}_*(X, \mathbb{C} \setminus \{0; 1\})$; this makes it possible to obtain the estimate $\gamma(X) \leqq (3r_1)^{r_1}$ (Sh.I. Kaliman). The case $\dim X = 2$ has been treated by T.M. Bandman ([5]; cf. further [31]).

Theorem 2. *Let X be a smooth irreducible affine algebraic surface. Then* $\gamma(X) \leqq (6r_1)^{2(r_1+r_2)}$.

For $\dim X > 2$ only little is known. It is clear that if $r_1 \leqq 1$ then $\gamma(X) = 0$ (cf. Chap. 1, Sec. 2.2, Corollary 2). Let $r_1 \geqq 2$. In Sec. 2.2, Chap. 1 (Example 1) it was shown that a nonconstant holomorphic function on X can not omit more than r_1 different values. It turns out that the following theorem holds.

Theorem 3 (V.Ya. Lin). *If on a smooth affine algebraic variety X with $r = \operatorname{rank} H^1(X, \mathbb{Z}) \geqq 2$ there exists a nonconstant holomorphic function which leaves out at least $r-1$ different values, then the nonconstant holomorphic functions on X omitting the values 0 and 1 are in number not more than the same functions on the complex line $\mathbb{C}$ with r punctures.* (For example, $\gamma(X) \leqq (3r)^r$.)

It follows readily from this theorem that if $\operatorname{rank} H^1(X, \mathbb{Z}) = 2$ then $\gamma(X) = 0$ or 6, and that if $\operatorname{rank} H^1(X, \mathbb{Z}) = 3$ then $\gamma(X) \leqq 36$. The first case, which has not been investigated, is the case of 3 dimensional manifolds with $\operatorname{rank} H^1(X, \mathbb{Z}) = 4$.

In many cases of importance for the applications one can not only estimate the number of functions $f \in \operatorname{Hol}_*(X, \mathbb{C} \setminus \{0; 1\})$ but also write down explicitly all such functions (cf. Chap. 1, Sec. 4.3).

§ 5. Other Finiteness Theorems

5.1. Finiteness of the Number of Targets. Let X, Y be algebraic varities. We say that Y is a target of X if there exists at least one dominant rational map $X \to Y$. The set of all Y which are targets of X will be denoted by $\operatorname{Targ}(X)$[30]. We denote further by $\mathscr{F}_m(X)$ the set of birational equivalence classes of projective algebraic varieties $Y \in \operatorname{Targ}(X)$ such that the m-th canonical map $\Phi_m: Y \to \Phi_m(Y)$ is birational. A projective algebraic variety Y is said to be semipositive of general type if it is birationally equivalent to a nonsingular projective manifold Y' of general type with semipositive canonical sheaf $K_{Y'}$ (the last thing means that the degree of the divisor $K_{Y'} | \Gamma$ on each curve $\Gamma \subset Y'$ is nonnegative or (equivalently) that $\langle c_1(K_{Y'} | \Gamma), \Gamma \rangle \geqq 0$). Let us denote by $\mathscr{E}(X)$ the set of all birational equivalence classes of semipositive projective varieties Y of general type with $Y \in \operatorname{Targ}(X)$.

Maehara [36] proved the following theorem.

---

[30] The elements of $\operatorname{Targ}(X)$ are really not the varieties themselves but rather classes of isomorphic (biregularly equivalent) varieties.

Theorem 1. *The sets* $\mathscr{F}_m(X)$ *and* $\mathscr{E}(X)$ *are finite for any projective algebraic variety.*[31]

For each smooth compact curve Γ of genus $g \geqq 2$ (respectively for each smooth projective surface Y of general type) the map $\Phi_3 : \Gamma \to \mathbb{CP}^{N_\Gamma}$ (respectively $\Phi_5 : Y \to \mathbb{CP}^{N_Y}$) is a projective imbedding (cf. [1]). Moreover, any two birationally equivalent smooth algebraic curves are isomorphic. Therefore from Theorem 1 one gets the following corollary.

Corollary 1 ([36]). *For a projective variety* X a) *the set of all smooth projective curves of genus* $g \geqq 2$ *belonging to* Targ(X) *is finite*; b) *the set of birational equivalence classes of projective surfaces* Y *of general type belonging to* Targ(X) *is finite.*

Let us remark that statement (b) is not true in characteristic $p > 0$.

In the noncompact situation, apparently, only one result on the finiteness of the number of targets is known.

Theorem 2 (Imayoshi [19]). *Let* X *be a Riemann surface of finite type. Then the set of all hyperbolic Riemann surfaces* Y *such that* $\mathrm{Hol}_*(X, Y) \neq \emptyset$ *is finite.*

A multidimension generalization of the de Franchis-Severi theorem (in its full extent, i.e., also with an estimate of the number of targets and the number of maps) was obtained in Howard and Sommese [16].

Theorem 3. *Let* X *be a connected compact complex manifold with negative first Chern class* $c_1(X)$ *(or, equivalently, with ample canonical bundle).* a) *The cardinality of the set* $T(X)$ *of all pairs* (Y, f), *where* Y *is a smooth projective curve of genus* $g \geqq 2$ *and* $f \in \mathrm{Hol}_*(X, Y)$, *admits the estimate* $\# T(X) \leqq A$, *where* A *depends only on the Chern numbers of* X. b) *If* $\dim X = 1$ *one has an effective estimate for* $\# T(X)$ *in terms of the genus of* X.

The proof of statement (a) amounts to a reduction to the case $\dim X = 1$, i.e., the classical de Franchis-Severi theorem. We give this reduction in a slightly more general situation using a method somewhat different from the one in [16] (cf. Sec. 3.1, Chap. 1).

Lemma 1. *Let* X *be as in Theorem 3 and let* $k \in \mathbb{N}$, $k \leqq n = \dim X$. *Then there exists a k-dimensional connected compact complex submanifold* X' *in* X *such that:* a) *for each irreducible k-dimensional C-hyperbolic complex space* Y *the restriction* $f \to f \mid X'$ *defines an imbedding* $\mathrm{Hol}(X, Y) \to \mathrm{Hol}(X' Y)$ *and* $f(X') = f(X)$ *for all* $f \in \mathrm{Hol}(X, Y)$; b) *the canonical sheaf* $K_{X'}$ *is ample*; c) *the Chern numbers of* X' *can be estimated from above by the Chern numbers of* X.

[31] The proof of this theorem reminds one of the proof of Bandman's theorem (Sec. 4.1, Theorem 1). The proof of the main lemma, based on the Kawamata-Fujita theory of semipositive manifolds, is also set forth in [88]).

Proof. If $n=k$ there is nothing to prove; so assume that $n>k$. By Matsusaka's theorem there exists an $m\in\mathbb{N}$ depending only on the Chern numbers of X (cf. Sec. 4.1, Remark 1) such that the m-th canonical map $\Phi_m: X \to \Phi_m(X)\subset\mathbb{CP}^N$ is biholomorphic. Let us identify X with $\Phi_m(X)$ and consider a generic (and therefore smooth) hyperplane section $Z\xrightarrow{i}X$. By Lefschetz's theorem Z is connected and the imbedding i induces an epimorphism of fundamental groups $i_*: \pi_1(Z)\to\pi_1(X)$ (this is an isomorphism if $n-1=\dim Z>1$). Therefore it follows from the Borel-Narasimhan theorem (Chap. 1, Sec. 3.1) that the restriction $f\to f\,|\,Z$ defines an imbedding $\mathrm{Hol}(X, Y) \to\mathrm{Hol}(Z, Y)$. Let $f\in\mathrm{Hol}_*(X, Y)$, $y_0\in f(X)$ and set $Z_{y_0}=f^{-1}(y_0)$. As $\dim Z_{y_0} \geqq n-k$ and Z is a hyperplane section of X, then $Z\cap Z_{y_0}\neq 0$. Therefore $f(Z) =f(X)$. By the adjunction formula $K_Z=(K_X+[Z])\,|\,Z$. As X is imbedded in $\mathbb{CP}^N$ with the aid of the m-th canonical map Φ_m, the bundle of the hyperplane section is $[Z]=mK_X$; therefore $K_Z=(m+1)K_X\,|\,Z$, which shows that K_Z is ample. By the Whitney product formula $c(TX\,|\,Z)=c(TZ)\,c(N_{Z|X})$. The normal bundle is $N_{Z|X}=[Z]\,|\,Z$ so that $c(N_{Z|X})=1-mi^*c_1(X)$. This means that $c(Z)=i^*[c(X)(1-mc_1(X))^{-1}]$. Finally, $\langle i^*(h), Z\rangle=\langle c_1([Z])\,h, X\rangle= \langle 1-mc_1(X)\,h, X\rangle$ for each $h\in H^{2m-2}(X,\mathbb{Z})$; therefore the Chern numbers of Z can be expressed in terms of m and the Chern numbers of X and, as a result, can be estimated by these Chern numbers. If $\dim Z>k$, we repeat the construction. The proof is complete.

For $k=1$ the genus g of the curve X' equals $1-c_1(X')/2$, so that Theorem 3(a) follows from Lemma 1 and the de Franchis-Severi theorem (recall that every smooth projective curve of genus $g\geqq 2$ is C-hyperbolic).

Remark 1. Let $n>k$, $X\subset\mathbb{CP}^N$ be a connected projective variety, Y_i k-dimensional complex spaces and $f_i\in\mathrm{Hol}(X, Y_i)$ $(i=1, 2, \ldots)$. Then one can find a hyperplane section Z of X such that $Y_i=Y_j$ and $f_i\,|\,Z=f_j\,|\,Z$ implies that $f_i=f_j$; moreover $f_i(X)=f_i(Z)$ for all i (the hyperplanes $H\subset\mathbb{CP}^N$ providing sections Z with these properties form an everywhere dense subset in $(\mathbb{CP}^N)^*$ with the complex topology). If X is connected and smooth and the canonical sheaf K_X ample, then one can assume that the section Z enjoys the same properties and that its Chern numbers can be estimated by the Chern numbers of X. Using this and the Kobayashi-Ochiai theorem (Theorem 1 of § 2), one can show that for compact X and Y, Theorem 1 in § 4 holds without the restriction $\dim Y=\dim X$, with a constant α depending on the Chern numbers of X (cf. Remark 1 and 2 in Sec. 4.1).

5.2. Families, Sections, Structures: Diverse Results. Here we formulate some finiteness theorems connected with the functional analogue of the Mordell conjecture and we give also some results on the finiteness of the number of holomorphic families of curves, complex structures etc.

a) Functional analogue of the Mordell conjecture. Let $p: X\to C$ be a proper regular map of a smooth algebraic surface X onto a smooth algebraic curve C. We assume that the fiber $X_c=p^{-1}(c)$ over a generic point $c\in C$ is a (compact) curve of genus $g \geqq 2$. Then either the set $\Gamma_{\text{rat}}(p)$ of all rational sections of the projection p is finite or else there exists over a suitable Zariski open subset $C' \subset C$ a fiber preserving biregular trivialization $\varphi: p^{-1}(C') \cong C' \times X_{c_0}$ ($c_0 \in C'$) and the set $\Gamma_{*\text{rat}}(p)$ of all nonconstant rational sections is finite (Yu.I. Manin [37]; other proofs and generalizations have been found by A.I. Parshin [43], Grauert [14], Samuel [49] and others). Multidimensional generalizations of the theorem have been given by Riebesehl [46] and Noguchi [41]; cf. also Martin-Deschamps [96] and J. Noguchi, Hyperbolic fiber spaces and Mordell's conjecture over function fields. Publ. Res. Inst. Math. Sci. *21*, 27–46 (1985). Riebesehl puts on the generic fibers of the projection $p: X \to Y$ restrictions of the type of negativity of the holomorphic sectional curvatures, while in the latter paper of Noguchi one puts restrictions of the type of negativity in the sense of Grauert of the tangent bundle.

b) Families of curves. Let us fix a curve C and a finite subset $S \subset C$. A family of projective curves over C is termed nondegenerate outside S if everywhere on $p^{-1}(C \setminus S)$ the differential dp has rank 1. A family which is nondegenerate outside S is called isotrivial if it admits a biregular trivialization after lifting to a finitely sheeted unbranched regular covering $\pi: B \to C \setminus S$ (an equivalent condition is that all curves $p^{-1}(c)$, $c \in C \setminus S$, are pairwise isomorphic). If C is a complete (i.e., smooth projective) curve, then the set of equivalence classes of non-isotrivial families of curves $p: X \to C$, with the generic fibers having fixed genus $g \geqq 2$ and the degeneracy set $S \subset C$, is finite (A.N. Parshin [43], S.Yu. Arakelov [2]).

c) Let M_g be the moduli space of smooth projective curves of given genus $g \geqq 2$. It is known that M_g is an irreducible quasiprojective algebraic variety (cf. [72]); let us fix one of its hyperplane sections H. Then for all complete curves $\Gamma_{g'} \subset M_g$ of genus g', the intersection index $\Gamma_{g'} \cdot H$ is bounded by a fixed number; there are no complete curves of genus 0 or 1 in M_g; the set of curves in M_g which are isomorphic to a given curve is finite (A.N. Parshin [43]). If $Z \subset M_g$ is a complete subvariety, then $\dim Z \leqq g-2$ ([74]).

d) Let X be a smooth projective surface of general type and C a curve of genus g on X. The number $2\delta = (K_X + C) \cdot C - 2g$ characterizes the "singularicity" of C. Set $T = TX$ and let $S^n T$ be the n-th symmetric power of X. Then the number of curves C on X with $C^2 < 0$ and $7 - 7g + 2\delta - C^2 > \dim H^1(X, S^5 T \otimes 2K_X)$ is finite. If $[c_1(X)]^2 > c_2(X)$ then there exists only a finite number of rational or elliptic curves on X (F.A. Bogomolov [7]).

e) Let $p: X \to C$ be a family of curves without degeneracy, the base and the fibers being hyperbolic. Then the universal covering of the surface X is biholomorphically equivalent to a bounded contractible domain of holomorphy (a Bergman domain) U in $\mathbb{C}^2$. If U is not homogeneous then the family $p: X \to C$ is not isotrivial. In this case U can serve as the universal covering only for a finite number of pairwise non-isomorphic complex-analytic surfaces Y, homeomorphic to X (G.B. Shabat [50]).

f) If X is a Kodaira surface then for each hyperbolic Riemann surface R of finite type, the family $\text{Hol}_*(R, X)$ is finite (Imayoshi [18]; see also

Y. Imayoshi: Holomorphic maps of compact Riemann surfaces into 2-dimensional compact C-hyperbolic manifolds. Math. Ann. *270*, 403–416 (1985)).

In the following example we give an explicit description of all holomorphic sections of some bundles connected with spaces of polynomials without multiple roots.

Example 1. Let $n>3$ and $k\in\mathbb{N}$. Let us consider the holomorphic bundle $\mathscr{E}_n^1=\{d_n^0\colon G_N^0\to\mathbb{C}^*\}$ with fiber SG_n (Chap. 1, Sec. 1.2, Example 2) and the bundles $\mathscr{E}_n^k=\{p_n^k\colon X_n^k\to\mathbb{C}^*\}$ with the same fiber, induced by the bundle $\mathscr{E}_n^1$ and the maps of the bases $e_k\colon\mathbb{C}^*\to\mathbb{C}^*$, $e_k(\zeta)=\zeta^k$. From the description of the holomorphic maps $\mathbb{C}^*\to G_n$ given in that example it follows that for each holomorphic map $f\colon\mathbb{C}^*\to G_n^0$ either $d_n^0(f(\zeta))\sim\zeta^{pn}$ or $d_n^0(f(\zeta))\sim\zeta^{q(n-1)}$ ($p,q\in\mathbb{Z}$), where the sign $\sim$ stands for homotopy within the class of holomorphic maps $\mathbb{C}^*\to\mathbb{C}^*$. If $s\colon\mathbb{C}^*\to\mathscr{E}_n^k$ is a holomorphic (meromorphic, continuous) section of the bundle $\mathscr{E}_n^k$, then $s(\zeta)=(\zeta,\sigma(\zeta))$ where $\sigma\colon\mathbb{C}^*\to G_n^0$ is a holomorphic (meromorphic, continuous) map such that $\zeta^k=d_n^0(\sigma(\zeta))$. By what was just said, for a holomorphic σ this is only possible if one of the conditions $k\equiv 0(\operatorname{mod} n)$ or $k\equiv 0(\operatorname{mod}(n-1))$ is fulfilled. If $k\equiv 0(\operatorname{mod} n)$ and $k\not\equiv 0(\operatorname{mod}(n-1))$, then $\mathscr{E}_n^k$ admits exactly n holomorphic sections ($\sigma_i(\zeta)=t^n+A_n\,\varepsilon_i\,\zeta^{k/n}\,t\in G_n^0$, where $\varepsilon_i^n=1$ and the constant A_n is chosen such that the discriminant of the polynomial $t^n+A_n t$ equals 1). If $k\equiv 0(\operatorname{mod}(n-1))$ and $k\not\equiv 0(\operatorname{mod} n)$, then $\mathscr{E}_n^k$ has exactly $n-1$ holomorphic sections ($\sigma_i(\zeta)=t^n+B_n\,\omega_i\,\zeta^{k/(n-1)}\in G_n^0$, where $(\omega_i)^{n-1}=1$ and the discriminant of the polynomial t^n+B_n equals 1). All these sections are regular. If $k\not\equiv 0(\operatorname{mod} n)$ and $k\not\equiv 0(\operatorname{mod}(n-1))$ (example: $k=1$), then $\mathscr{E}_n^k$ does not have any holomorphic sections. Finally, if $k\equiv 0(\operatorname{mod} n(n-1))$ then $\mathscr{E}_n^k$ is holomorphically trivial: $\mathscr{E}_n^k\cong\mathbb{C}^*\times SG_n$; all holomorphic sections become constant after the trivialization and the set of sections is in a natural way isomorphic to SG_n.

Thus, the bundle $\mathscr{E}_n^1=\{d_n^0\colon G_n^0\to\mathbb{C}^*\}$ has no holomorphic sections. The question of existence of meromorphic (respectively rational) sections of this bundle is equivalent to the existence of nonconstant meromorphic (respectively nonconstant rational) maps $\mathbb{C}^*\to SG_n$ satisfying the condition $\sigma(\varepsilon\zeta)=U(\varepsilon^{-1})\,\sigma(\zeta)$ for all $\zeta\in\mathbb{C}^*$ and all $n(n-1)$-th roots of unity ε. (If σ is such a map, then the formula $s(\zeta^{n(n-1)})=U(\zeta)\,\sigma(\zeta)$ defines a section $s\colon\mathbb{C}^*\to G_n^0$; conversely, a nonconstant map $\sigma\colon\mathbb{C}^*\to SG_n$, $\sigma(\zeta)=U(\zeta^{-1})\,s(\zeta^{n(n-1)})$, corresponding to a section $s\colon\mathbb{C}^*\to G_n^0$, satisfies the condition in question.) It is not hard to show that $\mathbb{C}^*$ does not admit any nonconstant rational maps into SG_3 and SG_4; therefore the bundles $\mathscr{E}_3^1$ and $\mathscr{E}_4^1$ have no rational sections. (It is not known whether there exists nonconstant rational maps $\mathbb{C}^*\to SG_n$ for $n>4$.) There exist nonconstant meromorphic maps $\mathbb{C}^*\to SG_3$ (and even maps $\mathbb{C}\to SG_3$) but it is not clear if there are among them any maps satisfying the above supplementary condition. Finally, let us remark that all bundles $\mathscr{E}_n^k$ have plenty of continuous sections.

References

For the convenience of the reader, references to reviews in Zentralblatt für Mathematik (Zbl.), compiled using the MATH database, and Jahrbuch über die Fortschritte der Mathematik (Jrb.) have been included as far as possible.

1. Algebraic surfaces. Tr. Mat. Inst. Steklova *75* (1965) [Russian]. Zbl. 154, 210. English transl.: Proc. Steklov Inst. Math. *75* (1965)
2. Arakelov, S.Yu.: Families of algebraic curves with fixed degeneracies. Izv. Akad. Nauk SSSR, Ser. Mat. *35*, 1269–1293 (1971) [Russian]. Zbl. 238.14012. English transl.: Math. USSR, Izv. *5* (1971), 1277–1302 (1972)
3. Bandman, T.M.: Surjective holomorphic maps of projective manifolds. Sib. Mat. Zh. *22*, No. 2, 48–56 (1981) [Russian]. Zbl. 462.32009. English transl.: Sib. Math. J. *22*, 204–210 (1981)
4. Bandman, T.M.: Surjective holomorphic maps of complex quasiprojective manifolds. Preprint no. 165. Novosibirsk: Institut Avtomatiki i Elektrometriĭ SO AN SSSR 1981 [Russian]
5. Bandman, T.: Estimates for the number of holomorphic maps of complex algebraic manifolds. Dissertation. Novosibirsk 1982 [Russian].
6. Bedford, E.: Invariant forms on complex manifolds with application to holomorphic mappings. Math. Ann. *265*, 377–397 (1983). Zbl. 532.32015
7. Bogomolov, F.A.: Families of curves on surfaces of general type. Dokl. Akad. Nauk SSSR *236*, 1041–1044 (1977) [Russian]. Zbl. 415.14013. English transl.: Sov. Math., Dokl. *18* (1977), 1294–1297 (1978)
8. Borel, A., Narasimhan, R.: Uniqueness conditions for certain holomorphic mappings. Invent. Math. *2*, 247–255 (1967). Zbl. 145, 318
9. Carathéodory, C.: Über die Abbildungen, die durch Systeme von analytischen Funktionen von mehreren Veränderlichen erzeugt werden. Math. Z. *34*, 758–792 (1932). Zbl. 3, 407
10. Drouilhet, S.J.: Criteria for algebraic dependence of meromorphic mappings into algebraic varieties. Ill. J. Math. *26*, 492–502 (1982). Zbl. 493.32023
11. Fujimoto, H.: Remarks to the uniqueness problem for meromorphic maps into $\mathbb{P}^N(\mathbb{C})$. Nagoya Math. J. *71*, 13–24, 25–41 (1978); *75*, 71–85 (1979), *83*, 153–181 (1981). Zbl. 358.32021. Zbl. 358.32022. Zbl. 431.32021. Zbl. 431.32022
12. Gorin, E.A., Lin, V.Ya.: Algebraic equations with continuous coefficients and some questions of algebraic braid theory. Mat. Sb., Nov. Ser. *78* (120), 579–610 (1969) [Russian]. Zbl. 211, 549
13. Gorin, E.A., Lin, V. Ya.: On separable polynomials over commutative Banach algebras. Dokl. Akad. Nauk SSSR *218*, 505–508 (1974) [Russian]. Zbl. 339.46037. English transl.: Sov. Math., Dokl. *15* (1974), 1357–1361 (1975)
14. Grauert, H.: Mordells Vermutung über rationale Punkte auf algebraischen Kurven und Funktionskörper. Publ. Math. Inst. Hautes Etudes Sci. *25*, 363–381 (1965). Zbl. 137, 405
15. Griffiths, P., Harris, J.: Principles of algebraic geometry. New York: Wiley 1978. Zbl. 408.14001
16. Howard, A., Sommese, A.J.: On the theorem of de Franchis. Ann. Sc. Norm. Super. Pisa, Cl. Sci., IV. Ser. *10*, 429–436 (1983). Zbl. 534.14016
17. Iitaka, S.: On logarithmic Kodaira dimension of algebraic varieties. In: Complex Analysis and Algebraic Geometry, pp. 178–189. Tokyo: Iwanami 1977. Zbl. 351.14016
18. Imayoshi, Y.: Generalizations of de Franchis theorem. Duke Math. J. 50, 393–408 (1983).
19. Imayoshi, Y.: An analytic proof of Severi's theorem. Complex Variables *2*, 151–155 (1983). Zbl. 585.32025
20. Kaliman, Sh.I.: Holomorphic endomorphisms of the manifold of complex polynomials of discriminant 1. Usp. Mat. Nauk *31*, No. 1 (187), 251–252 (1976) [Russian]. Zbl. 336.32004
21. Kalka, M., Shiffman, B., Wong, B.: Finiteness and rigidity theorems for holomorphic mappings. Mich. Math. J. *28*, 289–295 (1981). Zbl. 459.32011

22. Kato, T.: Analytic self-mapping reducing to the identity mapping. Comment. Math. Helv. *49*, 529–533 (1974). Zbl. 298.30018
23. Kato, T.: Analytic self-mapping inducing the identity on $H_1(W, \mathbb{Z}/m\mathbb{Z})$. Kodai Math. Sem. Reports *28*, 317–323 (1977). Zbl. 362.30016
24. Kaup, W.: Hyperbolische komplexe Räume. Ann. Inst. Fourier *18*, 303–330 (1968). Zbl. 174, 130
25. Kobayashi, S.: Hyperbolic manifolds and holomorphic mappings. New York: Dekker 1970. Zbl. 207, 379
26. Kobayashi, S.: Intrinsic distances, measures and geometric function theory. Bull. Am. Math. Soc. *82*, 357–416 (1976). Zbl. 346.32031
27. Kobayashi, S., Ochiai, T.: Meromorphic mappings onto compact complex spaces of general type. Invent. Math. *31*, 7–16 (1975). Zbl. 331.32020
28. Kodama, A.: On bimeromorphic automorphisms of hyperbolic complex spaces. Nagoya Math. J. *78*, 1–5 (1979). Zbl. 367.32002
29. Kurke, H.: An algebraic proof of a theorem of S. Kobayashi and T. Ochiai. Zürich: Forschungsinstitut für Mathematik ETH 1978.
30. Kwack, M.H.: Meromorphic mappings into compact complex manifolds with a Grauert positive bundle of q-forms. Proc. Am. Math. Soc. *87*, 699–703 (1983). Zbl. 532.32010
31. Lin, V.Ja. (= Lin, V.Ya.): Holomorphic mappings of some spaces connected with algebraic functions. In: Lect. Notes Math. 1043, 657–661. Berlin etc.: Springer-Verlag 1984. see Zbl. 545.30038
32. Lin, V.Ya.: Algebroid functions and holomorphic elements of homotopy groups of complex manifolds. Dokl. Akad. Nauk. SSSR *201*, 28–31 (1971) [Russian]. Zbl. 261.46054. English transl.: Sov. Math., Dokl. *12* (1971), 1608–1612 (1972)
33. Lin, V.Ya.: Algebraic functions with a universal discriminant manifold. Funkts. Anal. Prilozh. *6*, No. 1, 81–82 (1972) [Russian]. Zbl. 249.12101. English transl.: Funct. Anal. Appl. *6*, 73–75 (1972)
34. Lin, V.Ya.: On superpositions of algebraic functions. Funkts. Anal. Prilozh. *6*, No. 3, 77–78 (1972) [Russian]. Zbl. 272.12103. English transl.: Funct. Anal. Appl. *6* (1972), 240–241 (1973)
35. Lin, V.Ya.: Artin braids and groups and spaces connected with them. Itogi Nauki Tekh., Ser. Algebra Topologiya Geom. 17, 159–227 (1979) [Russian]. Zbl. 434.20020. English transl.: J. Sov. Math. *18*, 736–788 (1982)
36. Maehara, K.: A finiteness property of varieties of general type. Math. Ann. *262*, 101–123 (1983). Zbl. 438.14011
37. Manin, Yu.I.: Rational points of algebraic curves over function fields. Izv. Akad. Nauk. SSSR., Ser. Mat. *27*, 1395–1440 (1963). Zbl. 166, 169. English transl.: Transl., II. Ser., Am. Math. Soc. *50*, 189–234 (1966)
38. Martens, H.H.: Remarks on de Franchis' theorem. In: Lect. Notes Math. 1013, 160–163. Berlin etc.: Springer-Verlag 1983. Zbl. 527.14025
39. Martin-Deschamps, M., Lewin-Menegaux, R.: Applications rationelles séparables dominantes sur une variété du type général. Bull. Soc. Math. Fr. *106*, 279–287 (1978). Zbl. 417.14007
40. Minda, C.D.: Fixed point of analytic self-mappings of Riemann surfaces. Manuscr. Math. *27*, 391–399 (1979). Zbl. 405.30035
41. Noguchi, J.: A higher dimensional analogue of Mordell's conjecture over function fields. Math. Ann. *258*, 207–212 (1981). Zbl. 459.14002
42. Noguchi, J., Sunada, T.: Finiteness of the family of rational and meromorphic mappings into algebraic manifolds. Am. J. Math. *104*, 887–900 (1982). Zbl. 502.14002
43. Parshin, A.N.: Algebraic curves over function fields. I. Izv. Akad. Nauk SSSR, Ser. Mat. *32*, 1191–1219 (1968) [Russian]. Zbl. 181, 239. English transl.: Math. USSR, Izv. *2*, 1145–1170 (1968)
44. Rabotin, V.V.: Some remarks on the Borel-Narasimhan theorem. Sb. Nauchn. Tr., Tashk. Gos. Univ. 576, 77–80 (1979) [Russian]. R. Zh. Mat. 1980, 6A664

45. Rabotin, V.V.: The finiteness of the number of holomorphic maps in some algebraic varieties. In: Some questions of multidimensional complex analysis, 129–138. Krasnoyarsk 1980 [Russian]. Zbl. 486.32012
46. Riebesehl, D.: Hyperbolische komplexe Räume und die Vermutung von Mordell. Math. Ann. *257*, 99–110 (1981). Zbl. 451.32018
47. Sakai, F.: Degeneracy of holomorphic maps with ramification. Invent. Math. *26*, 213–229 (1974). Zbl. 276.32012
48. Sakai, F.: Kodaira dimension of complements of divisors. In: Complex Analysis and Algebraic Geometry, 239–257. Tokyo: Iwanami 1977. Zbl. 375.14009
49. Samuel, P.: Compléments a un article de Hans Grauert sur la conjecture de Mordell. Publ. Math. Inst. Hautes Etud. Sci. *29*, 311–318 (1966). Zbl. 144, 201
50. Shabat, G.B.: Local reconstruction of complex algebraic surfaces from universal coverings. Funkts. Anal. Prilozh. *17*, No. 2, 90–91 (1983) [Russian]. Zbl. 543.14007. English transl.: Funct. Anal. Appl. *17*, 157–159 (1983)
51. Shabat, G.B.: The Liouvilleness of maximal Abelian coverings of Riemann surfaces. Usp. Mat. Nauk *39*, No. 2, 131–132 (1984) [Russian]. Zbl. 552.14006. English transl.: Russ. Math. Surv. *39*, No. 2, 193–194 (1984)
52. Simha, R.R.: Holomorphic maps into compact complex analytic space. Arch. Math. *39*, 262–263 (1982). Zbl. 503.32011
53. Tsushima, R.: Rational maps to varieties of hyperbolic type. Proc. Japan Acad., Ser. A *55*, 95–100 (1979). Zbl. 443.14006
54. Urata, T.: Holomorphic mappings into taut complex analytic spaces. Tohoku Math. J., II. Ser *31*, 349–353 (1979). Zbl. 404.32013
55. Urata, T.: Holomorphic mappings onto certain compact complex analytic spaces. Tohoku Math. J., II. Ser *33*, 573–585 (1981). Zbl. 477.32025
56. Urata, T.: Holomorphic automorphisms and cancellation theorems. Nagoya Math. J. *81*, 91–103 (1981). Zbl. 416.32011
57. Wakabayashi, I.: Nonexistence of bounded functions on the homology covering surface of $\mathbb{P}^1\setminus\{3 \text{ points}\}$. J. Math. Soc. Japan *34*, 607–625 (1982). Zbl. 489.30031
58. Wong, B.: On the automorphism group of compact measure hyperbolic maifolds and complex analytic bundles with compact measure hyperbolic fibers. Proc. Am. Math. Soc. *62*, 54–56 (1977). Zbl. 358.32029
59. Zaĭdenberg, M.G.: The structure of finite degeneration of families of curves on affine surfaces and a characterization of the affine plane. Dokl. Akad. Nauk SSSR *287*, 272–276 (1986) [Russian]
60. Zaĭdenberg, M.G.: Rational $\mathbb{C}^*$-actions on $\mathbb{C}^2$, their quasi-invariants, and algebraic curves in $\mathbb{C}^2$ with Euler characteristic 1. Dokl. Akad. Nauk SSSR *280*, 277–280 (1985) [Russian]. Zbl. 595.14035. English transl.: Sov. Math., Dokl. *31*, 57–60 (1985)
61. Zaidenberg, M.G., Lin, V.Ya.: On holomorphically noncontractible bounded domains of holomorphy. Dokl. Akad. Nauk SSSR *249*, 282–285 (1979) [Russian]. Zbl. 444.32005. English transl.: Sov. Math., Dokl. *20*, 1262–1266 (1979)
62. Zaidenberg, M.G., Lin, V.Ya.: An irreducible simply connected algebraic curve in $\mathbb{C}^2$ is equivalent to a quasihomogeneous curve. Dokl. Akad. Nauk SSSR *271*, 1048–1052 (1983) [Russian]. Zbl. 564.14014. English transl.: Sov. Math., Dokl. *28*, 200–204 (1983)

Additional References

63. Accola, R.D.: Automorphisms of Riemann surfaces. J. Anal. Math. *18*, 1–5 (1967). Zbl. 158, 78
64. Bedford, E.: On the automorphism group of a Stein manifold. Math. Ann. *266*, 215–227 (1983). Zbl. 532.32014

65. Bochner, S., Martin, W.T.: Several complex variables. Princeton: Princeton University Press 1948. Zbl. 41, 52
66. Bochner, S., Montgomery, D.: Locally compact groups of differentiable transformations. Ann. Math. *47*, 639–653 (1946). Zbl. 61, 44
67. Bochner, S., Montgomery, D.: Groups on analytic manifolds. Ann. Math. *48*, 659–669 (1947). Zbl. 30, 75
68. Brieskorn, E.: Sur les groupes de tresses (d'apres V.I. Arnold). In: Lect Notes Math. 317, 21–44. Berlin etc.: Springer-Verlag 1973. Zbl. 277.55003
69. Brieskorn, E.: Die Fundamentalgruppe des Raumes der regulären Orbits einer endlichen komplexen Spiegelungsgruppe. Invent. Math. *12*, 57–61 (1971). Zbl. 204, 565
70. Cartan, H.: Sur les fonctions de plusieurs variables complexes, l'itération des transformations intérieures d'un domaine borné. Math. Z. *35*, 700–733 (1932). Zbl. 4, 406
71. Cartan, H.: Quotient d'un espace analytique par un groupe d'automorphismes. In: Algebraic geometry and topology, Symp. in honor of Lefschetz, 90–102. Princeton: Princeton University Press 1957. Zbl. 84, 72
72. Deligne, P., Mumford, D.: The irreducibility of the space of curves of given genus. Publ. Math. Hautes Etud. Sci. *36*, 75–109 (1969). Zbl. 181, 488
73. Demailly, J.P.: Fonctions holomorphes à croissance polynomiale sur la surface d'équation $e^x+e^y=1$. Bull. Sci. Math., II. Ser. *103*, 179–191 (1979). Zbl. 412.32007
74. Diaz, S.: A bound on the dimensions of complete subvarieties of M_g. Duke Math. J. *51*, 405–408 (1984). Zbl. 581.14017
75. Douady, A.: Le problème des modules des sous-espaces analytiques compacts d'un espace analytique. Ann. Inst. Fourier *16*, 1–95 (1966). Zbl. 146, 311
76. Fujimoto, H.: On meromorphic maps into a compact complex manifold. J. Math. Soc. Japan *34*, 527–539 (1982). Zbl. 497.32020
77. Grauert, H.: Über Modifikationen und exzeptionelle analytische Mengen. Math. Ann. *146*, 331–368 (1962). Zbl. 173, 330
78. Greenleaf, F.: Invariant means on topological groups and their applications. New York: Van Nostrand 1969. Zbl. 174, 190
79. Grothendieck, A.: Téchniques de construction en géométrie analytique. X. Construction de l'espace de Teichmüller. Appendice de J-P. Serre. In: Séminaire H. Cartan. Ecole norm. supér., 1960–1960, 1? année, fasc. 2, p. 17/1–17/20. Paris: 1962. Zbl. 142, 335
80. Hayman, W.K.: Meromorphic functions. Oxford: Clarendon 1964. Zbl. 115, 62
81. Hirschowitz, A.: A propos du principe d'Oka. C.R. Acad. Sci., Paris, Ser. A 272, 792–794 (1971). Zbl. 208, 351
82. Hurwitz, A.: Über algebraische Gebilde mit eindeutigen Transformationen in sich. Math. Ann. *41*, 403–443 (1893).
83. Inoue, M.: On surfaces of class VII_0. Invent. Math. *24*, 269–310 (1974). Zbl. 283.32019
84. Iskovskikh, V.A., Manin, Yu.A.: Three dimensional quartics and counterexamples to Lüroth's problem. Mat. Sb., Nov. Ser. *86* (128), 140–166 (1971) [Russian]. Zbl. 222.14009. English transl.: Math. USSR, Sb. *15* (1971), 141–166 (1972)
85. Kaĭmanovich, V.A.: Brownian motion and harmonic functions on covering manifolds, an entropy approach. Dokl. Akad. Nauk SSSR *288*, 1045–1049 (1986) [Russian]. Zbl. 615.60074. English transl.: Sov. Math., Dokl. *33*, 812–816 (1986)
86. Kaliman, Sh. I.: The holomorphic universal covering of the space of polynomials without multiple roots. Funkts. Anal. Prilozh. *9*, No. 1, 71 (1975) [Russian]. Zbl. 319.32012. English transl.: Funct. Anal. Appl. *9*, 67–68 (1975)
87. Kaliman, Sh.I.: The holomorphic universal covering of the space of polynomials without multiple roots. Teor. Funkts. Funkts. Anal. Prilozh. *28*, 25–35 (1977) [Russian]. Zbl. 449.32001
88. Kawamata, Y.: Hodge theory and Kodaira dimension. In: Algebraic varieties and analytic varieties, Proc. Symp., Tokyo, 1981, Adv. Stud. Pure Math. 1, 317–327. Amsterdam – New York: North Holland 1983. Zbl. 533.14017

89. Kodaira, K.: On holomorphic mappings of polydiscs into compact complex manifolds. J. Differ. Geom. *6*, 33–46 (1971). Zbl. 227.32008
90. Kondo, S.: Enriques surfaces with finite automorphism groups. Jap. J. Math., New Ser. *12*, 191–282 (1986). Zbl. 616.14031
91. Kra, I.: Automorphic forms and Kleinian groups. Reading: Benjamin 1972. Zbl. 253.30015
92. Krushkal', S.L., Apanasov, B.N., Gusevskiĭ, N.A.: Kleinian groups and uniformization with applications and problems. Novosibirsk: Nauka 1981 [Russian]. Zbl. 485.30001. English transl.: Providence: Am. Math. Soc. (1986)
93. Lin, V.Ya.: Representations of braids by permutations. Usp. Mat. Nauk *29*, No. 1 (175), 173–174 (1974) [Russian]. Zbl. 292.20039
94. Lin, V.Ya.: On representations of the braid group with permutations. Usp. Mat. Nauk *27*, No. 3 (165), 192 (1972) [Russian]. Zbl. 237.20012
95. Lyons, T., Sullivan, D.: Function theory, random path and covering spaces. J. Differ. Geom. *19*, 299–323 (1984). Zbl. 554.58022
96. Martin-Deschamps, M.: Propriétés de descente des variétés a fibre cotangent ample. Ann. Inst. Fourier *34*, No. 3, 39–64 (1984). Zbl. 535.14013
97. Maskit, B.: The conformal group of a plane domain. Am. J. Math. *90*, 718–722 (1968). Zbl. 172, 98
98. Matsumura, H.: On algebraic groups of birational transformations. Atti Accad. Naz. Lincei, Rend., Cl. Sci. Fis. Mat. Nat., VIII. Ser. *34*, 151–155 (1963). Zbl. 134, 166
99. Matsusaka, T.: On canonical polarized varieties II. Am. J. Math. *92*, 283–292 (1970). Zbl. 195, 228
100. Milnor, J.: Singular points of complex hypersurfaces. Princeton: Princeton University Press 1968. Zbl. 184, 484
101. Minda, C.D.: The hyperbolic metric and coverings of Riemann surfaces. Pac. J. Math. *84*, 171–182 (1979). Zbl. 388.30028
102. Mok, N.: Rigidity of holomorphic self-mappings and automorphism groups of hyperbolic Stein spaces. Math. Ann. *266*, 433–447 (1984). Zbl. 574.32021
103. Mori, A.: A note on unramified Abelian covering surfaces of closed Riemann surface. J. Math. Soc. Japan *6*, 162–176 (1954). Zbl. 59, 70
104. Murasugi, K.: Seifert fibre spaces and braid groups. Proc. Lond. Math. Soc., III. Ser. *44*, 71–84 (1982). Zbl. 489.57003
105. Nevanlinna, R.: Le théorème de Picard-Borel et la théorie des fonctions méromorphes. Paris: Gauthier-Villars 1929. Jrb. 55, 773
106. Nikulin, V.V.: On quotient groups of the automorphism groups of hyperbolic forms by subgroups generated by 2-reflections. Algebraic-geometric applications. In: Contemporary problems of mathematics. Itogi Nauki Tekh., Ser. Sovrem. Probl. Mat. *18*, 3–114 (1981) [Russian]. Zbl. 484.10021. English transl.: J. Sov. Math. *22*, 1401–14075 (1983)
107. Nikulin, V.V.: On describing the automorphism groups of the Enriques surfaces. Dokl. Akad. Nauk SSSR *277*, 1324–1327 (1984) [Russian]. Zbl. 604.14036. English transl.: Sov. Math., Dokl. *30*, 282–285 (1984)
108. Oikawa, K.: A supplement to the "Note on conformal mappings of a Riemann surface onto itself". Kodai Math. J. *8*, 115–116 (1956). Zbl. 73, 68
109. Peters, C.A.M.: Holomorphic automorphisms of compact Kähler surfaces and their induced actions in cohomology. Invent. Math. *52*, 143–148 (1979). Zbl. 415.32011
110. Pizer, A.K.: A problem on rational functions. Am. Math. Mon. *80*, 552–553 (1973). Zbl. 303.30001
111. Pham, P.: Singularités des processus de diffusion multiple. Ann. Inst. Henri Poincaré, Phys. Theor. *6*, 89–204 (1967). Zbl. 154, 461
112. Pólya, G.: Bestimmung einer ganzen Funktion endlichen Geschlechts durch viererlei Stellen. Math. Tidskrift B. 16–21 (1921). Jrb. 48, 354
113. Pyatetskiĭ-Shapiro, I.I., Shafarevich, I.R.: Torelli's theorem for algebraic surfaces of type K3. Izv. Akad. Nauk SSSR, Ser. Mat., *35*, 530–572 (1971) [Russian]. Zbl. 219.14021. English transl.: Math. USSR, Izv. 5, 547–588 (1971)

114. Ramspott, K.J.: Stetige und holomorphe Schnitte in Bündeln mit homogene Faser. Math. Z. *89*, 234–246 (1965). Zbl. 163, 321
115. Schmid, E.M.: Some theorems on value distributions of meromorphic functions. Math. Z. *120*, 61–92 (1971). Zbl. 196, 91
116. Shafarevich, I.R.: Basic algebraic geometry. Moscow: Nauka 1972 [Russian]. English transl.: Grundlehren 213. Berlin etc.: Springer 1974. Zbl. 258.14001
117. Shiffman, B.: Holomorphic and meromorphic mappings and curvature. Math. Ann. *222*, 171–194 (1976). Zbl. 329.32009. English transl.: Math. USSR, Izv. *6* (1972), 949–1008 (1973)
118. Siu, Y.T.: Every K3 surface is Kähler. Invent. Math. *73*, 139–150 (1983). Zbl. 557.32004
119. Varchenko, A.N.: Topological equisingularity theorems for families of algebraic mainfolds and families of polynomial maps. Izv. Akad. Nauk SSSR, Ser. Mat., *36*, 957–1019 (1972) [Russian]. Zbl. 251.14006
120. Wong, B.: Negative tangent bundles and hyperbolic manifolds. Proc. Am. Math. Soc. *61*, 90–92 (1976). Zbl. 346.32033
121. Zaĭdenberg, M.G.: Picard's theorems and hyperbolicity. Sib. Mat. Zh. *24*, No. 6 (142), 44–55 (1983) [Russian]. Zbl. 579.32039. English transl.: Sib. Math. J. *24*, 858–867 (1983)
122. Zaĭdenberg, M.G., Lin, V.Ya.: Holomorphic contractibility and quasihomogeneity. In: Abstracts from the conference "Theoretical and applied question in mathematics", 160–162. Tartu 1980 [Russian]
123. Yau, S.T.: On the Ricci curvature of a compact Kähler manifold and the complex Monge-Ampère equation. I. Commun. Pure Appl. Math. *31*, 339–411 (1978). Zbl. 362.53049

V. Holomorphic Maps in $\mathbb{C}^n$ and the Problem of Holomorphic Equivalence

S.I. Pinchuk

Translated from the Russian
by J. Peetre

Contents

Introduction

This Part is devoted to the problem of *holomorphic equivalence*. This problem consists of the following: given two domains D, $G \subset \mathbb{C}^n$, to determine whether there exists a *biholomorphic map* $f: D \to G$, or not.

In this Part we consider the problem of holomorphic equivalence within the class of bounded pseudoconvex domains with piecewise smooth boundaries. If $D \subset \mathbb{C}^n$ is a domain in this class and f a bounded holomorphic function in D, then by Fatou's theorem ([16], [63]) it has a.e. admissible limits in the sense of Stein-Chirka. Therefore its trace $\hat{f} \in L^\infty(\partial D)$ is well-defined and a CR-*function* on ∂D in a generalized sense (cf. [17]). Conversely, if a CR-function is given on ∂D then it extends to a holomorphic function on D. Thus, if there exists a biholomorphic map $f: D \to G$ then its trace $\hat{f}: \partial D \to \partial G$ is defined a.e. on ∂D and is a CR-*map*, so that D, G are biholomorphically equivalent iff their boundaries are "CR-equivalent". We must use quotes here, as without additional information on the boundary behavior of f and f^{-1}, it is, in general, not clear whether $\hat{f}$ and $\hat{f}^{-1}$ are each others inverses (this is, of course, the case when, for example, f extends to a homeomorphism of the closures $\bar{D}$ and $\bar{G}$). The above considerations indicate that questions of boundary correspondence must play an important rôle in problems of holomorphic equivalence. Questions of boundary behavior are also the most difficult ones in this direction.

Let us recall the well-known one dimensional results. Consider two domains D, $G \Subset C$ and a conformal mapping $f: G \to D$. Then

a) f extends to a homeomorphism $\hat{f}: \bar{D} \to \bar{G}$ provided the boundaries ∂D, ∂G are Jordan arcs (Carathéodory).

b) f extends to a C^{k-0}-diffeomorphism $\hat{f}: \bar{D} \to \bar{G}$ provided ∂D, $\partial G \in C^k$ $(1 \leqq k \leqq \infty)$ (cf. e.g. [18]).

c) f admits an analytic continuation to a neighborhood of $\bar{D}$ if ∂D, ∂G are real-analytic.

So far these results have not been extended completely to holomorphic mappings in $\mathbb{C}^n$. However, in additional restrictions on the domain analogues have been obtained.

A number of questions connected with holomorphic maps in $\mathbb{C}^n$ is discussed in the well-known paper by G.M. Khenkin and E.M. Chirka [34]. In the past years considerable progress has been made in this direction. In this Part we survey these new results. Some results, which will not be treated here, have been considered earlier in surveys by Burns and Shnider [10], Fefferman [26] and Wells [75]. In the last one of these there is also a historical sketch of the subject.

Chapter 1
The Problem of Holomorphic Equivalence and the Boundary Behavior of Maps

§ 1. General Notions

1.1. The Levi Form. Let Ω be a domain in $\mathbb{C}^n$ and $\rho \in C^2(\Omega)$ a real function. The *Levi form* of ρ at the point $\zeta \in \Omega$ is the Hermitean form

$$H_\rho(\zeta, a) = \sum_{\mu, \nu=1}^{n} \rho_{\mu\bar{\nu}}(\zeta)\, a_\mu \bar{a}_\nu, \quad a \in \mathbb{C}^n.$$

Here and in the sequel we employ the notation

$$\rho_\mu = \frac{\partial \rho}{\partial z_\mu}, \qquad \rho_{\bar{\nu}} = \frac{\partial \rho}{\partial \bar{z}_\nu}, \qquad \rho_{\mu\bar{\nu}} = \frac{\partial^2 \rho}{\partial z_\mu \partial \bar{z}_\nu}.$$

It follows easily from this definition that the Levi form is invariant under holomorphic transformations: if $f: \Omega' \to \Omega$ is a holomorphic map then

$$H_{\rho \circ f}(\zeta, a) = H_\rho(f(\zeta), df_\zeta(a)), \tag{1}$$

where df_ζ is the differential of f at the point ζ. In particular, let Ω' be a complex submanifold of Ω and denote the imbedding by $f: \Omega' \to \Omega$; then

$$H_\rho(\zeta, a) = H_{\rho'}(\zeta, a), \qquad \rho' = \rho \,|\, \Omega' \tag{2}$$

for $\zeta \in \Omega'$, $a \in T_\zeta(\Omega')$.

Let $D \subset \mathbb{C}^n$ be a domain with boundary of class C^k $(k \geqq 1)$ and Ω a suitable neighborhood of ∂D. A function $\rho \in C^k(\Omega)$ is called a *defining function* for D if $D \cap \Omega = \{z \in \Omega: \rho(z) < 0\}$ and $d\rho \neq 0$ on $\partial D \cap \Omega$.

It is well-known [32] that a *domain* $D \subset \mathbb{C}^n$ with boundary of class C^2 and defining function ρ is *pseudoconvex* iff $H_\zeta(\rho, a) \geqq 0$ for all $\zeta \in D$, $a \in T^c(\partial D)$, where $T^c_\zeta(\partial D) = \{a \in \mathbb{C}^n: \Sigma_\mu \rho_\mu(\zeta)\, a_\mu = 0\}$ is the complex tangent space to ∂D at the point ζ.

A *domain* $D \Subset \mathbb{C}^n$ with boundary of class C^2 and defining function ρ is called *strictly pseudoconvex* if $H_\rho(\zeta, a) > 0$ for $\zeta \in D$, $a \in T^c_\zeta(D) \setminus \{0\}$.

Strict (strong) and weak pseudoconvexity do not depend on the choice of the defining function.

The notions of defining function and pseudoconvexity, essentially, do not depend on the domain itself but only on the boundary. Therefore, without loss of generality, one can as well pass to the case of real hypersurfaces in $\mathbb{C}^n$. In the sequel, when speaking of the Levi form of the defining function of a hypersurface we will always intend its restriction to its complex tangent plane.

It is well-known that under biholomorphic maps pseudoconvex domains are mapped onto pseudoconvex domains. Let us consider the question of holomorphic invariance of the notion of strict pseudoconvexity. Let $f: D \to G$ be a biholomorphic mapping and D a strictly pseudoconvex domain. If f extends to a C^2-diffeomorphism[1] between the closures $\bar{D}$ and $\bar{G}$ then in view of (1) the domain G likewise must be strictly pseudoconvex. In the general case, this is not so, as the boundary may be non-differentiable. It is natural to ask whether G will be strictly pseudoconvex if we in addition assume that its boundary is sufficiently smooth. This question leads to a difficult questions concerning boundary correspondence. In § 2 of Chap. 2 we will give results quite close to its positive solution. In connection with this let us state the following result by Fornaess [27] which says that instead of requiring f to be a diffeomorphism it suffices to show the smoothness of f (or f^{-1}) up to the boundary.

Proposition 1. *Let D, $G \subset \mathbb{C}^n$ be bounded pseudoconvex domain with C^1-boundaries and $f: D \to G$ a biholomorphic mapping admitting a C^2-extension $\hat{f}: \bar{D} \to \bar{G}$. Then $\hat{f}$ is a C^2-diffeomorphism between $\bar{D}$ and $\bar{G}$.*

In problems of boundary behavior of maps an important rôle is played by plurisubharmonic defining functions, so we turn to their existence for some classes of domains.

First let $D \subset \mathbb{C}^n$ be a strictly pseudoconvex domain and $\rho \subset C^2(\Omega)$ an arbitrary defining function ($\Omega \supset \partial D$). The function $\tilde{\rho} = \rho + A\rho^2$ will then likewise be a defining function (possibly by decreasing Ω). Taking $A > 0$ sufficiently large, one can in view of (2) achieve that it becomes strictly pseudoconvex near ∂D. The defining function $\max(\tilde{\rho}, -\varepsilon)$ is, for $\varepsilon > 0$ sufficiently small, plurisubharmonic in D and strictly plurisubharmonic near ∂D.

Next let us consider the case of a (not necessarily strictly) pseudoconvex bounded domain $D \subset \mathbb{C}^n$ with C^2-boundary. In [36] it is shown that such a domain need not have plurisubharmonic defining functions. To some extent this can be compensated by the existence of a negative *exhausting plurisubharmonic function* in D. The following result is due to Diederich and Fornaess [19].

Lemma 2. *Let $D \subset C^n$ be a bounded pseudoconvex domain with boundary of class C^k ($2 \leqq k \leqq \infty$). Then in a suitable neighborhood $\Omega \supset \bar{D}$ there exist a defining function $\rho \in C^k(\Omega)$ for D and a number $\eta_0 > 0$ such that, for each $\eta \in (0, \eta_0)$, the exhausting function $\tilde{\rho} = -(-\rho)^\eta$ is strictly plurisubharmonic in D.*

The number η_0 fixed by Lemma 2 depends on D and in the general case there exists no constant $c > 0$ such that $\eta_0 > c$ for all domains. However, for arbitrary fixed $\zeta \in \partial D$ and $\eta \in (0, 1)$ there exist a neighborhood $\Omega \ni \zeta$ and

[1] Here in what follows C^k-diffeomorphisms means a diffeomorphism of class C^k. In analogous way we use words such as C^k-boundary, C^k-map(ping) etc.

a local defining function $\rho \in C^k(\Omega)$ such that $-(-\rho)^\eta$ is strictly plurisubharmonic in $D \cap \Omega$.

The application of plurisubharmonic defining (exhausting) functions to holomorphic maps is based on the following classical lemma due to Keldysh-Lavrent'ev-Hopf.

Lemma 3. *Let ρ be a negative plurisubharmonic function in a bounded domain $D \subset \mathbb{C}^n$ with C^2-boundary. Then there exists a constant $c>0$ such that $|\rho(z)| > c \cdot e(z, \partial D)$ for all $z \in D$* (here $e(z, \partial D)$ is the Euclidean distance between the point z and the boundary ∂D).

Combining Lemmata 2 and 3 one readily gets (cf., for example, [51], [55]).

Proposition 4. *Let D, $G \subset \mathbb{C}^n$ be bounded pseudoconvex domains with C^2-boundaries and let $f: D \to G$ be a biholomorphic map. Then there exist constants $c_1, c_2 \geqq 0$, $\alpha \geqq \beta > 0$ such that*

$$c_1 [e(z, \partial D)]^\alpha \leqq e(f(z), \partial D) \leqq c_2 [e(z, \partial D)]^\beta \tag{3}$$

for all $z \in D$. If $D(G)$ is strictly pseudoconvex the one can take $\beta = 1$ (respectively $\alpha = 1$).

Proof. Consider, for simplicity, the case when D and G are strictly pseudoconvex. Let φ be a defining function for G which is plurisubharmonic in the interior of G. The function $\varphi \circ f$ satisfies in D the estimate

$$\gamma_1 e(z, \partial D) \leqq |\varphi \circ f(z)| \leqq \gamma_2 e(f(z), \partial G). \tag{4}$$

The left hand part of this inequality follows from Lemma 3, whereas the right hand part is a consequence of the smoothness of φ near ∂G. From (4) the right hand part of (3), with $\alpha = \beta = 1$, follows at once. Repeating the reasoning with f^{-1} we get analogously the left hand side of (3). □

1.2. Proper Holomorphic Maps. Recall that a *map* $f: D \to G$ is said to be *proper* if the pre-image in D of every compact set $K \Subset G$ is compact. For a bounded domain this amounts to requiring that for every sequence $z^\nu \in D$ it follows from $e(z^\nu, \partial D) \to 0$ that $e(f(z^\nu), \partial G) \to 0$.

Proposition 5 ([46]). *Let D, $G \subset \mathbb{C}^n$ be strictly pseudoconvex domains and $f: D \to G$ a proper holomorphic mapping. If $f \in C^2(\bar{D})$ then f is locally biholomorphic.*

Proof. It suffices to show that the Jacobian $\det f'(z)$ is $\neq 0$ at the boundary ∂D. Let $\varphi \in C^2(\partial D)$ $(\Omega \supset \bar{G})$ be a defining function for G, plurisubharmonic there. Then $\varphi \circ f$ is negative and plurisubharmonic in D. In view of the hypothesis $f \in C^2(\bar{D})$, it extends to a neighborhood of ∂D as a function of class C^2. By Lemma 3 $d\varphi \circ f \neq 0$ on ∂D, i.e. $\varphi \circ f$ may be taken as a defining function for D. Consequently, $H_{\varphi \circ f}(\zeta, a) > 0$ for $\zeta \in \partial D$, $a \in T_\zeta^c(\partial D) \setminus \{0\}$.

Assume that $\det f'(\zeta) = 0$ for some point $\zeta \in \partial D$. Then we can find a complex line $l \subset \mathbb{C}^n$ such that $df_\zeta(l) = 0$. From $d\varphi \circ f \neq 0$ it follows that $l \subset T_\zeta^c(\partial D)$ and we get a contradiction, as for any vector $a \in l$ one must have, in view of (1), $H_{\varphi \circ f}(\zeta, a) = 0$. □

Remark. In the general case f need not be globally biholomorphic, as the following example shows:

$$D = \{z \in \mathbb{C}^2 : |z_1|^4 + |z_1|^{-4} + |z_2|^2 < 3\},$$
$$G = \{z \in \mathbb{C}^2 : |z_1|^2 + |z_1|^{-2} + |z_2|^2 < 3\},$$
$$f(z_1, z_2) = (z_1^2, z_2).$$

The biholomorphicity of f can be maintained in the following cases: a) G simply connected, b) $D = G$ (cf. [46]).

The following result sharpens Proposition 5.

Proposition 6 ([21]). *Let $D \subset \mathbb{C}^n$ be a strictly pseudoconvex domain, $G \subset \mathbb{C}^n$ a bounded domain with C^2-boundary and $f: D \to G$ a proper holomorphic map. If $f \in C^2(\bar{D})$ then f is locally biholomorphic and G is strictly pseudoconvex.*

The results of this Section are rather elementary. However, they are subordinated to a non-trivial and important question connected with the boundary behavior of maps: is it possible that Propositions 5 and 6 hold also without the assumption $f \in C^2(\bar{D})$? In § 2, Chap. 2 we give results close to a final solution of this problem.

1.3. Algebroidal Maps. Let $D \subset \mathbb{C}^n_z$, $G \subset \mathbb{C}^n_w$ be bounded domains and A an irreducible n-dimensional analytic set in $D \times G$ possessing the property that its boundary $bA = \bar{A} \setminus A$ is contained in the Shilov boundary $\partial D \times \partial G$. Consider the natural projection $\mathbb{C}^n_z \times \mathbb{C}^n_w \to \mathbb{C}^n_z$. It is evident that its restriction to A is a proper map $\pi_1: A \to D$ and therefore π_1 is a *branched analytic covering*[2]. This means in particular that π_1 is onto and that there exists an $(n-1)$-dimensional analytic subset $E \subset D$, called the *branch locus*, such that π_1 is an m-sheeted covering map of the set $A \setminus \pi_1^{-1}(E)$ onto $D \setminus E$. Each holomorphic function f on A can be represented as an *algebroidal function* of degree m on D, i.e. there exist functions $a_1(z), \ldots, a_m(z) \in \mathcal{O}(D)$ such that

$$f^m(p) + a_1(z) f^{m-1}(p) + \ldots + a_{m-1}(z) f(p) + a_m(z) = 0$$

for $p \in A$, $z = \pi_1(p)$. For f we may take, for instance, the coordinate functions w_ν; as a consequence we find that A is the graph of an m-valued proper *algebroidal* map $w: D \to G$, each component $w_\nu = w_\nu(z)$ of which being an algebroidal function,

$$w_\nu^m + a_{1\nu}(z) w_\nu^{m-1} + \ldots + a_{m\nu}(z) = 0 \tag{5}$$

(in fact, the degree of the algebroidal function w_ν can be less than m if the pseudopolynomial (5) is reducible).

Conversely, if $f: D \to G$ is an arbitrary proper algebroidal map, then its graph $A \subset D \times G$ is an analytic set and $bA \subset \partial D \times \partial G$. As the projection

[2] For the definition and properties of such coverings, see [30], Part IIIC, and Vol. 7, Part III.

$\pi_2: A \to G$ likewise is a branched analytic covering then the inverse $f^{-1}: G \to D$ of the map $f: D \to G$ must be proper and algebroidal. The degree of a proper analytic map $f: D \to G$ is defined to be the multiplicity of the covering $\pi_1: A \to D$. Let f be given by the equation (5), let $R_\nu(z)$ be the discriminant of $w_\nu(z)$ and set $S_\nu = \{z \dot{\in} D: R_\nu(z) = 0\}$. It is the evident that the branch locus of f is contained in the discriminant set $S = \bigcup_{\nu=1}^m S_\nu$.

Example. Let

$$D = \{z \in \mathbb{C}^2: |z_1|^2 + |z_2|^2 < 1\},$$
$$G = \{w \in \mathbb{C}^2: |w_1|^4 + |w_2|^4 < 1\}$$

and

$$A = \{(z, w) \in D \;\; G: z_1 = w_1^2, z_2 = w_2^2\}.$$

The projection $\pi_1: A \to D$ is an analytic covering of multiplicity 4 and A is the graph of the proper algebroidal map $w: D \to G$, $w_1^2 - z_1 = 0$, $w_2^2 - z_2 = 0$, the components of which have degree 2. The inverse map $w^{-1}: G \to D$ is proper and holomorphic.

§ 2. Survey of Some Results

2.1. Holomorphic Maps of Domains with Smooth Boundaries. The holomorphic image of a domain $D \subset \mathbb{C}^n$ with smooth boundary, of course, may have non-smooth boundary. It turns out that the smoothness cannot be lost in an arbitrary fashion. If the smoothness of the boundary is lost as the result of a biholomorphic map, then it is lost in a highly "irregular" manner. We will call $D \subset \mathbb{C}^n$ a domain with *piecewise C^k-smooth boundary* if ∂D is topologically a $(2n-1)$-dimensional manifold and if in some neighborhood U of the closure $\bar{D}$ there exist real functions $\rho_\nu \in C^k(U)$, $\nu = 1, \ldots, m$, such that

a) $\partial D \subset \bigcup_1^m S_\nu$ where $S_\nu = \{z \in U: \rho_\nu(z) = 0\}$;

b) for each subset $\{\nu_1, \ldots, \nu_r\} \subset \{1, \ldots, m\}$ one has $d\rho_{\nu_1} \wedge \ldots \wedge d\rho_{\nu_r} \neq 0$ in the intersection $S_{\nu_1} \cap \ldots \cap S_{\nu_r}$.

Theorem 7 ([50]). *Let $D \subset \mathbb{C}^n$ be a bounded pseudoconvex domain with piecewise C^2-smooth but not smooth boundary. Then there exists no biholomorphic map f from D onto a bounded domain $G \subset \mathbb{C}^n$ with C^2-boundary.*

The hypothesis of the boundedness of D and G in this theorem cannot be removed. In fact, take $f(z_1, z_2) = (z_1^2, z_2)$. Then f maps the domain $D = \{z \in \mathbb{C}^2: x_1 > 0, y_1 > 0\}$, with piecewise smooth but nonsmooth boundary, biholomorphically onto the unbounded domain $G = \{z \in \mathbb{C}^2: y_1 > 0\}$ with smooth boundary.

The proof of Theorem 7 is based on a comparison of the growth of the kernel functions $K_D(z)$ and $K_G(w)$ near the boundary. In G one can get an estimate for $K_G(w)$ in the following manner. As ∂G has C^2-smoothness, there exists an $\varepsilon > 0$, depending only on G, such that each point $w \in \partial G$ can be touched from the inside of G with a ball of radius ε. From this it follows readily that

$$K_G(w) \leqq c_1\, e(w, \partial G)^{-(n+1)}.$$

On the other hand, on ∂D one can find points near which the kernel function $K_D(z)$ has much higher growth: for some $k \geqq n$ on its k-dimensional skeleton Γ_k there exist points near which the estimate.

$$K_D(z) \geqq c_2\, e(z, \Gamma_k)^{-(n+2)}.$$

holds true. This shows that the Jacobian $\det f'(z)$ is $\equiv 0$.

Remarks. 1) In the additional assumption that D is an analytic polyhedron, Theorem 7 was obtained by G.M. Khenkin [33].

2) The question of holomorphic equivalence of analytic polyhedra is studied in detail in the paper [61] by S.E. Sharonov.

3) Theorem 7 can be generalized to the case of proper holomorphic maps [50], [33].

A result close to Theorem 7, but based on different methods, were obtained in [31].

Theorem 8 ([31]). *There exist no proper holomorphic map from a bounded domain $D \subset \mathbb{C}^n$ with smooth boundary onto the space of a holomorphic bundle whose basis and fiber have positive dimension.*

2.2. Domains with Noncompact Groups of Holomorphic Automorphisms. Each bounded *homogeneous domain* $D \subset \mathbb{C}$ (i.e. a domain admitting a transitive group of *holomorphic automorphisms*) is equivalent to a disk. In $\mathbb{C}^n$ with $n > 1$ the situation is more complicated. Already Poincaré [53], comparing the groups of holomorphic automorphisms of a bidisk and a ball in $\mathbb{C}^2$ (these are homogeneous domains), found that they were inequivalent. The systematic study of bounded homogeneous domains was initiated by É. Cartan [14] and finally lead to their complete classification (cf., for instance, [54]). It turned out that in $\mathbb{C}^n$ with $n < 7$ there exists up to biholomorphic equivalence only a finite number of different homogeneous domains (in $\mathbb{C}^2$ there is only the ball and the bidisk) but if $n \geqq 7$ the set of such a domain has the power of the continuum. We will not dwell in any detail on these results, because they have mainly an algebraic character, whereas we are here more interested in geometric conditions for holomorphic equivalence. Let us also mention that the methods based on a comparison of the group of holomorphic automorphisms have a limited applicability. The results of our Part show that holomorphic maps for $n > 1$ display a considerable rigidity: the equivalence of two domains picked up at random is more often an exception than the rule and the bulk of domains in $\mathbb{C}^n$ admit in general no automorphisms other than the identity map (cf. Chap. 3, § 1).

In this Section we set forth two results showing that the existence of a rich group of holomorphic automorphisms allows one to map biholomorphically a domain with a "nice" boundary onto a ball or a product of balls. The first result is due to B. Wong [76] and Rosay [56].

Theorem 9. *Let $D, G \subset \mathbb{C}^n$ be a bounded domain and assume that the boundary ∂G is strictly pseudoconvex in a neighborhood of the point $\zeta^0 \in \partial G$. Assume further that there exists a sequence of biholomorphic maps $f^\nu: D \to G$ such that for some point $z^0 \in D$ holds $f^0(z^0) \to \zeta^0 \in \partial G$ as $\nu \to \infty$. Then D and G are biholomorphically equivalent to a ball.*

Proof. We may assume that f^ν converges uniformly on compact subsets of D to a map $f: D \to \bar{G}$. Let $\varphi(w)$ be a strictly plurisubharmonic local defining function for G in a neighborhood of ζ^0. The function $\varphi \circ f(z)$ is defined, nonpositive and plurisubharmonic in a neighborhood of z^0. Also $\varphi \circ f(z^0) = 0$, as $f(z^0) = \zeta^0$. Therefore $\varphi \circ f(z) \equiv 0$ so that f must be a constant: $f(z) \equiv \zeta^0$.

Put $w^\nu = f^\nu(z^0)$ and let ζ^ν be the point on ∂G closest to w^ν. In order to construct a biholomorphism between D and G we will consider f^ν in various coordinate systems. First, let us consider, in a neighborhood $V \ni \zeta^0$, holomorphic coordinates such that $\zeta^0 = 0$, $\partial G \cap V$ is strictly convex and the Taylor expansion of φ takes the form $\varphi(w) = u_n + |w|^2 + o(|w|^2)$. The domain of definition of each map f^ν is thereby diminished but for each domain $\Omega \Subset D$ we can define f^ν in Ω from some index $\nu_0 = \nu(\Omega)$ on. Let h^ν be the composition of a shift and a unitary map sending ζ_ν onto 0 and $T^c_{\zeta_0}(\partial G)$ onto the plane $w_n = 0$. Then the defining functions of the domains $G^\nu = h^\nu(V \cap G\}$ have the form $\varphi^\nu(w) = c_\nu u_n + r^\nu(w)$ where

$$|r^\nu(w) - |w|^2| \leqq \varepsilon^\nu |w|^2 + \alpha(w), \tag{6}$$

and $c^\nu \to 1$, $\varepsilon^\nu \to 0$, $\alpha(w) = o(|w|^2)$. Put $F^\nu = h^\nu \circ f^\nu$. Then $F^\nu(z^0) = ('0, -\delta^\nu)$[3] and $\delta^\nu \to +0$. In the image let us make yet another linear transformation $('w, w_n) = (\sqrt{\delta^\nu}\,'\tilde{w}, \delta^\nu \tilde{w}_n)$ (the coordinate $'w$ in the complex tangent spaces is expanded $1/\sqrt{\delta^\nu}$ times and the normal coordinate w_n $1/\delta_\nu$ times; in the sequel we will for simplicity omit the twiddle). In terms of the new coordinates $F^\nu(z^0) = ('0, -1)$ and the defining function G^ν has the form

$$\psi^\nu(w) = \frac{1}{\delta^\nu} \varphi^\nu(\sqrt{\delta^\nu}\,'w, \delta^\nu w_n) = c^\nu u_n + \frac{1}{\delta^\nu} r^\nu(\sqrt{\delta^\nu}\,'w, \delta^\nu w_n)$$

(the factor $1/\delta^\nu$ in front of φ^ν is taken for the sake of normalization). The last term is positive and converges uniformly on compact sets to $|'w|^2$ as $\nu \to \infty$. We may assume that the sequence of functions $F^\nu_n(z)$, $\nu = 1, 2, \ldots$, converges uniformly on compact sets of D, because $\operatorname{Re} F^\nu_n(z) < 0$ and $F^\nu_n(z^0) = -1$. It follows from this that $'F^\nu(z)$ is uniformly bounded inside D. Consequently, we can pick from the sequence F^ν a subsequence converging to a map $F: D$

[3] Here we use the notation $w = ('w, w_n)$, $'w = (w_1, \ldots, w_n)$.

$\to\{u_n+|'w|^2<0\}$. The inverse maps $(F^\nu)^{-1}$ are uniformly bounded; therefore we can assume that they converges to a map H which is holomorphic in $\{u_n+|'w|^2<0\}$. As $F(z^0)=('0,-1)$, $H('0,-1)=z^0$, we see that F and H are each others inverses, so that F must be biholomorphic. It remains to make the observation that the domain $\{u_n+|w'|^2<0\}$ is biholomorphic to a ball. □

The basic idea in the preceding proof was the special expanding of the coordinates $('w, w_n)=(\sqrt{\delta^\nu}\,'\tilde{w}, \delta^\nu \tilde{w}_n)$, which allowed us to get from a strictly pseudoconvex piece of G the "ball" $\{u_n+|'w|^2<0\}$. The same method turns out to be useful also in several other problems. With the aid of it one can get the following result.

Theorem 10 ([52]). *Let $D\subset\mathbb{C}^n$ be a homogeneous bounded domain with piecewise C^2-smooth boundary. Then D is biholomorphically equivalent to a product of balls $B^{n_1}\times\ldots\times B^{n_k}$, where $B^{n_j}=\{z\in\mathbb{C}^{n_j}: |z|<1\}$, $n_1+\ldots+n_k=n$, $k\leqq p$ (p* being the largest codimension of the edges of the boundary ∂D).

2.3. Maps Onto. In this Section we set forth some results showing that for all domains D, $G\subset\mathbb{C}^n$ in a sufficiently large class there exists a surjective holomorphic map $f: D\to G$.

Theorem 11 ([28], [29]). *Let Ω be a connected paracompact n-dimensional complex manifold. Then there exist locally biholomorphic maps from the polydisk Δ^n and the ball B^n onto Ω and a number N such that the number of preimages of an arbitrary point $w\in\Omega$ is less than N.*

Løw [42] gave a simple construction of a domain $D\subset\mathbb{C}^n$ which can not be mapped holomorphically onto the ball. For D one can take a ball B^n with a sequence of spherical shells $K_\nu=\{z\in\mathbb{C}^n: r^\nu\leqq|z|\leqq s^\nu\}$ removed, having small "holes" so as to make D connected. Every holomorphic map $f: D\to B^n$ can be continued to B^n. Assuming, for simplicity, that $f(0)=0$, we get for the Jacobian the estimate $|\det f'(z)|\leqq n![e(z,\partial B^n)]^{-n}$. Therefore, if the K_ν lie sufficiently dense, then the volume of $f(D)$ does not suffice to cover the whole of B^n.

Theorem 12 ([42]). *Let B^n be a ball in $\mathbb{C}^n$ and $p\in\partial B^n$. There exists a holomorphic map $f: B^n\to\Delta^n$ which is surjective on each ball $\tilde{B}\subset B^n$ for which $p\in\partial\tilde{B}$.*

If $D\subset\mathbb{C}^n$ is an arbitrary bounded domain with C^2-boundary, then there exist balls $\tilde{B}^n$, B^n with a common boundary point such that $\tilde{B}^n\subset D\subset B^n$. It follows therefore from Theorems 11 and 12 that D can be mapped holomorphically onto an arbitrary domain $G\subset\mathbb{C}^n$.

Chapter 2
Boundary Correspondence and Holomorphic Equivalence

§ 1. Continuous Continuation

1.1. The Carathéodory Metric. Recall that the Carathéodory distance between two points p, q of a domain $D\subset\mathbb{C}^n$ is defined to be

$$c_D(p, q)=\sup_{h\in\mathcal{O}_p}\ln\frac{1+|h(q)|}{1-|h(q)|},$$

where $\mathcal{O}_p=\mathcal{O}_p(D, U)=\{h\in\mathcal{O}_p(D): |h|<1, h(p)=0\}$. It is well-known (cf., for example, [60]) that if D is bounded, then c_D is a metric which is invariant for biholomorphic maps and which diminishes under holomorphic maps. In particular, if $\Omega\subset D$ then $c_D(p, q)\leqq c_\Omega(p, q)$ for $p, q\in\Omega$.

It may happen that the Carathéodory metric is enlarged under algebroidal maps, but this does not happen in an arbitrary manner, as the following generalization of the classical Schwarz lemma holds.

Lemma 1 ([51]). *Let D be a domain in $\mathbb{C}^n$, $n\geqq 1$, and let $w=\varphi(z)$ be an algebroidal function of degree m in D. Let us assume that $|\omega|<1$ in D (that is, for $z\in D$ all values lie in the disk $|w|<1$) that $\varphi(p)\ni 0$ for some point $p\in D$ and that there is given a domain $\Omega\ni p$ relatively compact in D.*

Then there exist a constant $r=r(m, \Omega)<1$, not depending on φ, such that $|\varphi(z)|<r$ for all $z\in\Omega$ and all branches of the function $w=\varphi(z)$ obtained by analytic continuation inside Ω of the branch for which $\varphi(p)=0$.

Proof. We denote by $\mathcal{O}_p^m(D, U)=\mathcal{O}$ the set of all algebroidal functions φ satisfying the conditions of the lemma and by φ_p an irreducible germ (maybe, one of several) of φ at p such that $\varphi_p(p)=0$. Let $r(\varphi)=\sup_{z\in\Omega}|\varphi(z)|$ under the condition that one counts only those values of φ obtained by analytic continuation of φ_p along all possible paths inside Ω. We have to show that $r=\sup_{\varphi\in\mathcal{O}}r(\varphi)<1$. To this end let us consider a sequence $\varphi^\nu\in\mathcal{O}$ such that $r(\varphi_\nu)\to r$. Let

$$P_m^\nu(w, z)=w^m+a_{1\nu}(z)\,w^{m-1}+\ldots+a_{m\nu}(z)$$

be the Weierstrass polynomial of the functions φ^ν. As $|\varphi^\nu|<1$ the coefficients $a_{m\nu}$ will be uniformly bounded in D so that we may assume that they converge uniformly on compact sets to functions $a_k\in O(D)$. Then all roots of the polynomial $P_m(w, z)=w^m+a_1(z)\,w^{m-1}+\ldots+a_m(z)$ lie in the closed disk $|w|\leqq 1$. If we can for some $w_1\in\{|w|=1\}$ find a point $z_1\in D$ such that $P_m(w_1, z_1)=0$ then by the maximum principle $P_m(w_1, z)=0$ for all $z\in D$. This means that

P_m can be split into factors:

$$P_m(w, z) = P_{m-l}(w, z)(w - w_1)^{l_1} \dots (w - w_s)^{l_s},$$

where $w_1, \dots, w_s \in \{|w| = 1\}$, $l = l_1 + \dots + l_s$ and all roots of the polynomial P_{m-l} lie inside the disk $|w| < 1$. Therefore, for some $\varepsilon > 0$ and all $z \in \Omega$, the roots of P_{m-l} must satisfy the condition $|w| < 1 - 3\varepsilon$.

By the theorem on the continuous dependence of the roots of a polynomial on the coefficients, then for $z \in \Omega$ all roots of $P_m^\nu(w, z) = 0$, from some index ν_0 on, either lie in the disk $|w| \leqq 1 - 2\varepsilon$ or else satisfy the condition $|w| \geqq 1 - \varepsilon$. Therefore, under analytic continuation of the germs φ_p^ν within Ω all their values have to lie in the disk $|w| \leqq 1 - 2\varepsilon$ so that we may take $r(m, \Omega)$ to be $1 - 2\varepsilon$. □

Let $D \subset \mathbb{C}^n$ be a bounded domain, $p \in D$ and $d = e(p, \partial D)$. By $B(p, r)$ and $B_D^c(p, r)$ we then denote the ball with center at p and of radius r respectively in the usual (Euclidean) and in the Carathéodory metric.

Lemma 2 ([51]). *Let D, G be bounded domains and $f: D \to G$ an algebroidal map of degree m. Then for each $\varepsilon \in (0, 1)$ there exists $r > 0$, independent of $p \in D$, such that*

$$f(B(p, \varepsilon d)) \subset B_G^c(f(p), r)$$

(the inclusion is understood in the sense of Lemma 1).

Proof. As $B(p, \varepsilon d) \subset B(p, d) \subset D$ and $\varepsilon \in (0, 1)$ is fixed, then by Lemma 1 there exists a $c \in (0, 1)$ such that for all $p \in D$ and all $a \in \mathcal{O}_p^m(D, U)$ we have $|a(z)| < c$ for $z \in B(p, \varepsilon d)$. In particular, $|\beta \circ f(z)| < c$ for each $\beta \in \mathcal{O}_{f(p)}(G, U)$. It follows that $f(z) \in B_G^c(f(p), r)$ where $r = \ln(1 + c)/(1 - c)$. □

1.2. Continuous Extension. It is natural to call an algebroidal function (map) in a domain D, defined by the equation

$$w^m + a_1(z) w^{m-1} + \dots + a_m(z) = 0,$$

continuous in $\bar{D}$ if all the coefficients a_ν belong to $\mathcal{O}(D) \cap C(\bar{D})$.

Theorem 3 ([52]). *Let $G \subset \mathbb{C}^n$ be a strictly pseudoconvex domain, $D \subset \mathbb{C}^n$ a bounded domain with C^2-boundary and $f: D \to G$ a proper algebroidal map. Then f is continuous in $\bar{D}$.*

Proof. It is clear that D must be pseudoconvex. From each point $\zeta \in \partial D$ we trace the segment $[\zeta, p_0]$ ($p_0 = p_0(\zeta)$) of the inner normal to ∂D of length d_0, where $d_0 > 0$ is sufficiently small. For simplicity we consider only those points ζ for which the segments $[\zeta, p_0]$ do not intersect the set of branch points of f (they form a set $S \subset \partial D$ of full measure). We shall show that for each branch of f the limit $\lim_{p \to \zeta, p \in [p_0, \zeta]} f(p)$ exists uniformly for $\zeta \in S$. Apparently, the statement of the theorem will then follow.

Let p_k be the point of $[p_0, \zeta]$ such that $d_k = e(p_k, \zeta) = d_0 2^{-k}$ and put $q_k = f(p_k)$, $\delta_k = e(q_k, \partial G)$. We remark that Prop. 4 in Chap. 1 carries over unchanged to proper algebroidal maps, whence $\delta_k \leqq c d_k^\beta$. In view of Lemma 2 and taking into account that $p_k \in \operatorname{clos} B(p_{k-1}, d_{k-1}/2)$, we find that there exists $r>0$ such that $q_k \in B_k^c(q_{k-1}, r)$ for all q_k. Since G is strictly pseudoconvex, we must have $B_G^c(q_{k-1}, r) \subset B(q_{k-1}, R\sqrt{\delta_k})$, where $R>0$ depends only on r and G (cf. [48], [33]). Therefore $e(q_{k-1}, q_k) \leqq R c_1 (d_0 2^{-k})^{\beta/2}$ so that the sequence $f(p_k)$ must be convergent. The convergence of $f(p)$ for $p \to \zeta$, $p \in [p_0, \zeta]$ now follows from $f([p_{k-1}, p_k]) \subset B_G^c(q_{k-1}, r) \subset B(q_{k-1}, R\sqrt{\delta_k})$. This convergence is uniform in $\zeta \in S$, as all estimates have been uniform. □

Remarks. 1) The discriminants $R_\nu(z)$ of the components f_ν are symmetric polynomials in the coefficients of the Weierstrass polynomials. Therefore $R_\nu \in \mathcal{O}(D) \cap C(\bar{D})$. If E and E^* are respectively the branch locus and the discriminant set of f, then from the boundary uniqueness theorem in [47] it follows that $\bar{E} \cap \partial D$ and $\bar{E}^* \cap \partial D$ are nowhere dense subsets of the boundary.

2) If f is holomorphic it follows from the proof of Theorem 3 that f satisfies in $\bar{D}$ a Hölder condition with exponent $\alpha = \beta/2$. If, in addition, D is strictly pseudoconvex, then $\alpha = 1/2$. Various variants of Theorem 3 on the continuous continuation of biholomorphic and proper holomorphic maps were obtained in papers by G.A. Margulis [43], G.I. Khenkin [33], S.I. Pinchuk [46], Bedford, Diederich and Fornaess [3], [20], and others.

§ 2. Smooth and Analytic Continuation

2.1. Fefferman's Theorem. The following result is of major importance for the entire theory of holomorphic maps.

Theorem 4. *Every biholomorphic map $f: D \to G$ between strictly pseudoconvex domains $D, G \subset \mathbb{C}^n$ with C^∞ boundaries extends to a C^∞ diffeomorphism $\hat{F}: \bar{D} \to \bar{G}$.*

This theorem was obtained by Fefferman [24] in 1974. The original proof was quite delicate. It included a deep and subtle analysis of the asymptotics of the Bergman kernel and the boundary behavior of the geodesics in the Bergman metric. Today alternative approaches to the proof of this theorem are available (cf. Sec. 2.4).

2.2. Analytic Continuation. Fefferman's theorem is intimately connected with the following generalization of a classical theorem by Schwarz, obtained in [48] (cf. also [39], [72]).

Theorem 5. *Let $f: D \to G$ be a biholomorphic map between two strictly pseudoconvex domains $D, G \subset \mathbb{C}^n$ with real-analytic boundaries. Then f extends to a biholomorphic map of a suitable neighborhood of the closure $\bar{D}$ to a neighborhood of $\bar{G}$.*

For the problem of holomorphic equivalence the following corollary is of importance.

Corollary 6. *Two strictly pseudoconvex domains with real-analytic boundaries are biholomorphically equivalent iff their boundaries are biholomorphically equivalent.*

Remarks. 1) If in Theorem 5 the boundaries ∂D, ∂G are algebraic, then f is algebraic (Webster [71]).

2) Theorem 4 and 5 admit local formulations. Theorem 5 is a consequence of Theorem 4 and the following local statement.

Proposition 7 ([48]). *Let $\Gamma, \Lambda \subset \mathbb{C}^n$ be strictly pseudoconvex real-analytic hypersurfaces and let $f: \Gamma \to \Lambda$ be a C^1-smooth non-constant* CR-*map. Then f extends holomorphically to a neighborhood of Γ as a locally biholomorphic map.*

Let us write down a brief scheme for the proof of Proposition 7. Let ρ, φ be strictly plurisubharmonic defining functions for the surfaces Γ, Λ, respectively. In view of the strictly pseudoconvexity, the map f extends holomorphically to a suitable domain adjoining G from the side $\rho<0$. By Proposition 5, Chap. 1, we have $\det f' \neq 0$ on Γ. It remains, thus, to continue f analytically to the other side of Γ.

The real analyticity of ρ and φ means that they can be locally expanded in power series in powers of z, $\bar{z}$, i.e. $\rho=\rho(z,\bar{z})$, $\varphi=\varphi(z,\bar{z})$. Therefore we have on Γ

$$\varphi(f(z), \overline{f(z)})=0. \tag{1}$$

The main idea of the proof is to separate in this equation the holomorphic and the antiholomorphic constituents and, afterwards, to apply the symmetry principle. For $n=1$ the separation of the holomorphic and the antiholomorphic parts is easily achieved. By the implicit function theorem the equation (1) can be solved with respect to $\bar{f}$: $\bar{f}(z)=h(f(z))$. For $n>1$ just the single equation (1) does not suffice for this. In order to get the missing $n-1$ equations, we apply to (1) the *tangential Cauchy-Riemann operators*

$$T_\nu=\rho_n \frac{\partial}{\partial z_\nu}-\rho_\nu \frac{\partial}{\partial z_n}, \qquad \nu=1, \ldots, n-1$$

(we assume that in a neighborhood we have $\rho_n \neq 0$ so that the T_ν give at each point $\zeta \in \Gamma$ a basis for $T_\zeta^c(\Gamma)$). The system thus obtained can, in view of the implicit function theorem, be solved in the form $\bar{f}=h(f, Tf)$. The conditions in the latter theorem regarding the nondegeneracy of the corresponding Jacobian are fulfilled and are consequences of the strictly pseudoconvexity of Λ.

Let us consider in greater detail the case when Γ and Λ coincide with the sphere $S=\{|z|=1\}$ and the map f is defined locally in the intersection of Γ with a suitable neighborhood of Ω of the point $\zeta \in \Gamma$. For simplicity, we will assume that $\zeta=('0, 1)$. The equations $\varphi=0$, $T_\nu\,\varphi=0$ take the form

$$\begin{cases} \sum_{k=1}^{n} \overline{f_k(z)}\, f_k(z)=1, \\ \sum_{k=1}^{n} \overline{f_k(z)}\, T_\nu f_k(z)=1, \quad \nu=1, \ldots, n-1. \end{cases} \tag{2}$$

This system is linear in $\overline{f_k(z)}$ and therefore its solution has the form

$$\overline{f_k(z)}=R_k(f(z), Tf(z)), \quad z\in\Gamma\cap\Omega,$$

where the R_k are rational functions in $f(z)$ and $T_\nu f(z)$. Without loss of generality, we may assume that the map f extends holomorphically to $B^n\cap\Omega$. Let us consider the complex line $l=\{z\in\mathbb{C}^n\colon z_2=c_2, \ldots, z_n=c_n\}$, where $c_\nu=\text{const}$. If this line is sufficiently close to the point ζ, then it will intersect $B^n\cap\Omega$ in the disk $\Delta=\{|z_1|<r\}$, where $r^2=1-\sum_{\nu=2}^{n}|c_\nu|^2$. As in our case $T_\nu=\bar{z}_n\,\partial/\partial z_\nu-\bar{z}_\nu\,\partial/\partial z_n$, we have on the boundary $\partial\Delta$

$$T_1\, f_k(z)=\bar{c}_k\frac{\partial f_k(z)}{\partial z_1}-\bar{z}_1\frac{\partial f_k(z)}{\partial z_n}=\bar{c}_n\frac{\partial f_k(z)}{\partial z_1}-\frac{r^2}{z_1}\frac{\partial f_k(z)}{\partial z_n},$$

$$T_\nu\, f_k(z)=\bar{c}_n\frac{\partial f_k(z)}{\partial z_\nu}-\bar{c}_\nu\frac{\partial f_k(z)}{\partial z_n}, \quad \nu=2, \ldots, n-1.$$

As f is holomorphic in Δ, it follows from this that on $\partial\Delta$ the values of $T_\nu f_k(z)$ and, consequently, also those of $R_k(f(z), Tf(z))$ coincide with the boundary values of functions meromorphic in Δ. It follows now from the symmetry principle that for fixed $z_2=c_2, \ldots, z_n=c_n$ the functions f_k extend meromorphically in z_1 to the extended complex plane $\bar{\mathbb{C}}$. Therefore, the functions are rational in z_1. Analogously, one can show that f is rational in $z_2, \ldots, z_{n-1}$ and also in any direction close to them. By a theorem of Hurwitz and Weierstrass f must be rational in the entire collection of variables.

From this we readily obtain

Proposition 8. *Let Γ be connected open subset of the sphere $S\subset\mathbb{C}^n$, $n>1$, and $f\colon\Gamma\to S$ a non-constant* CR*-map of class C^1. Then f extends to a holomorphic automorphism of the ball $B^n=\{|z|<1\}$.*

This result illustrates a new typically multidimensional phenomenon of analytic continuation. It was first obtained by Poincaré and later rediscovered by Alexander [1]. Rudin [57] proved a variant of Proposition 8 in other assumptions on the map f. This again was extended to Siegel domains by A.M. Tumanov and G.M. Khenkin [66].

2.3. Applications. We may now free ourselves from the assumptions concerning the smoothness of f up to the boundary in Proposition 5 of Chapter 1. Let us first consider the case of the ball.

Theorem 9 ([2]). *Each proper holomorphic map $f\colon B^n \to B^n$ $(n>1)$ is biholomorphic.*

Proof. Let $E \subset B^n$ be the branch locus of the map f^{-1}. In view of Remark 1 to Theorem 3 we can find on the boundary ∂B^n a point $\omega \in \bar{E} \cap \partial B^n$. Let $\zeta \in f^{-1}(\omega)$. Then we can, apparently, find neighborhoods $U \ni \zeta$, $V \ni \omega$ such that f maps biholomorphically $U \cap B^n$ into $V \cap B^n$. By Fefferman's theorem (a local variant) $f \in C^\infty(U \cap \bar{B}^n)$ and, consequently (Proposition 8), f is an automorphism of B^n. □

Remark. As is seen from the proof, Theorem 9 remains in force if $f\colon B^n \to B^n$ is a proper algebroidal map.

Theorem 10 ([50]). *Let $D, G \subset \mathbb{C}^n$ be strictly pseudoconvex domains and $f\colon D \to G$ a proper holomorphic map. Then f is locally biholomorphic, i.e.* $\det f' \neq 0$.

Theorem 10 is obtained with the aid of the method of coordinate dilation used already in Sec. 2.2. of Chap. 1. For the proof of Theorem 10 one has to make dilations of coordinates in the image as well as in the preimage. This allows, after a passage to the limit, to reduce Theorem 10 to Theorem 9.

2.4. Generalizations. Recently one has found new approaches to the proof of Fefferman's theorem. One such proof is connected with papers by Webster [73], Ligocka [40], [41], [7], Bell [4], [5], [6], Diederich-Fornaess [21], [22] and others. This approach is based on a study of the Bergman projection and at the final stage reduces to subelliptic estimates for the $\bar{\partial}$-Neumann operator, which, however, cannot be considered as elementary. By contrast, the proof of Nirenberg, Webster and Yang [45] is more elementary. Finally, let us mention Lempert's approach [38], which is connected with extremal disks.

Let us consider some extensions of Fefferman's theorem. Following Bell, we say that a domain $D \Subset \mathbb{C}^n$ with C^∞-boundary satisfies condition R if $P(C^\infty(\bar{D})) \subset C^\infty(\bar{D})$, where $P\colon L^2(D) \to L^2(D) \cap \mathcal{O}(D)$ is the Bergman projection,

$$Pf(z) = \int_D K(z, w)\, f(w)\, dV_w.$$

It is well-known that condition R is satisfied by strictly pseudoconvex domains with C^∞-boundary [35] and pseudoconvex domains with real-analytic boundary [23].

Theorem 11 ([6], [22]). *Let $D, G \Subset \mathbb{C}^n$ be pseudoconvex domains with C^∞-boundary, where D satisfies condition R, and let $f\colon D \to G$ be a proper holomorphic map. Then f extends to a C^∞-map $\hat{f}\colon \bar{D} \to \bar{G}$.*

Taking account of Proposition 6 of Chap. 1 now follows

Proposition 12 ([21]). *Assume, in the hypothesis of Theorem 11, that D is strictly pseudoconvex. Then $\hat{f}$ is a local diffeomorphism and G is also strictly pseudoconvex.*

Theorem 12 gives a positive solution to the problems stated in Sections 1.1 and 1.2 of Chap. 1, in the case of domains with boundaries of class C^∞. In the case of boundaries with less smoothness we have the following partial result.

Theorem 13 ([50]). *Let $f: D \to G$ be a biholomorphic map of the bounded convex domain $D \subset \mathbb{C}^n$ with $C^{2+\varepsilon}$-boundary ($\varepsilon > 0$) onto a strictly pseudoconvex domain $G \subset \mathbb{C}^n$. Then D too is strictly pseudoconvex.*

A generalization of Fefferman's theorem to the case of finitely smooth boundaries was obtained by Lempert [38] and Ligocka [41].

Theorem 14. *Let $D, G \subset \mathbb{C}^n$ be strictly pseudoconvex domain with C^{k+4}-boundaries ($k \geqq 2$) and let $f: D \to G$ be a biholomorphic map. Then f extends to a $C^{k+1/2}$-diffeomorphism $\bar{f}: \bar{D} \to \bar{G}$.*

Chapter 3 Analytic Continuation and Holomorphic Equivalence

In this Chapter we consider questions of holomorphic equivalence for *real-analytic strictly pseudoconvex* (in brief: ASPS) *hypersurfaces* in $\mathbb{C}^n$, which already early attracted attention.[4] A number of invariants of such surfaces were obtained in the papers of Poincaré [53] and Segre [62]. Considerable progress in this direction is connected with the work of Cartan [13] and, especially, Tanaka [64], [65], Chern and Moser [15], where the problem of local holomorphic equivalence of germs of ASPS hypersurfaces is solved. Here we will only set forth the results by Moser on *normal forms* of ASPS hypersurface[5]. A survey of other results in [13], [15], [64], [65] can be found in the paper of Burns and Shneider [11] and in the following Part VI.

On the other hand, the results of Sec. 2.2 of Chap. 2 show that the holomorphic equivalence of strictly pseudoconvex domains with real-analytic boundaries is equivalent to the (global!) holomorphic equivalence of their boundaries, which are compact ASPS hypersurfaces. It turns out that for such surfaces one has a phenomenon of analytic continuation analogous to the one in Proposition 8 of Chap. 2 and so in many cases the local equivalence of ASPS hypersurfaces implies their global equivalence.

[4] For a more detailed treatment of these questions, see Vol. 7, Part IV.

[5] Moser introduced normal forms for a much larger class of hypersurfaces with nondegenerate Levi form.

§ 1. Local Transformation of Real Hypersurfaces into Normal Form

The simplest example of an ASPS surface is the sphere $S=\{z\in\mathbb{C}^n: |z|=1\}$. With the aid of the biholomorphic map

$$'z=\frac{2'\tilde{z}}{\tilde{z}_n+i}, \qquad z_n=\frac{\tilde{z}_n-i}{\tilde{z}_n+i}$$

it can be transformed into a surface given by the equation

$$\tilde{y}_n=|'\tilde{z}|^2. \tag{1}$$

Let now $\Gamma\subset\mathbb{C}^n$ be any ASPS hypersurface. We wish to write down an equation for Γ, thereby being guided by (1). The defining function ρ of Γ can be chosen in many ways. In order to get rid of this indeterminicy it is convenient to solve the equation $\rho=0$ with respect to some real variable. Without loss of generality, we may assume that it is the variable y_n. Then the equation for Γ takes the form

$$y_n=F('z, '\bar{z}, x_n),$$

where the right hand side is real analytic, that is, can be expanded in a power series in $'z, '\bar{z}, x_n$. Let us fix the point $\zeta\in\Gamma$ and a real analytic curve γ on Γ passing through ζ. We may assume that γ has the parametrization

$$'z='z(\tau), \qquad z_n=z_n(\tau), \qquad \tau\in(-\tau_0, \tau_0),$$

where $'z(\tau)$, $z_n(\tau)$ are real analytic and τ corresponds to ζ. We also assume that γ passes through ζ in a non-complex tangential direction, i.e. that

$$('\dot{z}(0), \dot{z}_n(0))\in T_\zeta(\Gamma)\setminus T_\zeta^c(\Gamma).$$

Then we have

Proposition 1. *In a neighborhood of $\zeta\in\Gamma$ there exists a biholomorphic map $h(z)$ mapping γ onto the line $'z=0$, $z_n=\tau$ (i.e. $h('z(\tau), z_n(\tau))=('0, \tau)$) such that, in the new local coordinates, Γ is given by the equation*

$$y_n=|'z|^2+\sum_{k,l\geqq 2} F_{kl}('z, '\bar{z}, x_n), \tag{2}$$

where the F_{kl} are homogeneous polynomials of degree k, l in $'z, '\bar{z}$ respectively, whose coefficients are analytic in x_n.

Let us notice that in (2) there are no terms of the form F_{k0}, F_{0l}, F_{k1}, F_{1l}. In order to simplify the equation of Γ even more, we use the nonuniqueness of the change of coordinates bringing the equation into the form (2) in a neighborhood of $\zeta\in\Gamma$. This nonuniqueness consists above all of that one has freedom in the choice of the curve γ and its parametrization. However, even after Γ has been brought into the form (2) one can a posteriori apply a transformation

$$('z, z_n) \to (U_{z_n}\,'z, z_n), \tag{3}$$

where U_{z_n} is an arbitrary nondegenerate $(n-1)\times(n-1)$ matrix function, holomorphic in z_n and unitary for $z_n=\tau: |U_\tau'|=|'z|$. This transformation fixes points of γ and preserves the form of the equation (2).

Using tensor notation, let us define the trace $\operatorname{tr} F_{kl}=G_{k-1,l-1}$ of the polynomial

$$F_{k,l}=\sum_{1\leqq\alpha_\nu,\beta_\mu\leqq n-1} a_{\alpha_1,\dots,\alpha_k,\beta_1,\dots,\beta_l}\, z_{\alpha_1}\cdot\ldots\cdot z_{\alpha_k}\cdot\bar z_{\beta_1}\cdot\ldots\cdot\bar z_{\beta_l}$$

(it is assumed that the coefficient $a_{\alpha\beta}(x_n)$ do not change under arbitrary permutations of the indices $\alpha_1,\dots,\alpha_k$ or the indices $\beta_1,\dots,\beta_l$). For $k, l\geqq 1$ let us set

$$\operatorname{tr} F_{k,l}=\sum b_{\alpha_1,\dots,\alpha_{k-1},\beta_1,\dots,\beta_{l-1}}\, z_{\alpha_1}\cdot\ldots\cdot z_{\alpha_{k-1}}\cdot\bar z_{\beta_1}\cdot\ldots\cdot\bar z_{\beta_{l-1}},$$

where

$$b_{\alpha_1,\dots,\alpha_{k-1},\beta_1,\dots,\beta_{l-1}}=\sum_{\alpha_k=\beta_l} a_{\alpha_1,\dots,\alpha_k,\beta_1,\dots,\beta_l}.$$

This definition is very simple and transparent if $m=2$. In this case $F_{kl}=a(x_2)\,z_1^k\,\bar z_1^l$ so that $\operatorname{tr} F_{kl}=a(x_2)\,z_1^{k-1}\,\bar z_1^{l-1}$, i.e. $F_{kl}=|z_1|^2\operatorname{tr} F_{kl}$.

It turns out that in the neighborhood of an arbitrary point $\zeta\in\Gamma$ there exists a system of local coordinates such that $\zeta=0$ and Γ has an equation of the form (2) with

$$\operatorname{tr} F_{22}=0,\quad (\operatorname{tr})^2 F_{32}=0,\quad (\operatorname{tr})^3 F_{33}=0. \tag{4}$$

in this case we say that the equation of Γ near the point ζ is in *normal form*. Notice that if $n=2$ the condition (4) means that $F_{22}=0$, $F_{32}=0$, $F_{33}=0$.

Let us explain the geometric meaning of (4). Consider the transformation of coordinates in Proposition 7 and take x_n to be the parameter on γ. Then every admissible curve is given by an equation $'z={}'z(x_n)$ (in this case $\operatorname{Re} z_n(x_n)=x_n$ and $\operatorname{Im} z_n(x_n)$ is defined by the equation of Γ). It turns out that the equation $(\operatorname{tr})^2 F_{32}=0$ is precisely equivalent to the condition that γ satisfies a certain second order differential equation

$$'\ddot z=Q(x_n, 'z, '\bar z, '\dot z, '\dot{\bar z}). \tag{5}$$

This equation is rather complicated already for the sphere (1). One knows however that in the general case Q is a rational (vector-)function in $'\dot z, '\dot{\bar z}$ with analytic coefficients in $x_n, 'z, '\bar z$; its denominator vanishes on complex tangential directions and only in such directions. Let us notice that the condition $(\operatorname{tr})^2 F_{32}=0$ does not change if we take another parametrization of γ or if we make (2) subject to a transformation of the form (3). Thus, (5) defines on Γ a biholomorphically invariant family of curves termed *chains*. Through each point $\zeta\in\Gamma$ in a non-complex tangential direction there passes exactly one chain and if the surface Γ is given by an equation of the form (2) then the line $'z=0$, $y_n=0$ is a chain iff $(\operatorname{tr})^2 F_{32}=0$.

Let A be an arbitrary unitary $(n-1)\times(n-1)$ matrix. It turns out that there exists a unique map of type (3) such that $U_0=A$ and such that in the new coordinates $\operatorname{tr} F_{22}=0$. The matrix A can here be interpreted as a new choice of orthonormal basis in $T_0^c(\Gamma)$. We still have freedom in the choice of the parametrization of γ. A new parametrization corresponds to a change of coordinates

$$('z, z_n)\to(\sqrt{\dot{q}(z_n)}\,'z, q(z_n)),$$

where $q(0)=0$, $q(\bar{z}_n)=\overline{q(z_n)}$. Such a transformation preserves the form of the equation (2) and the relations $\operatorname{tr} F_{22}=0$, $(\operatorname{tr})^2 F_{32}=0$. The function $q(z_n)$ can be chosen such that one also has the condition $(\operatorname{tr})^3 F_{33}=0$. Then $q(x_n)$ must satisfy a certain third order differential equation. As we as before require that $q(0)=0$, the solution of this equation is determined by two initial conditions $\dot{q}(0)=\alpha$, $\ddot{q}(0)=\beta$.

Thus, the reduction of Γ into normal form is not unique. It depends on the following data (initial conditions):

a) the point $\zeta\in\Gamma$ corresponding to the origin in the normal form;

b) a direction $v\in T_\zeta(\Gamma)\setminus T_\zeta^c(\Gamma)$ defining the chain γ, which in the normal form has the equation $'z=0$, $y_n=0$.

c) the choice of an orthonormal basis in $T_\zeta^c(\Gamma)$;

d) two real parameters fixing the parametrization of γ.

The following fundamental result holds true.

Theorem 2 ([9], [15]). *Let $\Gamma\subset\mathbb{C}^n$ be an* ASPS *hypersurface given by the equation $y_n=F('z,'\bar{z},x_n)$. To each choice of initial conditions* a)–d) *there corresponds a unique holomorphic map $h=(h_1,\dots,h_n)$ transforming the equation of Γ into normal form* (2), (4). *The map h and the coefficients F_{kl} in* (2) *depend continuously (and also analytically) on these initial conditions and the function* $F('z,'\bar{z},x_n)$ (this implies, for instance, for h that each coefficient of the Taylor expansion of h_k at the point ζ depends only on a finite number of the Taylor coefficients of $F('z,'\bar{z},x_n)$ at this point and that this dependence is analytic) *and the radius of convergence of the series defining h and the corresponding normal equation* (2) *can be estimated from below by a suitable positive function depending continuously on the parameters in* a)–d).

A simple consequence of Theorem 2 is the Poincaré-Alexander theorem (Proposition 8, Chap. 2). In fact, the normal form of a sphere is $y_n=|'z|^2$. It is not hard to see that each choice of the initial conditions a)–d) defines a unique fractional-linear automorphism of the surface $y_n=|'z|^2$. Therefore, each local automorphism must be global.

In the case of a general ASPS hypersurface $\Gamma\subset\mathbb{C}^n$, the map h, giving the passage from one normal form to another, need not be fractional-linear in general. However, one has

Proposition 3 ([15]). *Suppose that the* ASPS *hypersurface $\Gamma\subset\mathbb{C}^n$ has near the origin the normal form* (2), (4) *and let $z\to h(z)$ be a biholomorphic map*

of a neighborhood $\Omega \ni 0$ mapping the line $'z=0$, $y_n=0$ into itself. Then the surface $h(\Gamma)$ has normal form iff h is a fractional-linear map of the form

$$('z, z_n) \to \left(\frac{\sqrt{|a|}}{b z_n + 1} U('z), \frac{a z_n}{b z_n + 1} \right), \tag{6}$$

where $U: \mathbb{C}^{n-1} \to \mathbb{C}^{n-1}$ is unitary and $a, b \in \mathbb{R}$, $a \neq 0$.

Thus, near any point the local equation of an ASPS hypersurface can be brought into a canonical form, the normal form (2), (4). This transformation is not unique and depends on a finite number of parameters b)–d). As shown in [12], [74], this nonuniqueness can be eliminated at points in "general position" – so-called non-umbilical points – if one requires some auxiliary conditions on the coefficients F_{22}, F_{32} if $n>2$ and on F_{42}, F_{43}, F_{52} if $n=2$. In this "strengthened" normal form the coefficients F_{kl} generate a complete system of invariants for Γ. Therefore the set of all equivalence classes (under biholomorphic maps) of germs of ASPS surfaces depend on a infinite number of parameters.

The basic results of the theory of normal forms carry over from real-analytic surfaces to surfaces of class C^∞, but we have to state them in this case in terms of formal series without any assertions regarding their convergence. From this it follows that also the hypersurfaces of class C^∞ have a rather large supply of CR-invariants, which depend on infinitely many variables. Therefore, the results of the paper [12] (we will not give their exact formulation here) look completely natural and, in some sense, "evident": strictly pseudoconvex domains with C^∞ boundary "in general position" do not admit other holomorphic automorphisms than the identity map and "almost all" sufficiently small deformations lead to domains inequivalent to the given one.

If a connected ASPS surface Γ is, near a point $\zeta \in \Gamma$, locally biholomorphically equivalent to a sphere then it is, apparently, locally equivalent to a sphere near any other point. Such *surfaces* are called *spherical*; they form a rather special but still quite important class of ASPS surfaces. For them we have

Proposition 4 ([48]). *Let $\Gamma \subset \mathbb{C}^n$ be a connected* ASPS *surface, U a neighborhood of a point $\zeta \in \Gamma$, $f: U \to \mathbb{C}^n$ a nonconstant holomorphic map and assume that $f(U \cap \Gamma) \subset S$ (where $S = \{|z|=1\}$). Then f can be continued along each path on Γ as a locally biholomorphic map.*

Let us mention some further results pertaining to normal forms. A.G. Vitushkin [67] has proved:

Theorem 5. *Let D, $G \subset \mathbb{C}^n$ be strictly pseudoconvex domains with real-analytic non-spherical boundaries. Then there exists a neighborhood $\Omega \supset \bar{D}$ such that all biholomorphic maps from D onto G can be continued holomorphically to Ω and are there equicontinuous.*

Theorem 6 (N.G. Kruzhilin, A.V. Loboda [37]). *Let Γ be a non-spherical ASPS hypersurface with $0 \in \Gamma$. Then there exists a local system of coordinates such that in terms of these coordinates the group of local automorphisms of Γ fixing the origin 0 is a subgroup of the unitary group $U(n)$.*

In many questions of analytic continuation a variant of the normal form (2), (4), due to A.G. Vitushkin [68], is often useful. It is gotten from (2), (4) with the aid of the map

$$('z, z_n) \to (2U('z)/(i+z_n);\ (i-z_n)/(i+z_n)), \tag{7}$$

sending (2) into a surface defined near the point $'z=0$, $\rho=1$, $\theta=0$ by the equation

$$1-\rho^2 = |'z|^2 + \sum_{k,l \geqq 2} \Phi_{kl}('z, '\bar{z}, x_n), \tag{8}$$

where $z_n = \rho e^{i\theta}$. Then (4) is translated into the condition

$$\operatorname{tr} \Phi_{22} = 0, \qquad (\operatorname{tr})^2 \Phi_{32} = 0, \qquad (\operatorname{tr})^3 \Phi_{33} = 0, \tag{9}$$

and the line $'z=0$, $y_n=0$ into the circumference $'z=0$, $\rho=1$.

Let γ^* be an arc on the chain γ of the surface Γ. A *normalization* of Γ with respect to the arc γ^* is a locally biholomorphic map h in a neighborhood of γ^* which maps Γ into a surface of the form (8), (9) and γ^* into the circumference $\rho=1$, $'z=0$.

For the normal forms (8), (9) Proposition 3 remains in force with the only change that instead of the maps (6) we now have to consider transformations of the type

$$('z, z_n) \to (\sqrt{q'(z_n)}\, U('z), q(z_n)), \tag{10}$$

where $q(z_n) = e^{i\alpha}(z_n - a)/(1-\bar{a} z_n)$ ($|a|<1$, $\alpha \in \mathbb{R}$) is an arbitrary fractional-linear automorphism of the disc $|z_n|<1$. From this we obtain the following property of the normal forms (8), (9), giving them a preference as compared to (2), (4).

Proposition 7 ([67]). *Each normalization of an ASPS surface Γ with respect to an arc γ^* of a chain γ extends unlimitedly along γ and gives a normalization of Γ with respect to entire chain γ. If, moreover, the increment of the normal parameter θ along γ^* equals $2\pi k$ (k = integer) then the increment of any other parameter $\tilde{\theta}$ likewise equals $2\pi k$.*

§ 2. Analytic Continuation of Local Maps

Let Γ and Λ be connected ASPS hypersurfaces in $\mathbb{C}^n$ and assume that Λ is compact. In a neighborhood U of the point $\zeta \in \Gamma$ let there be defined a nonconstant holomorphic maps $f: U \to \mathbb{C}^n$ such that $f(U \cap \Lambda) \subset \Lambda$. If Λ is the sphere $|z|=1$ then f can be continued analytically along each path on Γ (Prop. 4). It is natural to conjecture that this is true also in the general case. However, this is not so, as the following example due to Burns and Shnider [10] shows. Let

$$\Gamma=\{z\in\mathbb{C}^2: y_2=|z_1|^2\},$$
$$\Lambda=\{z\in\mathbb{C}^2: \sin\ln|z_2|+|z_1|^2=0,\ e^{-\pi}\leqq|z_2|\leqq 1\}.$$

Then the map $f(z_1, z_2)=(z_1/\sqrt{z_2}, \exp(i\ln z_2))$, with a suitable choice of a branch on $\ln z_2$, maps $\Gamma\setminus\{0\}$ into Λ but does not extend to the point $z=0$.

The surface Γ in the above example is curious, as it is a non simply connected compact spherical hypersurface. Indeed, this circumstance is the main obstacle to continuing f to the point $z=0$. This may seem surprising, but the case of non simply connected spherical surfaces in $\mathbb{C}^n$ is exceptional in the sense that for them the above conjecture on analytic continuation cannot be true. The following theorem holds:

Theorem 8 ([49]). *Let Γ, $\Lambda\subset\mathbb{C}^n$ be nonspherical connected* ASPS *hypersurfaces, assuming that Λ is compact. Let f be a nonconstant holomorphic map of some open subset $\Gamma^*\subset\Gamma$ into Λ. Then f can be continued analytically along each path on Γ as a locally biholomorphic map.*

The complete proof of this result is rather long, so we cannot give it here. We will just try to further clarify why the spherical case is excluded in the theorem. The spherical surface $y_n=|'z|^2$ admits automorphisms of the form

$$('z, z_n)\to\left(\frac{1}{\sqrt{\delta}}U('z), \frac{1}{\delta}z_n\right), \tag{11}$$

where $\delta>0$ and $U:\mathbb{C}^{n-1}\to\mathbb{C}^{n-1}$ unitary. For $\delta\to 0$ the map (11) gives an expansion as large as we wish. An analogous situation holds also in the Burns-Shnider example in a neighborhood of the point $z=0$. In the nonspherical case such dilations are not possible, which essentially was proved in Sec. 2.2, Chap. 1.

All the basic results on Moser's normal forms also hold true for real analytic hypersurfaces with a nondegenerate Levi form. Therefore, it might be interesting to try to extend Theorem 8 to the case of such surfaces (on complex manifolds). However, under maps of nonspherical surfaces with nondegenerate Levi form of indefinite sign arbitrary large dilations, analogous to (11), are possible. A corresponding example has been supplied by V.K. Beloshapka [8]. The reason for this is that the map (11) preserves the normal form of such surfaces if U is a pseudounitary map preserving the corresponding indefinite Hermitean form. But such a map U, as well as its inverse, can have arbitrary large norm.

Recently A.G. Vitushkin, V.V. Ezhov and N.G. Kruzhilin [69] obtained a different proof of Theorem 8 in the more general case when Γ, Λ are ASPS hypersurfaces in general n-dimensional complex manifolds ($n\geqq 2$). It is obtained with the aid of the normal forms of type (8), (9).

From Theorem 8 we readily derive

Theorem 9 [49]. *A necessary and sufficient condition for two strictly pseudoconvex domains D, $G \subset \mathbb{C}^n$ with simply connected real analytic boundaries to be holomorphically equivalent is that there exist points $p \in \partial D$, $q \in \partial G$, sufficiently small neighborhoods $U \ni p$, $V \ni q$ and a biholomorphic map $f\colon U \cap \partial D \to V \cap \partial G$.*

In the case of non simply connected ASPS boundaries the holomorphic equivalence of their pieces does not imply the holomorphic equivalence of the boundaries, as under analytic continuation along the boundary one can now get a multivalued map. It is easy to find the corresponding examples, but these are also the single exceptions from the rule.

Let us call an ASPS *hypersurface Γ completely inhomogeneous* if the neighborhoods (relative to Γ) of two arbitrary disjoint points $p_1, p_2 \in \Gamma$ are not holomorphically equivalent, i.e. there exists no biholomorphic map $h\colon U_1 \cap \Gamma \to U_2 \cap \Gamma$ (where $U_1 \ni p_1$, $U_2 \ni p_2$) such that $h(p_1) = p_2$. Complete inhomogeneity of ASPS hypersurfaces is a property of general position. Theorem 9 remains in force for arbitrary strictly pseudoconvex domains D, $G \subset \mathbb{C}^n$ with real analytic and completely inhomogeneous boundaries. In fact, it is clear that such boundaries are nonspherical. Therefore, in view of Theorem 8, each biholomorphic map $f\colon U \cap \partial D \to V \cap \partial G$ can be continued analytically along any path on ∂D and, if ∂D, ∂G are completely inhomogeneous, this continuation is well-defined.

References

For the convenience oı the reader, references to reviews in Zentralblatt für Mathematik (Zbl.), compiled using the MATH database, have been included as far as possible.

1. Alexander, H.: Holomorphic mappings from the ball and polydisc. Math. Ann. *209*, 249–256. Zbl. 272.32006
2. Alexander, H.: Proper holomorphic mappings in $\mathbb{C}^n$. Indiana Univ. Math. J. *26*, 137–146 (1977). Zbl. 391.32015
3. Bedford, E., Fornaess, J.E.: Biholomorphic maps of weakly pseudoconvex domains. Duke Math. J. *45*, 711–719 (1978). Zbl. 401.32006
4. Bell, S.: Biholomorphic mappings and the $\bar{\partial}$-problem. Ann. Math., II. Ser. *114*, 103–113 (1981). Zbl. 423.32009
5. Bell, S.: Proper holomorphic mappings and the Bergman projection. Duke Math. J. *48*, 167–175 (1981). Zbl. 465.32014
6. Bell, S., Catlin, D.: Proper holomorphic mappings extended smoothly to the boundary. Bull. Am. Math. Soc. *7*, 269–272 (1982). Zbl. 491.32018
7. Bell, S., Ligocka, E.: A simplification and extension of Fefferman's theorem on biholomorphic mappings. Invent. Math. *57*, 283–289 (1980). Zbl. 411.32010
8. Beloshapka, V.K.: Example of a noncontinuable holomorphic map of an analytic hypersurface. Mat. Zametki *32*, 121–123 (1982) [Russian]. Zbl. 518.32011. English transl.: Math. Notes *32*, 540–541 (1983)
9. Vitushkin, A.G.: Holomorphic continuation of maps of compact hypersurface. Izv. Akad. Nauk SSSR, Ser. Mat. *46*, 28–35 (1982) [Russian]. Zbl. 571.32011. English transl.: Math. USSR, Izv. *20*, 27–33 (1983)

10. Burns, D., jr., Shnider, S.: Spherical hypersurfaces in complex manifolds. Invent. Mat. *33*, 223–246 (1976). Zbl. 357.32012
11. Burns, D., jr., Shnider, S.: Real hypersurfaces in complex manifolds. In: Several complex variables. Proc. Symp. Pure Math. 30, part 2, 141–167. Providence: Am. Math. Soc. 1977. Zbl. 422.32016
12. Burns, D., jr., Wells, R.O., jr.: Deformations of strictly pseudoconvex domains. Invent. Math. *46*, 237–253 (1978). Zbl. 412.32022
13. Cartan, É.: Sur la géometrie pseudo-conforme des hypersurfaces de deux variables complexes. I, II. Ann. Mat. Pura Appl., IV. Ser. *11*, 17–90 (1932). Zbl. 5, 373; Ann. Scuola Norm. Sup. Pisa, II. Ser. *1*, 333–354 (1932). Zbl. 5, 374 (≡Oeuvres, partie II, vol. II, pp. 1231–1306. Zbl. 58, 83; partie III, vol. II, pp. 1217–1238. Zbl. 59, 153)
14. Cartan, É.: Sur les domaines bornés homogènes de l'espace de n variables complexes. Abh. Math. Semin. Univ. Hamb. *11*, 116–162 (1936) (≡Oeuvres (Paris 1952), partie I, vol. II pp. 1295–1307). Zbl. 11, 123
15. Chern, S.S., Moser, J.K.: Real hypersurfaces in complex manifolds. Acta Math. *133*, 219–271 (1974). Zbl. 302.32015
16. Chirka, E.M.: The theorems of Lindelöf and Fatou in $\mathbb{C}^n$. Mat. Sb. *92* (134), 622–644 (1973) [Russian]. Zbl. 285.32005. English transl.: Math. USSR, Sb. *21* (1973), 619–639 (1975)
17. Chirka, E.M.: Analytic representation of CR-functions. Mat. Sb. *98* (140), 591–623 (1975) [Russian]. Zbl. 321.32004. English transl.: Math. USSR, Sb. *27* (1975), 526–553 (1977)
18. Chirka, E.M.: Boundary regularity of analytic sets. Mat. Sb., Nov. Ser. *117* (159), 291–336 (1982) [Russian]. Zbl. 525.32005. English transl.: Math. USSR, Sb. *45*, 291–336 (1983)
19. Diederich, K., Fornaess, J.E.: Pseudoconvex domains: bounded strictly plurisubharmonic exhaustion functions. Invent. Math. *39*, 129–141 (1977). Zbl. 353.32025
20. Diederich, K., Fornaess, J.E.: Proper holomorphic maps onto pseudoconvex domains with real-analytic boundary. Ann. Math. II. Ser. *110*, 575–592 (1979). Zbl. 394.32012
21. Diederich, K., Fornaess, J.E.: Proper holomorphic images of strictly pseudoconvex domains. Math. Ann. *259*, 279–286 (1982). Zbl. 486.32013
22. Diederich, K., Fornaess, J.E.: Smooth extendability of proper holomorphic mappings. Bull. Am. Math. Soc., New Ser. *7*, 264–268 (1982). Zbl. 521.32014
23. Diederich, K., Fornaess, J.E.: Pseudoconvex domains with real-analytic boundaries. Ann. Math., II. Ser. *107*, 371–384 (1978). Zbl. 378.32014
24. Fefferman, C.: The Bergman kernel and biholomorphic mappings of pseudoconvex domains. Invent. Math. *26*, 1–65 (1974). Zbl. 289.32012
25. Fefferman, C.: Monge-Ampère equations, the Bergman kernel, and geometry of pseudoconvex domains. Ann. Math., II. Ser. *103*, 395–416 (1976). Zbl. 322.32012
26. Fefferman, C., Beals, M., Grossman, R.: Strictly pseudoconvex domains in $\mathbb{C}^n$. Bull. Am. Math. Soc., New Ser. *8*, 125–322 (1983). Zbl. 546.32008
27. Fornaess, J.E.: Biholomorphic mappings between weakly pseudoconvex domains. Pac. J. Math. *74*, 63–65 (1978). Zbl. 353.32026
28. Fornaess, J.E., Stout, E.L.: Polydiscs in complex manifolds. Math. Ann. *227*, 145–153 (1977). Zbl. 331.32007
29. Fornaess, J.E., Stout, E.L.: Regular holomorphic images of balls. Ann. Inst. Fourier *32*, No. 2, 23–36 (1982). Zbl. 452.32008
30. Gunning, R., Rossi, H.: Analytic functions of several variables. Englewood Cliffs: Prentice Hall 1965. Zbl. 141, 86
31. Hörmander, L.: An introduction to complex analysis in several variables. Princeton: Van Nostrand 1966. Zbl. 138, 62
32. Huckleberry, A., Ormsby, E.: Non-existence of proper holomorphic mappings between certain complex manifolds. Manuscr. Math. *26*, 371–379 (1979). Zbl. 422.32009
33. Khenkin, G.M.: An analytic polyhedron holomorphically nonequivalent to a strictly pseudoconvex domain. Dokl. Akad. Nauk SSSR *210*, 1026–1029 (1973) [Russian]. Zbl. 288.32015. English transl.: Sov. Math., Dokl. *14*, 858–862 (1973)
34. Khenkin, G.M., Chirka, E.M.: Boundary properties of holomorphic functions in several complex variables. In: Contemporary problems in mathematics (Itogi Nauki Tekhn., Ser.

Sovrem. Probl. Mat.) 13–142. Moscow: VINITI 1975 [Russian]. Zbl. 335.32001. English transl.: J. Sov. Math. *5*, 612–687 (1976)
35. Kohn, J.J.: Harmonic integrals on strictly pseudoconvex manifolds, I, II. Ann. Math., II. Ser. *78*, 112–148 (1963). Zbl. 161, 93; *79*, 450–472 (1964). Zbl. 178, 113
36. Kohn, J.J., Nirenberg, L.: A pseudoconvex domain non-admitting a holomorphic support function. Math. Ann. *201*, 265–268 (1973). Zbl. 248.32013
37. Kruzhilin, N.G., Loboda, A.V.: Linearization of local automorphisms of pseudoconvex domains. Dokl. Akad. Nauk SSSR *271*, 280–282 (1983) [Russian]. Zbl. 582.32040. English transl.: Sov. Math., Dokl. *28*, 70–72 (1983)
38. Lempert, L.: La métrique de Kobayashi et la représentation des domaine sur la boule. Bull. Soc. Math. Fr. *109*, 427–474 (1981). Zbl. 492.32025
39. Lewy, H.: On the boundary behavior of holomorphic mappings. Accad. Naz. Linzei. *35*, 1–8 (1977).
40. Ligocka, E.: How to prove Fefferman's theorem without use of differential geometry. Ann. Pol. Math. *39*, 117–130 (1981). Zbl. 489.32016
41. Ligocka, E.: The Hölder continuity of the Bergman projection and proper holomorphic mappings. Stud. Math. *80*, 89–107 (1984). Zbl. 566.32017
42. Løw, E.: An explicit holomorphic map of bounded domains in $\mathbb{C}^n$ with C^2-boundary onto the polydisc. Manuscr. Math. *42*, 105–113 (1983). Zbl. 545.32011
43. Margulis, G.A.: Boundary correspondence under biholomorphic mappings of multivariate domains. In: Abstract of the all union conf. of the theory of functions of complex variables, 137–138. Kharkov 1971 [Russian].
44. Moser, J., Webster, S.: Normal forms for real surfaces in $\mathbb{C}^2$ near complex tangents and hyperbolic surface transformations. Acta Math. *150*, 255–296 (1983). Zbl. 519.32015
45. Nirenberg, L., Webster, S., Yang, P.: Local boundary regularity of holomorphic mappings. Commun. Pure Appl. Math. *33*, 305–338 (1980). Zbl. 436.32018
46. Pinchuk, S.I.: On proper holomorphic maps of strictly pseudoconvex domains. Sib. Mat. Zh. *15*, 909–917 (1974) [Russian]. Zbl. 289.32011. English transl.: Sib. Math. J. *15*, 644–649 (1974)
47. Pinchuk, S.I.: A boundary uniqueness theorem for holomorphic functions of several complex variables. Mat. Zametki *15*, 205–212 (1974) [Russian]. Zbl. 285.32002. English transl.: Math. Notes *15*, 116–120 (1974)
48. Pinchuk, S.I.: On analytic continuation of holomorphic maps. Mat. Sb. *98* (140), 416–435 (1975) [Russian]. Zbl. 366.32010. English transl.: Math. USSR, Sb. *27*, 375–392 (1975)
49. Pinchuk, S.I.: On holomorphic maps of real-analytic hypersurfaces. Mat. Sb., Nov. Ser. *105* (147), 574–593 (1978) [Russian]. Zbl. 389.32008. English transl.: Math. USSR, Sb. *34*, 503–519 (1978)
50. Pinchuk, S.I.: Holomorphic nonequivalence for some classes of domains in $\mathbb{C}^n$. Mat. Sb., Nov. Ser. *111* (153), 67–94 (1980) [Russian]. Zbl. 442.32005. English transl.: Math. USSR, Sb. *39*, 61–86 (1981)
51. Pinchuk, S.I.: On the boundary behavior of analytic sets and algebroidal maps. Dokl. Akad. Nauk SSSR *268*, 296–298 (1982) [Russian]. Zbl. 577.32008. English transl.: Sov. Math., Dokl. *27*, 82–85 (1983)
52. Pinchuk, S.I.: Homogeneous domains with piecewise smooth boundaries. Mat. Zametki *32*, No. 5, 729–735 (1982) [Russian]. Zbl. 576.32041. English transl.: Math. Notes *32*, 849–852 (1983)
53. Poincaré, H.: Sur les fonctions analytiques de deux variables et la représentation conforme. Rend. Circ. Mat. Palermo *23*, 185–220 (1907) (≡Oeuvres, t. 4, 244–289)
54. Pyatetskiĭ-Shapiro, I.I.: Geometry of classical domains and the theory of automorphic functions. Moscow: Nauka 1961 [Russian]. French transl.: Paris: Dunod 1966. English transl.: New York ete: Gordon and Breach (1969). Zbl. 137, 275
55. Range, R.M.: The Carathéodory metric and holomorphic maps on a class of weakly pseudoconvex domains. Pac. J. Math. *78*, 173–189 (1978). Zbl. 396.32005
56. Rosay, J.P.: Sur une caractérisation de la boule parmi les domaines de $\mathbb{C}^n$ par son groupe d'automorphismes. Ann. Inst. Fourier *29*, 91–97 (1979). Zbl. 402.32001

57. Rudin, W.: Holomorphic maps that extend to automorphisms of a ball. Proc. Am. Math. Soc. *81*, 429–432 (1981). Zbl. 497.32011
58. Rudin, W.: Function theory in the unit ball of $\mathbb{C}^n$. Grundlehren 241. Berlin etc.: Springer 1980. Zbl. 495.32001
59. Sadullaev, A.: A Schwarz lemma for circular domains and its applications. Mat. Zametki *27*, 245–253 (1980) [Russian]. Zbl. 431.32018. English transl.: Math. Notes *27*, 120–125 (1980)
60. Shabat, B.V.: Introduction to complex analysis. Vol. 2. Moscow: Nauka 1976 [Russian]. 3rd ed. (1985). Zbl. 578.32001
61. Sharonov, S.E.: On holomorphic maps of polyhedra. Mat. Sb., Nov. Ser. *116* (158), 128–135 (1981) [Russian]. Zbl. 492.32027. English transl.: Math. USSR, Sb. *44*, 117–123 (1983)
62. Segre, B.: Intorno al problema di Poincaré della rappresentazione pseudoconforme. Atti. Accad. Naz. Lincei, VI. Ser. *13*, 676–683 (1931). Zbl. 3, 213
63. Stein, E.: Boundary behavior of holomorphic functions of several complex variables. Princeton: Princeton University Press 1972. Zbl. 242.32005
64. Tanaka, N.: On the pseudo-conformal geometry of hypersurface of the space of n complex variables. J. Math. Soc. Japan *14*, 397–429 (1962). Zbl. 113, 63
65. Tanaka, N.: On generalized graded Lie algebras and geometric structures. J. Math. Soc. Japan *19*, 215–254 (1967). Zbl. 165, 560
66. Tumanov, A.E., Khenkin, G.M.: Local characterizations of analytic automorphisms of classical domains. Dokl. Akad. Nauk SSSR *267*, 796–799 (1982) [Russian]. Zbl. 529.32014. English transl.: Sov. Math., Dokl. *26*, 702–705 (1982)
67. Vitushkin, A.G.: Holomorphic continuation of maps of compact hypersurfaces. Izv. Akad. Nauk SSSR, Ser. Mat. *46*, No. 1, 28–35 (1982) [Russian]. Zbl. 571.32011. English transl.: Math. USSR, Izv. *20*, 27–33 (1983)
68. Vitushkin, A.G.: Global normalization of a real-analytic surface along chains. Dokl. Akad. Nauk SSSR *269*, 15–18 (1983) [Russian]. Zbl. 543.32003. English transl.: Sov. Math., Dokl. *27*, 270–273 (1983)
69. Vitushkin, A.G., Ezhov, V.V., Kruzhilin, N.G.: Continuation of local maps of pseudoconvex surfaces. Dokl. Akad. Nauk SSSR *270*, 271–274 (1983). Zbl. 558.32003. English transl.: Sov. Math., Dokl. *27*, 580–583 (1983)
70. Vladimirov, V.S.: Methods of the theory of functions of several complex variables. Moscow: Nauka 1964 [Russian]. English transl.: Cambridge, London: M.I.T. Press 1966. Zbl. 125, 319
71. Webster, S.: On the mapping problem for algebraic real hypersurfaces. Invent. Math. *43*, 53–68 (1977). Zbl. 348.32005
72. Webster, S.: On the reflection principle in several complex variables. Proc. Am. Math. Soc. *71*, 26–28 (1978)
73. Webster, S.: Biholomorphic mappings and the Bergman kernel off the diagonal. Invent. Math. *51*, 155–163 (1979). Zbl. 385.32019
74. Webster, S.: Moser normal form at a non-umbilic point. Math. Ann. *233*, 97–102 (1978). Zbl. 358.32013
75. Wells, R.O., jr.: The Cauchy-Riemann equations and differential geometry. Bull. Am. Math. Soc., New Ser. *6*, 187–199 (1982). Zbl. 496.32012
76. Wong, B.: Characterizations of the unit ball in $\mathbb{C}^n$ by its automorphism group. Invent. Math. *41*, 253–257 (1977). Zbl. 385.32016

VI. The Geometry of CR-Manifolds

A.E. Tumanov

Translated from the Russian
by J. Peetre

Contents

Introduction

The geometry of CR-manifolds goes back to Poincaré and received a great attention in the works of É. Cartan, Tanaka, Moser, Chern and others (cf. [44]). In this chapter we consider results connected with the equivalence problem for CR-manifolds in its differential geometric aspect and some applications of this.

§ 1. CR-Manifolds

A real manifold M[1] is called an (abstract) Cauchy-Riemann manifold or a CR-*manifold* if at each point $x \in M$ in the tangent space $T_x M$ there is a distinguished subspace $T_x^c M$ and a complex structure on it, both depending

[1] In this Part, we assume, unless we do not say the contrary, that all object encountered are smooth of class C^∞.

smoothly on x. The complex dimension of the subspace $T_x^c(M)$ is called the CR-dimension of M and will be denoted by CR dim M.

The collection of subspaces $T_x^c(M)$ generates a complex tangent distribution $T^c(M)$ or a CR-*structure*. The complexified distribution $T^c(M)\otimes_{\mathbb{R}}\mathbb{C}$ decomposes into a sum

$$T^c(M)\otimes_{\mathbb{R}} C = T^{(0,1)}(M) + T^{(1,0)}(M).$$

If J_x: $T_x^c(M)\to T_x^c(M)$ is the operator giving the complex structure in $T_x^c(M)$ $(J_x^2 = -\mathrm{id})$ then

$$T_x^{(0,1)}(M) = \{X + iJ_x X : X \in T_x^c(M)\},$$
$$T_x^{(1,0)}(M) = \{X - iJ_x X : X \in T_x^c(M)\}.$$

A function f on a open subset $U\subset M$ is called a CR-*function* if it satisfies the Cauchy-Riemann equations

$$Xf = 0, \qquad X \in \Gamma_U(T^{(0,1)}(M)), \tag{1.1}$$

where $\Gamma_U(T^{(0,1)}(M))$ the set of vector fields on U (sections of the bundle $T^{(0,1)}M$ on the set U) and Xf stands for differentiation of f along X.

A map between CR-manifolds f: $M_1\to M_2$ is termed a CR-*map* if it induces a homomorphism of complex vector bundles $T^c(M_1)\to T^c(M_2)$.

The notion of CR-manifold arose in the study of the tangential Cauchy-Riemann equations on real submanifolds of $\mathbb{C}^n$ (cf. Vol. 7, Chap. II).

Let M be a real submanifold of $\mathbb{C}^n$ with coordinates $(z_1, \ldots, z_N)$. Set

$$T_x^c(M) = T_x(M)\cap JT_x(M), \quad x\in M,$$

where J is the operation of multiplication by the imaginary unit in $\mathbb{C}^N$. On the space $T_X^c(M)$ we have an induced complex structure. Generally speaking, the dimension of $T_x^c(M)$ may depend on $x\in M$; if it is constant then M is a CR-manifold. In this case we say that the CR-structure on M is induced by the complex structure of the ambient space. For manifolds $M\subset\mathbb{C}^N$ with an induced CR-structure, the bundle $T^{(0,1)}(M)$ consists of complex tangent vectors which can be expressed linearly in $\partial/\partial\bar{z}_1, \ldots, \partial/\partial\bar{z}_N$, and equation (1.1) reduces to the tangential Cauchy-Riemann equation on M.

It follows from this description of $T^{(0,1)}(M)$ for $M\subset\mathbb{C}^N$ that

$$[X, Y]\in\Gamma_U(T^{(0,1)}(M)), \quad \text{provided } X, Y\in\Gamma_U(T^{(0,1)}(M)). \tag{1.2}$$

Condition (1.2) for an abstract CR-manifold will be called the *integrability condition for the* CR-*structure* [20]. It is a necessary condition for the existence of an imbedding of M into $\mathbb{C}^N$ such that the CR-structure on M is induced by the complex structure on $\mathbb{C}^N$. In the real-analytic case this condition is also sufficient for the existence of a local imbedding of M into $\mathbb{C}^N$ (cf. [3], [30]). As Nirenberg [27] has demonstrated, for smooth CR-manifolds the last statement is in general not true. From now on we will only consider integrable CR-structures.

An important concept for CR-structures is the *Levi form*. Let M be a CR-manifold. The (Hermitian) Levi form $L_M(x)$ at the point $x \in M$ is defined on $T_x^{(1,0)}(M)$ (and likewise on $T_x^{(0,1)}(M)$ and $T_x^c(M)$) and takes its values in $N_x = T_x(M)/T_x^c(M) \otimes_{\mathbb{R}} \mathbb{C}$. It is given by the expression

$$L_M(x)(Z(x), W(x)) = i[Z, W](x) \quad (\text{mod } T_x^c(M) \otimes_{\mathbb{R}} \mathbb{C}), \tag{1.3}$$

where $Z, W \in \Gamma_U(T^{(1,0)}(M))$. This definition is consistent and agrees with the conventional definition of the Levi form on surfaces (cf., for example, [28]). In fact, let $M \subset \Omega \subset \mathbb{C}^N$ be given in the form

$$M = \{z \in \Omega : r(z) = 0\},$$

where $r = (r_1, \ldots, r_m)$ is a real vector function such that $\bar{\partial} r_1 \wedge \ldots \wedge \bar{\partial} r_m \neq 0$. Then, setting $\theta = i\bar{\partial} r \mid M$ (θ is a real vectorvalued 1-form) we have

$$T^c(M) = \{X \in T(M) : \theta(X) = 0\}. \tag{1.4}$$

By a well-known formula in differential calculus (cf. [34]) we have for Z, $W \in \Gamma(T^{(1,0)}(M))$

$$2d\theta(Z, \bar{W}) = Z\theta(\bar{W}) - \bar{W}\theta(Z) - \theta([Z, \bar{W}]), \tag{1.5}$$

where in the right hand side, in view of (1.4), only the last term is nonvanishing. If

$$Z(x) = \sum \xi_i \frac{\partial}{\partial z_i}, \qquad W(x) = \sum \eta_i \frac{\partial}{\partial z_i},$$

then, choosing coordinates in N_x with the aid of the form θ, we get from (1.5):

$$L_M(x)(Z(x), W(x)) = \sum_{i,j=1}^{N} \frac{\partial^2 r}{\partial z_i \partial \bar{z}_j} \xi_i \bar{\eta}_j,$$

from which it follows, in particular, that the value of the Levi form depends only on the values of the fields Z, W at the point x.

Let M be a CR-manifold. If the Levi form vanishes identically, then M is fibered locally by complex submanifolds of dimension $n = \text{CR dim } M$ (Sommer [32], [33]). In fact if $L_M \equiv 0$ then it follows from (1.3) that the differential system $T^c(M) \subset T(M)$ is involutive, i.e. that $[T^c(M), T^c(M)] \subset T^c(M)$. The integral manifolds of this system are almost complex manifolds. From the integrability conditions of a CR-structure (1.2) it follows that the almost complex structure is integrable, and so by the Newlander-Nirenberg theorem [26] they must be complex submanifolds of M. Conversely, if M contains a complex manifold $V \subset M$, $\dim_{\mathbb{C}} V = \text{CR dim } M$, then it follows from the definition (1.3) of the Levi form that it vanishes at each point $x \in V$. Further results on fibrations by complex submanifolds may be found in Freeman [18], [19], Bedford and Kalka [4], Bryant [7].

The generalized Levi forms lead to the Levi-Tanaka algebra (cf. [20], [24]). Let us assume that on the CR-manifold M there exists a sequence of subbundles (differential systems)

$$T^c(M)=D^1\subset D^2\subset\ldots\subset D^\mu=T(M)$$

such that

$$\Gamma(D^{i+1})=[\Gamma(D^i),\Gamma(D^1)]+\Gamma(D^i),\quad i\geqq 1.$$

Then we say that the CR-*manifold* M (or the CR-*structure*) is *regular*.

As examples of regular CR-structure we have hyperplanes in $\mathbb{C}^N$ with nowhere vanishing Levi form. For them holds $\mu=2$. The first examples of regular manifolds with $\mu>2$ appeared in papers by H. Lewy (cf., for example, [22]), in connection with questions pertaining to holomorphy hulls of CR-manifolds. The following example is a modification of one of Lewy's examples.

Example 1.1. Let $M\subset\mathbb{C}^3$ be given by the equations

$$|z_1|^2-|z_2|^2=1,\quad |z_3|^2-|z_2|^2=2,\quad z_1\neq 0,\quad z_2\neq 0.$$

Here $T^{(1,0)}(M)$ is spanned by the vector field

$$\xi=\frac{\bar z_2}{\bar z_1}\frac{\partial}{\partial z_1}+\frac{\partial}{\partial z_2}+\frac{\bar z_2}{\bar z_3}\frac{\partial}{\partial z_3}.$$

One can show that the fields

$$X,\ JX,\ [X,JX],\ [X,[X,JX]],$$

where $X=\xi+\bar\xi$, are linearly independent.

Thus M is regular and $\mu=3$, $\operatorname{CR}\dim M=1$, $\dim D^1=2$, $\dim D^2=3$, $\dim D^3=\dim M=4$.

Let M be a regular CR-manifold. Set

$$\mathfrak{g}^1=D^1,\quad \mathfrak{g}^{i+1}=D^{i+1}/D^i,\quad i\geqq 1,$$
$$\mathfrak{m}=\mathfrak{g}^1\oplus\ldots\oplus\mathfrak{g}^\mu.$$

Then the operation of commutation of vector fields induces on $\Gamma(\mathfrak{m})$ a structure of Lie algebra. It is easy to see that

$$[fX,gY]=fg[X,Y]\quad\text{for } f,g\in C^\infty(M),\ X,Y\in\Gamma(\mathfrak{m}),$$

from which it follows that the Lie algebra structure is defined pointwise, i.e.

$$\mathfrak{m}_x=\mathfrak{g}_x^1\oplus\ldots\oplus\mathfrak{g}_x^\mu,\quad x\in M,$$

is a Lie algebra and

$$[X(x),Y(x)]=[X,Y](x),\quad X,Y\in\Gamma(\mathfrak{m}).$$

The algebra $\mathfrak{m}_x$ is termed the *Levi-Tanaka algebra* of the manifold M at the point x. It enjoys the following properties. Setting $\mathfrak{a}=\mathfrak{m}_x$, $\mathfrak{a}^j=\mathfrak{a}_x^j$ for $x\in M$ and $1\leqq j\leqq\mu$; $\mathfrak{a}^k=0$ for $k>\mu$ we have:

1) $\mathfrak{a}^1$ spans $\mathfrak{a}$ as a Lie algebra; (1.6)
2) $[\mathfrak{a}^j,\mathfrak{a}^k]\subset\mathfrak{a}^{j+k}$; (1.7)
3) $[JX,JY]=[X,Y]$, $X,Y\in\mathfrak{a}^1$. (1.8)

Property (1.7) means that $\mathfrak{m}_x$ is a (generalized) *graded Lie algebra* (cf. [35]). Property (1.8) follows from the integrability conditions for a CR-structure (1.2).

The operation of commutation in $\mathfrak{g}_x^1$ actually coincides with the imaginary part of the Levi form.

If the Levi-Tanakà algebra is the same at all points, then the *manifold* is said to be *strongly regular* [24]. More exactly, a CR-manifold M is strongly regular if for any two points $x, y \in M$ there exists an isomorphism of Lie algebras $\sigma\colon M_x \to M_y$ such that

$$\sigma(\mathfrak{g}_x^j) = \mathfrak{g}_y^j,$$
$$\sigma(J_x X) = J_y\, \sigma(X), \quad X \in \mathfrak{g}_x^1.$$

The hypersurfaces in $\mathbb{C}^N$ with a nondegenerate Levi form constitute an example of strongly regular CR-manifolds.

It is natural to ask if every finite dimensional graded Lie algebra

$$\mathfrak{a} = \sum_{k=1}^{\mu} \mathfrak{a}^k$$

(where $\mathfrak{a}^1$ is a complex space), enjoying the properties (1.6)–(1.8), is the Levi-Tanaka algebra of some strongly regular CR-manifold. The answer to this question is positive. In [24] there is given a natural construction of a strongly regular CR-manifold $M = M(\mathfrak{a})$ in the space

$$\mathfrak{a}^1 \oplus \sum_{k=2}^{\mu} \mathfrak{a}^k \otimes_{\mathbb{R}} \mathbb{C}$$

with the Levi-Tanaka algebra $\mathfrak{a}$. The manifold $M(\mathfrak{a})$ is called the *standard* CR-*manifold* corresponding to $\mathfrak{a}$.

In what follows we restrict attention to the case $\mu = 2$. Then the Levi-Tanaka algebra is defined by the Levi form and the manifold $M(\mathfrak{a})$ has a simple description.

Set $\mathfrak{a}^1 = \mathbb{C}^n$, $\mathfrak{a}^2 = \mathbb{R}^m$, $n, m \geqq 1$. In view of (1.8) there exists an Hermitian vectorial quadratic form F on $\mathbb{C}^n$ such that

$$[u, v] = \operatorname{Im} F(u, v), \quad u, v \in \mathfrak{a}^1.$$

Then $M = M(\mathfrak{a})$ is given in the form

$$M = \{(z, w) \in \mathbb{C}^n \oplus \mathbb{C}^m \colon \operatorname{Im} w = F(z, z)\}. \tag{1.9}$$

Standard manifolds are homogeneous. For example, on the manifold (1.9) there acts transitively the group of "parallel transports", the affine maps of the form

$$z \to z + a, \quad a \in \mathbb{C}^n,$$
$$w \to w + b + 2iF(z, a) + iF(a, a), \quad b \in \mathbb{R}^m. \tag{1.10}$$

Standard manifolds are the simplest ones and also most important. As a result of a comparison of general CR-manifolds with standard ones one obtains invariants; these will be discussed in the following Sections.

§ 2. The Equivalence Problem for CR-Manifolds

A diffeomorphic CR-map $f\colon M_1\, M_2$ of two CR-manifolds will be referred to as a CR-*isomorphism* or a CR-equivalence.

The *equivalence problem for* CR-manifolds consists of finding necessary and sufficient conditions for two manifolds to be CR-equivalent. This problem may be considered as a special case of the general *equivalence problem for G-structures* (cf. [34]), first considered by É. Cartan. If G is subgroup of the full linear group $GL(N, \mathbb{R})$ then a G-structure on an N-dimensional manifold M is defined by a reduction of the bundle of frames $\mathscr{F}(M)$ (or coframes) of M to the group G. In other words, a G-structure on M is nothing but a subbundle $B\subset\mathscr{F}(M)$ which is a principle G-bundle. For example, a CR-structure of CR-dimension n on a real manifold of dimension $N=2n+m$ is a G-structure with G the group of linear automorphisms of $\mathbb{R}^N=\mathbb{C}^n\oplus\mathbb{R}^m$ which preserve the first component $\mathbb{C}^n$ and are complex linear on it.

Two G-structure B_1 and B_2 on manifolds M_1 and M_2 are termed equivalent if there exists a diffeomorphism $f\colon M_1\to M_2$ such that for the induced map $f_*\colon\mathscr{F}(M_1)\to\mathscr{F}(M_2;$ holds $f_*(B_1)=B_2$. Such a diffeomorphism f is called an *isomorphism of the G-structures* B_1 and B_2.

É. Cartan developed a general approach towards the solution of the equivalence problem for G-structures (cf. [34]), in which the group G, is replaced, step by step, by a simpler "smaller" group. The goal is to reduce the equivalence problem for G-structures to the equivalence problem of $\{e\}$-structures, where $\{e\}$ is the group consisting of one element.

On a manifold P an $\{e\}$-structure defines a global section of the frame bundle $\mathscr{F}(P)$ and, consequently, a global trivialization of the tangent bundle $T(P)$. Therefore, an $\{e\}$-structure is called an *absolute parallelism* (or, for brevity, just a parallelism). A parallelism is given by a vectorial 1-form $\omega=(\omega_1, \ldots, \omega_N)$ such that at each point $x\in P$ the map $\omega_x\colon T_x(P)\to\mathbb{R}^N$ is nondegenerate.

The reduction of the equivalence problem for G-structures to the one for $\{e\}$-structures amounts to constructing for each manifold M equipped with a G-structure in a canonical way a bundle $P\to M$ with a parallelism ω on P. It is required that with each isomorphism of G-structures $f\colon M_1\to M_2$ one can associate on the corresponding bundles a map $F\colon P_1\to P_2$ which preserves the parallelism, i.e. $F^*\,\omega_2=\omega_1$, in such a way that the diagram

$$\begin{array}{ccc} P_1 & \xrightarrow{F} & P_2 \\ \downarrow & & \downarrow \\ M_1 & \xrightarrow{f} & M_2 \end{array}$$

is commutative. Conversely, each map $F: P_1 \to P_2$ which preserves the parallelism is by necessity a lifting of an isomorphism of G-structures $f: M_1 \to M_2$.

Examples of G-structures which can be reduced to parallelisms are Riemannian and conformal structures. For instance, a Riemannian structure on a N-dimensional manifold M is an $O(N)$-structure, where $O(N)$ is the group of orthogonal matrices of order N. The Levi-Civita connection on the bundle O_M of orthogonal frames on M together with the canonical $\mathbb{R}^N$-valued form on the frame bundle (cf. [34]) defines a parallelism on O_M.

The reduction of G-structure to parallelism has several advantages, because for $\{e\}$-structures the equivalence problem is well understood (cf. [34]). In particular, by a theorem of Kobayashi's the automorphism group of an $\{e\}$-structure on a manifold P is a Lie group of dimension not exceeding the dimension of P, and each such automorphism is uniquely defined by the image of a single point. Therefore, if a G-structure can be reduced to an $\{e\}$-structure its automorphism group too must be a Lie group, the same as the one for the corresponding $\{e\}$-structure.

Let us make still another observation regarding the use of $\{e\}$-structures, concerning the regularity of isomorphisms.

Proposition 2.1. *Let P_1 and P_2 be manifolds of smoothness C^k (C^∞ or $\mathbb{R}$-analytic) with parallelisms ω_1 and ω_2 of smoothness C^{k-1} (respectively C^∞ and $\mathbb{R}$-analytic). Then each C^1-smooth map $F: P_1 \to P_2$ such that $F^*\omega_2 = \omega_1$ is in fact smooth of class C^k (respectively C^∞ and $\mathbb{R}$-analytic).*

The proof of Proposition 2.1 follows immediately from the fact that the equation $F^*\omega_2 = \omega_1$ is a first order quasilinear elliptic equation. Let us remark that the equation corresponding to an isomorphism of G-structures reducing it to a $\{e\}$-structure is not at all elliptic. For instance, as we will see below, CR-structures can in many cases be reduced to $\{e\}$-structures but then the corresponding equations defining the CR-map are elliptic only in the trivial case $T^c(M) = T(M)$, that is, for complex manifolds.

In connection with CR-structures the procedure of É. Cartan in its original form does not lead to the goal. Nevertheless, Cartan [11] solved the equivalence problem for three dimensional CR-manifolds while in 1967 Tanaka [35] obtained the solution of the equivalence problem for the class of strongly regular CR-manifolds. Let us remark that this result by Tanaka became widely known only in 1974 when it was obtained independently by Chern [13] for hypersurfaces in $\mathbb{C}^{n+1}$ (cf. [36]). Let us state the main results of Tanaka's paper [35].

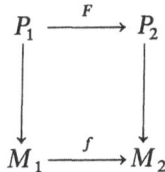

Let $\mathfrak{a}=\mathfrak{a}^1 \oplus \mathfrak{a}^2$ be a graded Lie algebra satisfying (1.6)–(1.8). Let us set $\mathfrak{g}_{-k}=\mathfrak{a}^k$, $k=1, 2$. It is not hard to see that there exists a maximal graded Lie algebra

$$\mathfrak{g}=\sum_{k=-2}^{\infty} \mathfrak{g}_k,$$

satisfying the following conditions:

1) if $X \in \mathfrak{g}_k$, $k \geqq 0$, and $[X, \mathfrak{g}_{-1}]=0$ then $X=0$;

2) $[X, JY]=J[X, Y]$, $X \in \mathfrak{g}_0$, $Y \in \mathfrak{g}_{-1}$, where J is the operation of multiplication with the imaginary unit in $\mathfrak{g}_{-1}=\mathfrak{a}^1$.

The algebra $\mathfrak{g}$ will be called the extension of $\mathfrak{a}$.

Example 2.2. Let $\mathfrak{a}=\mathfrak{a}^1 \oplus \mathfrak{a}^2$ where $\mathfrak{a}^1$ is the space of complex matrices of order $q \times s$ and $\mathfrak{a}^2$ the space of Hermitian matrices of order q. Let us set

$$F(U, V)=UV^*, \quad U, V \in a^1 \tag{2.1}$$

(the star stands for Hermitian conjugation) and define the commutation operation in $\mathfrak{a}^1$ by the relation

$$[U, V]=\operatorname{Im} F(U, V)=\frac{1}{2i}(UV^*-VU^*).$$

One can show that the extension $\mathfrak{g}$ is finite dimensional and consists of five components

$$g=\sum_{k=-2}^{2} \mathfrak{g}_k, \qquad \dim \mathfrak{g}_k=\dim \mathfrak{g}_{-k}, \qquad \dim \mathfrak{g}=4q^2+s^2+4qs-1.$$

For $q=1$ this example will be studied in some greater detail in the next Section.

Theorem 2.3 (Tanaka [35]). *Let* $\mathfrak{g}$ *be the extension of the graded Lie algebra* $\mathfrak{a}=\mathfrak{a}^1 \oplus \mathfrak{a}^2$. *If* $\dim \mathfrak{g}<\infty$ [2] *then the strongly regular* CR-*structures whose Lewy-Tanaka algebra is isomorphic to* $\mathfrak{a}$ *reduce in a natural way to absolute parallelisms in spaces whose dimension equals* $\dim \mathfrak{g}$.

Corollary 2.4. *Let* M *be a strongly regular* CR-*manifold with Lewy-Tanaka algebra* $\mathfrak{a}$ *and let* $\mathfrak{g}$ *be the extension of* $\mathfrak{a}=\mathfrak{a}^1 \oplus \mathfrak{a}^2$. *If* $\dim \mathfrak{g}<\infty$ *then the group of* CR-*automorphisms* $\operatorname{Aut}_{CR}(M)$ *is a Lie group of dimension not exceeding* $\dim \mathfrak{g}$.

For example, the group $\operatorname{Aut}_{CR}(M)$ is finite dimensional for the standard manifold M of the form (1.9) corresponding to the form F in example 2.2. Also the group $\operatorname{Aut}_{CR}(S)$, where S is a domain in M, is finite dimensional. Let us remark that M is equivalent to the skeleton of a classical domain of the first type (cf. [38] and Vol. 7, Chap. II).

---

[2] A necessary and sufficient condition for this condition to hold is that the operation of taking commutators is nondegenerate, i.e. that $x \in \mathfrak{a}^1$ and $[\mathfrak{a}^1, x]=0$ implies $x=0$.

The structure of the Lie algebra $\mathfrak{a}=\mathfrak{a}^1 \oplus \mathfrak{a}^2$ defined by (1.6)–(1.8) determines an Hermitian quadratic form F on $\mathfrak{a}^1$ with values in $\mathfrak{a}^2$. This form F is called stable if every form $\tilde{F}$ sufficiently close to it has the form

$$\tilde{F}(u, v)=A_2^{-1} F(A_1 u, A_1 v)$$

where A_1 and A_2 are nondegenerate $\mathbb{C}$-linear respectively $\mathbb{R}$-linear selfmaps of the spaces $\mathfrak{a}^1$ and $\mathfrak{a}^2$. If F is stable and the Levi-Tanaka algebra of the CR-manifold M at the point x is isomorphic to $\mathfrak{a}$ then M is strongly regular in a neighborhood of this point. However, stable forms are rather rare and, therefore, strong regularity usually is not generic property. For instance, a form F of the type (2.1) is stable only if $q=1$ or $s=1$.

In [35] Tanaka gave three cases when strong regularity is a property of general position (a generic property):

$$m=1, \qquad m=n^2-1, \qquad m=n^2,$$

where $n=\mathrm{CR}\dim M$ and m is the real codimension of the structure. This list can be complemented also with the case $m=n=2$.

It is would be interesting to try to eliminate the requirement of strong regularity in Theorem 2.3 and, in particular, answer the question when $\mathrm{Aut}_{\mathrm{CR}}(M)$ is a finite dimensional Lie group. If $m>1$ a positive answer to this question is known only for manifolds satisfying the conditions of theorem 2.3, and for standard manifolds (cf. § 4).

§ 3. Hypersurfaces in $\mathbb{C}^{n+1}$

Let us apply the main theorem in the case $m=\dim \mathfrak{a}^2=1$, corresponding to hypersurfaces in $\mathbb{C}^{n+1}$ [13], [36]. Other approaches to the solution of the equivalence problem for hypersurfaces in $\mathbb{C}^{n+1}$ can be found in [28], [12], [13], [14], [17], [31], [42], [44]. If $m=1$ and the Levi form of a connected CR-manifold is nondegenerate at each point then it must be strongly regular and its Levi-Tanaka algebra determines the signature of the Levi form.

Let us set $\mathfrak{a}=\mathfrak{g}_{-1}\oplus\mathfrak{g}_{-2}$, $\mathfrak{g}_{-1}=\mathbb{C}^n$, $\mathfrak{g}_{-2}=\mathbb{R}$. For $x, y\in\mathfrak{g}_{-1}$, $x=(x^1, \ldots, x^n)$, $y=(y^1, \ldots, y^n)$,

$$[x, y]=-2 \operatorname{Im} F(x, y),$$
$$F(x, y)=x^\alpha g_{\alpha\bar{\beta}}\, y^{\bar{\beta}}, \quad y^{\bar{\beta}}=\overline{y^\beta},$$

F is a nondegenerate Hermitian quadratic form with p plus signs and q minus signs. Consider the extended matrix

$$g=(g_{i\bar{j}})=\begin{pmatrix} 0 & 0 & -i \\ 0 & g_{\alpha\bar{\beta}} & 0 \\ i & 0 & 0 \end{pmatrix},$$

where the Greek indices run from 1 to n and the Latin ones from 0 to $n+1$. The corresponding Hermitian quadratic form has $p+1$ plus signs and $q+1$ minus signs. Let us denote by $\mathfrak{g}$ the Lie algebra $su(p+1, q+1)$ of matrices with zero traces which preserve the form $\tilde{F}(x, y)=x^i g_{ij} y^j$. The matrix $a=(a_i^j)$ belongs to $\mathfrak{g}$ presicely when $ag+ga^*=0$, tr $a=0$, or

$$a_i^k g_{k\bar{j}}+g_{i\bar{k}} a_{\bar{j}}^{\bar{k}}=0 \qquad (a_{\bar{j}}^{\bar{k}}=\overline{a_j^k}),\ a_i^i=0.$$

The algebra $\mathfrak{a}$ is realized as a subalgebra of $\mathfrak{g}$: an element $(x^\alpha, t)\in\mathfrak{g}_{-1}\oplus\mathfrak{g}_{-2}$ corresponds to the matrix

$$\begin{pmatrix} 0 & x^\alpha & t \\ 0 & 0 & 2ix_\alpha \\ 0 & 0 & 0 \end{pmatrix},$$

where $x_\alpha=g_{\alpha\bar{\beta}} x^{\bar{\beta}}$, $x^{\bar{\beta}}=\overline{x^\beta}$.

The algebra g is a graded Lie algebra

$$\mathfrak{g}=\sum_{k=-2}^{2} \mathfrak{g}_k,$$

where

$$\mathfrak{g}_0=\left\{\begin{pmatrix} a_0^0 & 0 & 0 \\ 0 & a_\beta^\alpha & 0 \\ 0 & 0 & a_{n+1}^{n+1} \end{pmatrix} : \begin{array}{l} a_0^0+a_{n+1}^{n+1}=0 \\ a_\beta^\alpha g_{\alpha\bar{j}}+g_{\beta\bar{\alpha}} a_{\bar{j}}^{\bar{\alpha}}=0 \\ a_0^0+a_\alpha^\alpha+a_{n+1}^{n+1}=0 \end{array}\right\},$$

$$\mathfrak{g}_1=\left\{\begin{pmatrix} 0 & 0 & 0 \\ -iy & 0 & 0 \\ 0 & y & 0 \end{pmatrix}\right\}$$

$$\mathfrak{g}_2=\left\{\begin{pmatrix} 0 & 0 & 0 \\ 0 & 0 & 0 \\ t & 0 & 0 \end{pmatrix} : t\in\mathbb{R}\right\}.$$

One can show that g is an extension of $\mathfrak{a}$ and, as g is finite dimensional, Theorem 2.3 is applicable to it.

In order to formulate the theorem more completely in the case at hand, let us consider in $\mathbb{C}^{n+1}$ the standard manifold

$$Q=\{(z, w): \operatorname{Im} w=\tfrac{1}{2} F(z, z)\}, \tag{3.1}$$

where $F(z, z)=z^\alpha g_{\alpha\bar{\beta}} z^{\bar{\beta}}$. Passing to the projective space $\mathbb{CP}^{n+1}$ we obtain in homogeneous coordinates a quadric

$$\frac{1}{2} g_{\alpha\bar{\beta}} z^\alpha z^{\bar{\beta}}+\frac{i}{2}(z^{n+1} z^{\bar{0}}-z^0 z^{\overline{n+1}})=0,$$

where $\mathbb{C}^{n+1}$ is realized in $\mathbb{CP}^{n+1}$ as $z^0=1$, $z^{n+1}=w$. The group $G=SU(p+1, q+1)/Z$, where $SU(p+1, q+1)$ is the group of linear transformations in $\mathbb{C}^{n+2}$ with determinant 1, which preserve the form with matrix $g=(g_{ij})$, and Z its center, acts transitively on this quadric. $\mathfrak{g}$ is the Lie algebra of G.

It is well-known that G is the group of all CR-automorphisms of the quadric Q. Let H be the subgroup of G stabilizing the point $0 \in Q$, that is, the point with homogeneous coordinates $(1, 0, \ldots, 0)$. The Lie algebra of H is the subalgebra $\mathfrak{g}_0 \oplus \mathfrak{g}_1 \oplus \mathfrak{g}_2$.

Theorem 3.1 (Cartan [11] – Tanaka [36] – Chern [13]). *Let M be a* CR-*manifold with integrable* CR-*structure,* $\dim M = 2n+1$, CR $\dim M = n$, *whose Levi form has type (p, q), $p+q=n$. Then there exists a principal H-bundle $P \to M$ and a $\mathfrak{g}$-valued 1-form ω on M such that*

$$\omega(p)\colon T_p(P) \to g \text{ is an isomorphism, } p \in P, \tag{3.2}$$

$$R_a^* \omega = \mathrm{Ad}\, a^{-1} \omega, \quad a \in H,$$

where $R_a\colon P \to P$ is the operation of right translation and $\mathrm{Ad}\colon G \to GL(g)$ *the adjoint representation. The correspondence $M \to (P, \omega)$ is a natural one; in particular, a map $\varphi\colon P_1 \to P_2$ is the lift of a* CR-*isomorphism $f\colon M_1 \to M_2$ iff $\varphi^* \omega_2 = \omega_1$.*

A form ω subject to the conditions in Theorem 3.1 is termed a *Cartan connection*. Property (3.2) expresses the fact that ω defines a parallelism on P.

Let us write down the construction of the bundle P. Let θ be a real 1-form on M defining $T^c(M)$, i.e. $T_x^c(M) = \{X \in T_x(M)\colon \theta(X) = 0\}$[3]. Let E be the one dimensional subbundle on $T^*(M)$ such that

$$E_x = \{u\theta(x)\colon u > 0\}.$$

As on subbundles of the cotangent bundle, there exists on E a canonical 1-form (cf. [34], Chap. 3) which we denote $\theta^{n+1} = u\theta$. At each point $e = u\theta(x)$ we consider coframes of the form $p = (\theta^0, \theta^\alpha, \theta^{n+1})$, where θ^{n+1} is as defined above, θ^0 is a real covector and the θ^α are obtained by lifting from M to E of suitable covectors, complex linear on $T_x^c(M)$. The bundle P is defined as the set of those coframes which satisfy

$$d\theta^{n+1} = -i g_{\alpha\bar\beta}\, \theta^\alpha \wedge \theta^{\bar\beta} + \theta^{n+1} \wedge \theta^0. \tag{3.3}$$

The left hand side of this equation makes sense, as the form θ^{n+1} is defined on the whole of E. The existence of such coframes is guaranteed by the type of the Levi form, along with the integrability conditions.

Studying the rule for change of coframes subject to (3.3) one can show that P is a principal H-bundle on M.

On P there exist canonical (basis) 1-forms $\tilde\theta^0, \tilde\theta^\alpha, \tilde\theta^{n+1}$:

$$\tilde\theta^j(p)(X) = \theta^j(\pi_* X),$$

[3] For $p \neq q$ the Levi form determines the orientation of M, while for $p = q$ one has to assume in Theorem 3.1 that M is an oriented CR-manifold.

where $\pi: P \to E$ is the projection, $X \in T_p(P)$. For simplicity we shall omit the "twiddle" over θ^j. These forms satisfy the same equation (3.3).

We can define on P a connection ω such that

$$\omega_0^\alpha = \theta^\alpha, \qquad \omega_0^{n+1} = \theta^{n+1}, \qquad 2\,\mathrm{Re}\,\omega_0^0 = \theta^0. \tag{3.4}$$

Let us consider the *curvature form* of the connection ω:

$$\Omega = d\omega + \tfrac{1}{2}[\omega, \omega]$$

or, in terms of coordinates,

$$\Omega_i^j = d\omega_i^j + \omega_i^k \wedge \omega_k^j. \tag{3.6}$$

Thanks to the integrability conditions, one can pick ω in such a way as

$$\Omega_i^j = S_{ik\bar{l}}^j\, \omega_0^k \wedge \omega_{\bar{0}}^{\bar{l}},$$

where S_{ikl}^j are functions on P such that

$$S_{i\bar{j}k\bar{l}} = g_{s\bar{j}}\, S_{ik\bar{l}}^s = 0, \tag{3.7}$$

provided any of the indices i, j, k, l equals zero, and

$$S_{i\bar{j}k\bar{l}} = S_{k\bar{j}i\bar{l}} = S_{k\bar{l}i\bar{j}} = \bar{S}_{j\bar{i}l\bar{k}}. \tag{3.8}$$

Chern [13] proved that ω is defined uniquely by the requirement

$$S_{ik\bar{l}}^i = 0. \tag{3.9}$$

The form ω is a CR-invariant, because it is uniquely defined and satisfies condition (3.4).

Let us remark that for the standard manifold (3.1) the construction described leads to the bundle $G \to G/H \simeq Q$ and to the Maurer-Cartan form, that is, the 1-form ω on G with values in $\mathfrak{g} = T_e(G)$ such that the map $\omega(e)$: $\mathfrak{g} \to \mathfrak{g}$ is the identity. The Maurer-Cartan form satisfies

$$d\omega + \tfrac{1}{2}[\omega, \omega] = 0. \tag{3.10}$$

This means that the CR-curvature of the quadric Q vanishes.

Also the converse is true: if the curvature form of a CR-manifold vanishes then it is locally equivalent to a quadric. In fact, if (3.10) is fulfilled then in a neighborhood on any point $p \in P$ one can introduce the structure of local Lie group, isomorphic to G. By Theorem 3.1 this isomorphism is subordinated to a CR-isomorphism in a suitable neighborhood of M and Q.

From the transformation rule for the tensor $S_{\alpha\sigma\bar{\tau}}^\beta$ under the action of H it follows that $S_{\alpha\sigma\bar{\tau}}^\beta = 0$ on each fiber of the bundle P, provided it vanishes at least one point of the fiber. Therefore, one can make the following definition: the *point* $x \in M$ is said to be *umbilical* ($n \geqq 2$) if the tensor $S_{\alpha\sigma\bar{\tau}}^\beta$ vanishes on the fiber over x. In the case $n = 1$ one has $S_{\alpha\sigma\bar{\tau}}^\beta \equiv 0$, in view of (3.9), and in the preceding definition one has to take the function $S_{22\bar{1}}^1$ instead of $S_{\alpha\sigma\bar{\tau}}^\beta$.

Proposition 3.2. *If all points of a* CR-*manifold are umbilical, then the manifold must be locally equivalent to a quadric.*

Proof. Differentiating (3.5) we obtain

$$d\Omega_j = \sum_{k=-2}^{2} [\Omega_{j+k}, \omega_{-k}], \tag{3.11}$$

where $\Omega = \Sigma\Omega_k$, $\omega = \Sigma\omega_k$ are the expansions corresponding to the grading of $\mathfrak{g}$.

By the assumption of the proposition, $\Omega_0 = 0$ (if $n=1$ we have $\Omega_1 = 0$) whereas the identities $\Omega_{-1}=0$, $\Omega_{-2}=0$ are always fulfilled, in view of (3.7). Therefore, invoking (3.11), we get $\Omega=0$, from which follows that the manifold is locally equivalent to a quadric.

It follows from Theorem 3.1 that on a CR-manifold there exists a family of invariant curves, termed "chains" in [13]. Let us consider on P the differential system defined by the equations

$$\omega_0^\alpha = 0, \qquad \omega_{n+1}^\alpha = 0. \tag{3.12}$$

From (3.6), (3.7) it is clear that the system is involutive. The projections of its integral manifolds are curves on M. They are called *chains*. Chains are curves which are "straightened" if one reduces the equations of the hypersurface to normal form [13]. Through each point $x \in M$ there passes in each tangential direction not belonging to $T_x^c(M)$ precisely one chain.

On chains there is a natural parameter defined up to fractional linear transformations. This follows from the fact that if we to (3.12) also adjoin the equation

$$\omega_\beta^\alpha = 0, \tag{3.13}$$

then the forms θ^{n+1}, θ^0 and $-2\omega_{n+1}^0$ satisfy on the integral manifolds of the differential system thus obtained precisely the structure equations of the group of projective transformations of the line.

Let us also point out that, as to each point $p \in P$ there corresponds a frame in $T^c(M)$, the equations (3.12), (3.13) give a parallel transport of complex tangent vectors along chains.

An interesting approach to the study of the properties of chains on the boundary of a strongly pseudoconvex domain $D \subset \mathbb{C}^{n+1}$ was given by Fefferman [16] (cf. also [8]). The chains appear as projections onto M of light rays of a certain Lorentz metric on the manifold $\partial D \times S^1$, the product of ∂D with the unit circumference. The conformal class of this metric is a CR-invariant. The Fefferman metric is defined with the aid of the Bergman kernel function of D or using an approximate solution of the Monge-Ampère equation on D.

The application of the invariants (P, ω) to problems of analysis and geometry is sometimes complicated by the fact that the structure group H is noncompact. However, as proved by Webster, for *strictly pseudoconvex* CR-*manifolds* (with a Levi form of definite sign) on the set of non-umbilical points

$M^* \subset M$ the bundle $P|M^*$ reduces to the compact group $U(n)$, whereas on M^* there arises a CR-*invariant Riemannian metric* (cf. [8], [41]). For $n=1$ more is known. In this case $P|M^*$ reduced to a group with two elements and on M^* one can find 9 CR-invariant functions, the scalar CR-invariants of É. Cartan [11]. Using these Wells, Burns and Shnider [10] showed that the majority of deformations of a strongly pseudoconvex domain in $\mathbb{C}^2$ leads to rigid domains which are biholomorphically nonequivalent with each other and the original domain.

Let us indicate the construction of a CR-invariant metric on M^*. Set $s = S_{\alpha\bar{\beta}\sigma\bar{\tau}} S^{\alpha\bar{\beta}\sigma\bar{\tau}}$. The umbical points of a strongly pseudoconvex domain are characterized by the condition $s=0$. Consider the set $P_1 \subset P$,

$$P_1 = \{p \in P \colon s(p) = 1\}.$$

One can show that the image of P_1 under the projection $P \to E$ provides a section of the bundle E over the set M^*, which we denote by θ^*. This is a CR-invariant 1-form on M^* defining $T^c(M^*)$. There exists a unique vector field X^* on M^* satisfying the conditions:

$$\langle X^*, \theta^* \rangle = 1, \qquad X^* \lrcorner\, d\theta = 0.$$

where $\lrcorner$ is the sign for inner multiplication (cf. [34]).

Now we can define the Riemannian metric on M^* putting

$$\begin{aligned} &(X^*, X^*) = 1, \\ &(X^*, X) = 0, \quad X \in T^c(M), \\ &(X, Y) = d\theta(X, JY), \quad X, Y \in T^c(M). \end{aligned}$$

Using this CR-invariant metric we get the following

Theorem 3.3 (Webster [41]). *If a compact strictly pseudoconvex* CR-*manifold M admits a connected non-compact group of* CR-*automorphisms, then M is locally equivalent to the standard sphere $S^{2n+1} \subset \mathbb{C}^{n+1}$.*

Proof. Assuming the contrary, then in view of Proposition 3.2 the set M^* of nonumbilical points must be non-empty.

Let Y be a vector field on M generating a one parameter group of CR-automorphisms with non-compact closure. Write Y in the form

$$Y = fX^* + \tilde{Y},$$

where $\tilde{Y} \in \Gamma(T^c(M))$, $f = \theta^*(Y)$. Let us first show that f can not vanish identically on any open set $U \subset M$. Indeed, as Y is an infinitesimal CR-automorphism, we have

$$[Y, X] \in \Gamma(T^c(M)) \quad \text{for } X \in \Gamma(T^c(M)). \tag{3.14}$$

If now $f \equiv 0$, then it follows from (3.14) that the Levi form vanishes on all pairs of vectors (Y, X), $X \in T^c(M)$, which is impossible by virtue of its nondegeneracy.

Let K_ε be a connected component of the set

$$\{x \in M : |f(x)| \geqq \varepsilon\},$$

where $\varepsilon > 0$ is sufficiently small. One can prove that at umbilical points f must vanishes. Therefore a CR-invariant metric is defined in K_ε.

It follows from (3.14) and the CR-invariance of X^* that $Yf \equiv 0$. Therefore K_ε is invariant with respect to G_1.

As we have already remarked, a Riemannian structure defines a parallelism on the bundle of orthogonal frames, and each isometry is lifted to an automorphism of this parallelism. Each such automorphism is defined by the image of one point, while the space of orthogonal frames is compact. Therefore, the group G_1 is contained in a compact space (the group of isometries of K_ε), which contradicts the assumption made on G_1. The proof is complete.

As a complement to Theorem 3.3 one can show (cf. [41]) that if the fundamental group $\pi_1(M)$ is finite then M is globally equivalent to the sphere S^{2n+1}. In the case when M bounds a relatively compact domain D in a complex manifold, Burns and Shnider, Klembeck, B. Wong have shown that the group of CR-automorphism $\operatorname{Aut}_{CR}(M)$ is compact, with the exception for manifolds globally equivalent to S^{2n+1} (cf. [9]).

Theorem 3.3 was applied by Burns and Shnider to the classification problem for homogeneous CR-manifolds [10].

In conclusion let us state some results concerning holomorphic curves in CR-manifolds. On CR-manifolds there can exist complex curves, for example, owing to the degeneracy of the Levi form (cf. [18], [33]). Strictly pseudoconvex manifolds do not contain complex curves. However, for an indefinite Levi form the situation is different. In this case Sommer [33] gave examples both of hypersurfaces in $\mathbb{C}^3$ containing complex curves and of such not containing complex curves. An example of a hypersurface containing complex curves is the quadric. The following result is due to Bryant [7].

Proposition 3.4. *Let M be a hypersurface in $\mathbb{C}^3$ with a nondegenerate Levi form on indefinite sign. If through each point $x \in M$ in each direction annihilating the Levi form there passes a complex curve, then M must be equivalent to a quadric Q.*

Proof. Let us fix a nonzero vector c^α, $\alpha = 1, 2$, such that

$$g_{\alpha\bar\beta}\, c^\alpha c^{\bar\beta} = 0, \tag{3.15}$$

and let us consider a differential system on P defined by the equations

$$c^1\theta^2 - c^2\theta^1 = 0,$$
$$\theta^{n+1} = 0, \quad (n=2),$$
$$c_\alpha c^\beta \omega^\alpha_{\bar\beta} = 0, \quad (c_\alpha = g_{\alpha\bar\beta}\, c^{\bar\beta}).$$

One can show that the projections to M of the integral manifolds of this system are the only holomorphic curves on M. Applying (3.6) we find that the integrability condition for this system has the form

$$S_{\alpha\bar{\beta}\sigma\bar{\tau}}\, c^\alpha c^{\bar{\beta}} c^\sigma c^{\bar{\tau}} = 0.$$

Taking account of (3.9), it follows from this formula, which holds for all c^α satisfying (3.15), that $S_{\alpha\bar{\beta}\sigma\bar{\tau}}=0$, and so, in view of Proposition 3.2, the hypersurface M must be locally equivalent to a quadric.

For $n>2$ Proposition 3.4 does not hold true. Bryant [7] has proved that a hypersurface $M \subset \mathbb{C}^{n+1}$ with a Lorentzian Levi form can have both more holomorphic curves than a quadric and fewer ones. The family of holomorphic curves on M depends on at most n^2 real parameters, whereas a quadric contains a $(4n-4)$-parameter family of complex curves.

§ 4. Forced Smoothness and the Continuation Principle for CR-Maps

Let $f\colon M_1 \to M_2$ be a CR-isomorphism. There arises the question: must f be infinitely smooth (analytic), provided M_1 and M_2 enjoy the same property. It is clear that is not always the case. For instance, this is not so if M_1 and M_2 are fibered by complex submanifolds of dimension $n = \mathrm{CR}\dim M$, that is, in the case when the Levi form vanishes identically.

If M_2 and M_2 are hypersurfaces with a nondegenerate Levi form, the answer is positive. In fact, for Levi forms of indefinite sign the situation is trivial, because in this case the CR-map can be continued holomorphically to a neighborhood of M_1. For strictly pseudoconvex hypersurfaces by a theorem of Lewy's [22] a CR-map f (or its inverse) can be continued to one side of the hypersurface M_1 (respectively M_2), and the smoothness of f follows from the local variant (cf. [2]) of a famous theorem by Fefferman [15]:

Theorem 4.1. *Let D_1 and D_2 be bounded strictly pseudoconvex domains in $\mathbb{C}^N$ with C^∞-smooth boundaries. Then each biholomorphic map $f\colon D_1 \to D_2$ extends to a C^∞-smooth map $\bar{D}_1 \to \bar{D}_2$.*

Besides the rather difficult original proof [15] of Theorem 4.1, there are now several somewhat simpler proofs and likewise generalizations of the theorem (cf. [28]). Here we display the interesting proof by Naruki [25] based on the Cartan-Tanaka CR-invariants.

Let D be a bounded strictly pseudoconvex domain in $\mathbb{C}^N$ with C^∞-smooth boundary. Let us consider in $\mathbb{C}^{N+1}$ the domain

$$\tilde{D} = \{(z_0, z)\colon z_0 \in \mathbb{C},\ z \in D,\ |z_0|^{2(N+1)} K(z) < 1\},$$

where $K(z)$ is the Bergman kernel function for D.

From relatively simple facts about the boundary behavior of the kernel function [15] it follows that $\tilde{D}$ is strictly pseudoconvex with a smooth boundary of class C^N.

Let now $f: D_1 \to D_2$ be a biholomorphic map and $J(z)$ its Jacobian. For simplicity, let us assume that D is simply connected and let us pick a holomorphic function $\chi(z)$ such that $\chi^{N+1} = J$.

Put $\tilde{f}(z_0, z) = (\chi(z) z_0, f(z))$. From the transformation rule for the kernel function under biholomorphic maps it follows that $\tilde{f}(\tilde{D}_1) = \tilde{D}_2$ and that everywhere, except on $\partial D_1 \subset \partial \tilde{D}_1$, $\tilde{f}$ has a smooth continuation to the boundary.

If the boundaries $\partial \tilde{D}_1$ and $\partial \tilde{D}_2$ are sufficiently smooth ($N \geqq 7$), there exist bundles P_1 and P_2 with Cartan connections ω_1 and ω_2, corresponding to the CR-manifolds $\partial \tilde{D}_1$ and $\partial \tilde{D}_2$. The map $g = \tilde{f} \,|\, \partial \tilde{D}_1 \setminus \partial D_1$ is lifted on $P_1 \,|\, \partial \tilde{D}_1 \setminus \partial D_1$ to a map G such that $G^* \omega_2 = \omega_1$.

Naruki's main argument consists of the observation that the map G, which preserves the parallelism (the Cartan connection), can not have singularities on the manifold $P_1 \,|\, \partial D_1$, of codimension 2 in P_1. Therefore G has a C^N-smooth continuation to the whole of P_1. This means that g and, consequently, f too extend to ∂D_1. The infinite smoothness of f follows the fact that the Cartan connections corresponding to the boundaries of the domains D_1 and D_2 themselves are preserved. This completes the proof of Theorem 4.1 ($N \geqq 7$).

Let us return to the question raised in the beginning of this Section. For strongly regular CR-manifolds, satisfying the condition of Theorem 2.3, as for hypersurfaces, such forced smoothness holds true. However, in these cases it suffices to assume that the map is so smooth that it lifts on a suitable bundle to a parallelism preserving map. Then, in view of Proposition 2.1, the lifted map and, consequently, also f itself will be C^∞.

If M_1 and M_2 are $\mathbb{R}$-analytic, then by the same argument it follows that each sufficiently smooth CR-isomorphism $f: M_1 \times M_2$ must be $\mathbb{R}$-analytic. For manifolds M_1 and M_2, imbedded in C^N, again it follows from this that f admits a holomorphic continuation to suitable neighborhoods.

There is another proof of the analyticity of CR-manifolds, based on the multivariate symmetry principle due to S.I. Pinchuk [28] and H. Lewy [23] and the analytic continuation of CR-functions with "the edge of the wedge" [1].

A bilinear, possibly vector valued, form $F(x, y)$ is said to be nondegenerate if $F(x, y) = 0$ for all y entails that $x = 0$.

Theorem 4.2 ([37], [43]). *Let M_1 and M_2 be $\mathbb{R}$-analytic* CR-*manifolds in $\mathbb{C}^N$. Assume also that at each point $x \in M_i$ the Levi form is nondegenerate and that its values span the entire space $N_x = T_x(M_i)/T_x^c(M_i)$. Then each* CR-*isomorphism $f: M_1 \to M_2$ of class C^1 extends to a biholomorphic map of suitable neighborhoods of M_1 and M_2.*

It would be interesting to have a generalization of this result to regular CR-manifolds with $\mu > 2$.

Interesting questions about CR-maps arise in connection with a well-known result by Poincaré-Alexander stating that local CR-automorphism of a sphere a global. In the Parts by A.G. Vitushkin (Vol. 7, Part IV) and

S.I. Pinchuk (this Vol., Part V) there are considered generalizations of this result to strictly pseudoconvex $\mathbb{R}$-analytic hypersurfaces. Here we just mention that if there are no umbilical points on the hypersurfaces then the continuation of local CR-maps is a straightforward application of Webster's CR-invariant metric (cf. § 3).

Proposition 4.3. *Let M_1 and M_2 be connected $\mathbb{R}$-analytic strictly pseudoconvex $(2n+1)$-dimensional* CR-*manifolds of* CR-*dimension n without umbilical points. Then each* CR-*isomorphism $f: U_1 \to U_2$ of relatively compact domains $U_i \subset M_i$ admits an analytic continuation to a suitable neighborhood of $\bar{U}_1$. If M_1 is simply connected and M_2 compact, then f extends to a CR-isomorphism $M_1 \to M_2$.*

Proof. The map f must be an isometry for the CR-invariant metrics on M_1 and M_2. Local isometries of analytic manifolds extend along geodesics (cf. [21], chap. VI, § 6). This gives the desired continuation.

For an $\mathbb{R}$-analytic CR-structures of codimension $m>1$ the phenomenon of continuation of local CR-maps is so far only known for standard manifolds of the form

$$M=\{(z, w) \in \mathbb{C}^n \oplus \mathbb{C}^m : \operatorname{Im} w = F(z, z)\}, \tag{4.1}$$

where F is a $\mathbb{C}^m$-valued Hermitian quadratic form on $\mathbb{C}^n$, $n, m \geqq 1$. The cone of values of the form F is defined to be the convex cone $C \subset \mathbb{R}^m$ which is the set of inner points of the convex hull of the set

$$\{F(z, z) : z \in \mathbb{C}^n\}.$$

As Naruki [24] (cf. also [1]) has shown the holomorphy hull of a manifold M of the type (4.1) is the domain

$$D=\{(z, w) \in \mathbb{C}^n + \mathbb{C}^m : \operatorname{Im} w - F(z, z) \in C\}.$$

Theorem 4.4 ([37]). *Let M_i $(i=1, 2)$ be standard manifolds on the type* (4.1) *such that the corresponding Hermitian forms F_i are nondegenerate and their cones of values non-empty. Then each C^1* CR-*map $f: U_1 \to U_2$ of domains $U_i \subset M_i$ extends to a rational map $D_1 \to D_2$ of the holomorphy hulls of the manifolds M_i.*

Let us remark that if the cones of values of the forms F_i are cuspidal (do not contain any entire lines) then the extended map $D_1 \to D_2$ is biholomorphic and in the formulation of Theorem 4.1 one can dispense of the smoothness of the map f, provided f is a homeomorphism satisfying the tangential Cauchy-Riemann equations (1.1) in the sense of distributions.

Besides the exact description of the hull of holomorphy for standard manifolds, the proof of Theorem 4.1 also utilizes the multivariate symmetry principle. For this it is very essential that a standard manifold of type (4.1) contains plenty of three dimensional standard manifolds, for example,

$$M_{z^*}=\{(z, w)\in M: z=\lambda z^*, w=\mu F(z^*, z^*), \lambda, \mu\in\mathbb{C}\},$$

where $F(z^*, z^*)\neq 0$.

It would be interesting to clarify when an arbitrary CR-manifold contains three dimensional CR-manifolds of CR-dimension one.

It follows from Theorem 4.1 that the group $\mathrm{Aut}_{CR}(S)$, where S is a domain in the standard manifold M, satisfying the assumptions of the theorem, must be a finite dimensional Lie group.

Theorem 4.4 can be used to the description of the proper holomorphic maps of *Siegel domains*. A Siegel domain (of the second kind, cf. [1]) is a domain of the form

$$D=\{(z, w)\in\mathbb{C}^n\oplus\mathbb{C}^m: \operatorname{Im} w-F(z, z)\in V\}, \tag{4.2}$$

where V is an open cone in $\mathbb{R}^m$ containing the cone of values C of the Hermitian form F. A Siegel domain (4.2) is called nondegenerate if $C\neq\emptyset$.

The map $f: D_1\to D_2$ is called proper, if the preimage $f^{-1}(K)$ of any compact set $K\subset D_2$ is compact.

Theorem 4.5 ([39]). *Proper holomorphic maps of nondegenerate Siegel domains are biholomorphic and rational.*

Theorem 4.5 is a generalization of a corresponding result of Alexander's concerning proper holomorphic maps of a ball. Theorem 4.5 reduces to Theorem 4.4 if we take account of the transformation rule of Bergman projection under proper holomorphic maps, as obtained by Bell [5]. Let us remark that the conclusion of Theorem 4.5 holds true for all classical domains which do not contain the unit disk as a direct factor (cf. [29]). This follows at once from the results in [38] and [6].

References

For the convenience of the reader, references to reviews in Zentralblatt für Mathematik (Zbl.), compiled using the MATH database, have been included as far as possible.

1. Airapetyan, R.A., Khenkin, G.M.: Integral representations of differential forms on Cauchy-Riemann manifolds and the theory of CR-functions. Usp. Mat. Nauk *39*, No. 3 (237), 39–106 (1984) [Russian]. Zbl. 589.32035. English transl.: Russ. Math. Surv. *39*, No. 3, 41–118 (1984)
2. Alexander, H.: Proper holomorphic mappings in $\mathbb{C}^n$. Indiana Math. J. *26*, 137–146 (1977). Zbl. 391.32015
3. Andreotti, A., Hill, C.D.: Complex characteristic coordinates and tangential Cauchy-Riemann equations. Ann. Sc. Norm. Sup. Pisa, Sci. Fis. Mat., III. Ser. *26*, 299–324 (1972). Zbl. 256.32006
4. Bedford, E., Kalka, M.: Foliations and complex Monge-Ampère equations. Comm. Pure Appl. Math. *30*, 543–571 (1977). Zbl. 351.35063
5. Bell, S.: Proper holomorphic mappings and the Bergman projection. Duke Math. J. *48*, 167–175 (1981). Zbl. 465.32014

6. Bell, S.: Proper holomorphic mappings between circular domains. Comment. Math. Helv. *57*, 532–538 (1982). Zbl. 511.32013
7. Bryant, R.L.: Holomorphic curves in Lorentzian CR-manifolds. Trans. Am. Math. Soc. *272*, 203–220 (1982). Zbl. 517.32008
8. Burns, D., Shnider, S.: Real hypersurfaces in complex manifolds. Proc. Symp. Pure Math. *30*, Part 2, 141–168 (1977). Zbl. 422.32016
9. Burns, D., Shnider, S.: Geometry of hypersurfaces and mapping theorems in $\mathbb{C}^n$. Comment. Math. Helv. *54*, 199–217 (1979). Zbl. 444.32012
10. Burns, D., Shnider, S., Wells, R.O.: Deformations of strongly pseudoconvex domains. Invent. Math. *46*, 237–253 (1978). Zbl. 412.32022
11. Cartan, É.: Sur la géométrie pseudo-conforme des hypersurfaces de deux variables complexes, I–II. Ann. Mat. Pura Appl. IV. Ser. *11*, 17–90 (1932). Zbl. 5, 373; Ann. Scuola Norm. Sup. Pisa II. Ser. *1*, 33–354 (1932). Zbl. 5, 374. Also: Oeuvres Partie II, Vol. 2, 1231–1307; Partie III, Vol. 1, pp. 333–354
12. Chern, S.: On the projective structure of a real hypersurface in $\mathbb{C}^n$. Math. Scand. *36*, 74–82 (1975). Zbl. 305.53019
13. Chern, S., Moser, J.: Real hypersurfaces in complex manifolds. Acta Math. *133*, 219–271 (1974). Zbl. 302.32015
14. Faran, J.: Segre families and hypersurfaces. Invent. Math. *60*, 135–172 (1980). Zbl. 464.32011
15. Fefferman, C.: The Bergman kernel and biholomorphic mappings of pseudoconvex domains. Invent. Math. *26*, 1–65 (1974). Zbl. 289.32012
16. Fefferman, C.: Monge-Ampère equations, the Bergman kernel, and geometry of pseudoconvex domains. Ann. Math., II. Ser. *103*, 395–416 (1976). Zbl. 322.32012
17. Fefferman, C.: Parabolic invariant theory in complex analysis. Adv. Math. *31*, 131–262 (1979). Zbl. 444.32013
18. Freeman, M.: The Levi form and local complex foliations. Proc. Am. Math. Soc. *57*, 369–370 (1976). Zbl. 328.32006
19. Freeman, M.: Local biholomorphic straightening of real submanifolds. Ann. Math., II. Ser. *106*, 319–352 (1977). Zbl. 372.32005
20. Greenfield, S.J.: Cauchy-Riemann equations in several variables. Ann. Scuola Norm. Sup. Pisa, Sci. Fis. Mat., III. Ser. *22*, 275–314 (1968). Zbl. 159, 375
21. Kobayashi, S., Nomizu, K.: Foundations of differential geometry, I. New York and London: Interscience 1963. Zbl. 119, 375
22. Lewy, H.: On hulls of holomorphy. Commun. Pure Appl. Math. *13*, 587–591 (1960). Zbl. 113, 61
23. Lewy, H.: On the boundary behavior of holomorphic mappings. Acad. Naz. Lincei *35:1*, 1–8 (1977).
24. Naruki, I.: Holomorphic extension problem for standard real submanifolds of second kind. Publ. Res. Inst. Math. Sci., Kyoto Univ. *6*, 113–187 (1970). Zbl. 225.32008
25. Naruki, I.: On extendibility of isomorphisms of Cartan connections and biholomorphic mappings of bounded domains. Tohoku Math. J., II. Ser. *28*, 117–122 (1976). Zbl. 346.32003
26. Newlander, A., Nirenberg, L.: Complex analytic coordinates in almost complex manifolds. Ann. Math., II. Ser. *65*, 391–404 (1957). Zbl. 79, 161
27. Nirenberg, L.: Lectures on linear partial differential operators. Providence: Am. Math. Soc. 1973. Zbl. 267.35001
28. Pinchuk, S.I.: Holomorphic maps in $\mathbb{C}^n$ and the problem of holomorphic equivalence. In: Contemporary problems of mathematics. Fundamental directions. Itogi Nauki Tekh., Ser. Mat. Anal. 9, 195–222. Moscow: VINITI 1986 [Russian]. English transl.: this Volume pp. 173–199
29. Pyatetskiĭ-Shapiro, I.I.: The geometry of classical domains and the theory of automorphic functions. Moscow: Nauka 1961 [Russian]. French transl.: Paris: Dunod 1966. Zbl. 137, 275. English transl.: New York etc: Gordon and Breach (1969)
30. Rossi, H.: Differentiable manifolds in complex Euclidean space. In: Proc. internat. congr. math., Moscow, 1966, 512–516. Moscow: MIR 1968. Zbl. 192, 440

31. Segre, B.: Intorno al problema di Poincaré della rappresentazione pseudoconforme, I. Atti Acad. Naz. Lincei, VI. Ser. *13*, 676–683 (1931). Zbl. 3, 213
32. Sommer, F.: Komplex-analytische Blätterung reeller Mannigfaltigkeiten im $\mathbb{C}^n$. Math. Ann. *136*, 111–113 (1958). Zbl. 92, 299
33. Sommer, F.: Komplex-analytische Blätterung reeller Hyperflächen im $\mathbb{C}^n$. Math. Ann. *137*, 393–411 (1959). Zbl. 92, 299
34. Sternberg, S.: Lectures on differential geometry. Englewood Cliffs: Prentice Hall 1964. Zbl. 129, 131
35. Tanaka, N.: On generalized graded Lie algebras and geometric structures. J. Math. Soc. Japan *19*, 215–254 (1967). Zbl. 165, 560
36. Tanaka, N.: Graded Lie algebras and geometric structures. In: Proc. U.S.-Japan Seminar Diff. Geom., Kyoto, 1965, 147–150. Tokyo: Nippon Hyoronsha 1966. Zbl. 163, 439
37. Tumanov, A.E., Khenkin, G.M.: Local characterization of holomorphic automorphism of Siegel domains. Funkts. Anal. Prilozh. *17*, No. 4, 49–61 (1983) [Russian]. Zbl. 572.32018. English transl.: Funct. Anal. Appl. *17*, 285–294 (1983)
38. Tumanov, A.E., Khenkin, G.M.: Local characterization of automorphisms of classical domains. Dokl. Akad. Nauk SSSR *267*, 796–799 (1982) [Russian]. Zbl. 529.32014. English transl.: Sov. Math., Dokl. *26*, 702–705 (1982)
39. Tumanov, A.E., Khenkin, G.M.: Proper maps of Siegel domains. In: Complex methods in mathematical physics. All union school of young scientists, Donetsk, 1984, p. 185 [Russian].
40. Vitushkin, S.I.: Holomorphic maps and the geometry of hypersurfaces. In: Contemporary problems of mathematics. Fundamental directions. Itogi Nauki Tekh. 7, 167–226. Moscow: VINITI 1985 [Russian].
41. Webster, S.M.: On the transformation group of a real hypersurface. Trans. Am. Math. Soc. *231*, 179–190 (1977). Zbl. 368.57013
42. Webster, S.M.: Some birational invariants for algebraic real hypersurfaces. Duke Math. J. *45*, 39–46 (1978). Zbl. 373.32012
43. Webster, S.M.: Analytic discs and regularity of CR-mappings of real submanifolds in $\mathbb{C}^n$. In: Complex analysis of several variables, Madison, 1982. Proc. Symp. Pure Math. *41*, 199–208. Providence: Am. Math. Soc. 1984. Zbl. 568.32013
44. Wells, R.O.: The Cauchy-Riemann equations and differential geometry. Bull. Am. Math. Soc., New Ser. *6*, 187–199 (1982). Zbl. 496.32012

VII. Supersymmetry and Complex Geometry

A.A. Roslyĭ, O.M. Khudaverdyan, A.S. Schwarz

Translated from the Russian
by J. Peetre

Contents

Introduction

In the past years supersymmetric theories have gained great importance in physics. By this one intends field theoretical models based on a new form of symmetry dubbed supersymmetry. Supersymmetry connects boson and fermion fields with each other [13], [46], [48], [26], [6]. The observed properties of particles cannot satisfy the demands of supersymmetry (for instance, supersymmetry would lead to the equality of mass for the boson and the corresponding fermion). However, an increasing number of physicists have arrived at the conviction that the action functional of interactions encountered in nature must be supersymmetrical (although for the ground state (the physical vacuum) and, consequently, for the observed spectra of particles supersymmetry is broken). Perhaps the most weighty foundation for such a belief is the mathematical beauty of the supersymmetric theories and the remarkable property of cancellation of the divergencies appearing in these theories. It is question of the circumstance that in quantum field theories one encounters divergencies arising from the integration over large momenta (ultraviolet divergencies). In supersymmetry the most dangerous of these divergencies cancel. Moreover, there exist models completely free of ultraviolet divergencies. Presently great hopes are put on such supersymmetric theories which take account of the presence of gravitational interactions. Thus an important constituent part of these theories is played by *supergravity*, a supersymmetric theory containing Einstein's theory of gravity. Many people think that on the basis of such supersymmetric models it will be possible to build a theory unifying all known interactions – the strong, the electromagnetic, the weak and the gravitational interactions.

The mathematical basis for supersymmetric theories is a branch of mathematics which is now usually called supermathematics. In the creation of supermathematics a prominent rôle belongs to F.A. Berezin, who on the basis of his studies of second quantization arrived at the view that there exists a counterpart of analysis where elements of the Grassman algebra are considered as functions of anticommuting variables. Berezin invented a theory of functions of commuting and anticommuting variables and constructed objects now called supergroups and Lie superalgebras [2]. Later on, supergroups appeared in the physics literature in the construction of symmetries interchanging fermion and boson fields and many examples of supersymmetric theories were given. Here the notion of superspace introduced by Salam and Strathdee turned out to be most useful, a space where coordinates may be both commutative and anticommutative objects [36]. (The definition given in [36] is not rigorous but can be remedied.) The reading of Berezin's papers does not leave any doubts that he mastered the notion of superspace long before Salam and Strathdee (in the "physicists" standard of rigour). This may be seen, for instance, from the fact that in [3] the term "group with anticommuting parameters" is used for the nowadays accepted term "Lie supergroup". However, it really was the paper [36] that opened up the road for the application of the notion of superspace to the construction of supersymmetric field theories in terms of superfields (fields on superspace).

In physics one often considers the superspace $\mathbb{R}^{4|4}$ with four commuting real coordinates x^a and four anticommuting real coordinates. Instead of the four real anticommuting coordinates it is convenient to use two complex anticommuting ones θ^α and their conjugates $\bar{\theta}^{\dot{\alpha}}$. The Poincaré group[1] acts in this space in such a way that the x^a transform as a vector[2], whereas

[1] The *Poincaré group* is the group of affine maps of the space $\mathbb{R}^4$ preserving an invariant quadratic form of signature $(+---)$ (that is, the group of motions in Minkowski space).

[2] That is, as the coordinates of a point in $\mathbb{R}^4$ under the natural action of the Poincaré group on this space.

the θ^α transform as a two component spinor[3]. The Poincaré group may be extended by adding the transformations

$$x^a \to x^a + i\sigma^a_{\alpha\dot\beta}(\theta^\alpha \bar\varepsilon^{\dot\beta} - \varepsilon^\alpha \bar\theta^{\dot\beta}),\\ \theta^\alpha \to \theta^\alpha + \varepsilon^\alpha, \tag{0.1}$$

where ε^α are two anticommuting parameters and $\sigma^a_{\alpha\dot\beta}$ are the elements of the 2×2 matrices σ^a *(Pauli matrices)* having the form[4]:

$$\sigma^0 = \begin{pmatrix}1 & 0\\ 0 & 1\end{pmatrix}, \quad \sigma^1 = \begin{pmatrix}0 & 1\\ 1 & 0\end{pmatrix}, \quad \sigma^2 = \begin{pmatrix}0 & -i\\ i & 0\end{pmatrix}, \quad \sigma^3 = \begin{pmatrix}1 & 0\\ 0 & -1\end{pmatrix}. \tag{0.2}$$

The supergroup (a group with anticommuting parameters) generated by the transformations of the Poincaré group together with the transformations (0.2) is called the *Poincaré supergroup*. The space $\mathbb{R}^{4|4}$ equipped with the Poincaré supergroup is called *Minkovski superspace*.

A *superfield* on $\mathbb{R}^{4/4}$ is defined to be an expression of the type

$$\begin{aligned}\Phi(x, \theta, \bar\theta) = {} & A(x) + B_\alpha(x)\theta^\alpha + C_{\dot\beta}(x)\bar\theta^{\dot\beta} + D_{\alpha\beta}(x)\theta^\alpha\theta^\beta + \\ & + E_{\alpha\dot\beta}(x)\theta^\alpha\theta^{\dot\beta} + F_{\dot\alpha\dot\beta}(x)\bar\theta^{\dot\alpha}\bar\theta^{\dot\beta} + G_{\alpha\beta\dot\gamma}(x)\theta^\alpha\theta^\beta\bar\theta^{\dot\gamma} + H_{\dot\alpha\dot\beta\gamma}(x)\bar\theta^{\dot\alpha}\bar\theta^{\dot\beta}\theta^\gamma + \\ & + K_{\alpha\beta\dot\gamma\dot\delta}(x)\theta^\alpha\theta^\beta\bar\theta^{\dot\gamma}\bar\theta^{\dot\delta}.\end{aligned}$$

The coefficients in products of an even number of anticommuting coordinates correspond to boson fields and coefficients with an odd number of products of anticommuting coordinates to fermion fields. In quantum field theory fermion fields take anticommuting values. One can also consider superfields taking values in an arbitrary representation of the Lorentz group, for instance, vector superfields. In order to construct a supersymmetric theory one has to write down an action functional for the superfields which is invariant with respect to the Poincaré supergroup.

In the sequel we will mainly be occupied with an investigation of the theory of supergravity [44]. In the construction of supergravity it is natural to use a vector superfield $H^a(x, \theta, \bar\theta)$ (a minimal superfield containing a spin 2 field, which can be identified with gravity), cf. [27], [28], [11]. The action functional for supergravity must be invariant with respect to a supergroup containing the Poincaré supergroup as well as the group of diffeomorphisms of four dimensional space, which is the invariance group of conventional gravitation theory. Such a supergroup can in fact be constructed, although

[3] The subgroup of the Poincaré group consisting of linear maps is called the *Lorentz group*. We will also have occasion to employ the spin representation of the Lorentz group, which is a two dimensional representation (*two component spinors*) of the group $SL(2, \mathbb{C})$, locally isomorphic to the Lorentz group, and likewise with its conjugate representation (dotted spinors). Let us remark that in the sequel $SL(2, \mathbb{C})$ too often will be referred to as the Lorentz group. (Moreover, by the Poincaré group we actually then will understand the group obtained from it upon substituting $SL(2, \mathbb{C})$ for the Lorentz group.)

[4] The Pauli matrices define a map from the tensor product of the two dimensional spaces of the spinor representation of the Lorentz group and its conjugate to the four dimensional space of the vector representation. This map commutes with the action of the Lorentz group.

its construction is nontrivial. It turns out that the transformations of this supergroup "mix" field and space variables. More precisely, one has to associate with the vector superfield $H^a(x, \theta, \bar{\theta})$ its graph, a surface in the space $\mathbb{R}^{8|4}$ with eight commuting and four anticommuting coordinates. In $\mathbb{R}^{8|4}$ it is possible to introduce a complex structure, converting it into a complex superspace $\mathbb{C}^{4|2}$. Then the group of symmetries for supergravity can be realized as the group of volume preserving analytic transformation of the superspace $\mathbb{C}^{4|2}$. (This group will in the sequel be denoted by $\mathscr{L}$.) A transformation in $\mathscr{L}$ carries a surface in $\mathbb{C}^{4|2}$ into another surface; considering the transformation of graphs of vector superfields we obtain an action of the group $\mathscr{L}$ on vector superfields.

Supergravity in terms of vector superfields and an action functional which is invariant with respect to $\mathscr{L}$ were constructed in papers by V.I. Ogievetskiĭ and E.S. Sokachev [27], [28][5]. The formulation of supergravity given by Ogievetskiĭ and Sokachev suggests that as the basic object of supergravity it is convenient to take the surface of real dimension $(4|4)$ in complex superspace $\mathbb{C}^{4|2}$. From this it is clear that in the analysis of supergravity methods developed in mathematics in the study of real surfaces in complex space may be useful. In the present Part we show that in fact this is the case. In particular, we make it clear that the action functional in supergravity may be written in a simple way using the Levi form of the surface. The results set forth, thus obtained in supergravity using methods of complex geometry, constitute the main objective of the present Part.[6]

However, before passing to the analysis of supergravity we must state a few notions and results which here play an auxiliary rôle but have an interest of their own right. First we discuss a field theoretic formalism in which the rôle of field is played by surfaces (as follows from what we said above, it is precisely the formalism in which field and space variables are treated on equal footing, which is useful for supergravity). Then we briefly develop the main concepts concerning supermathematics (§ 2) and discuss the notion of supersymmetry (§ 3). The following two sections are devoted to supergravity and to the applications to supergravity of the methods of complex geometry. In § 6 we indicate how the theory of G-structures may be used in supergravity.

In physics literature there is presently given a considerable attention also to super gauge theories, that is, supersymmetric generalizations of gauge theory. It turns out that if we write these theories in terms of superfields the language of complex geometry[6] still is most useful. § 7 is devoted to some such applications of the methods of complex geometry. Finally, in § 8 the geometry of the superspace under consideration, corresponding thus

---

[5] An analogous approach is developed also in Gates-Siegel [25].

[6] Many interesting applications of the methods of complex geometry to the study of supersymmetric theories can be found in Yu.I. Manin's monograph [25], which, in particular, contains an introduction (mainly for mathematicians) to a circle of ideas connected with supergravity and super gauge theories (including the point of view of the twistor approach).

to supergravity, is studied from the point of view of the geometry of a distinguished family of submanifolds in superspace, so-called super light like geodesic surfaces.

§ 1. Field and Space Variables on Equal Footing

In the study of the structure of physical theories it is often useful to make a formal unification of space and field variables. Such an approach is given e.g. in [45], [27], [28].

Let us develop the fundamentals of this subject only in the case of the simplest example. Let there be given an action functional S, the basic object defining the content of the physical theory:

$$S(\varphi)=\int \mathscr{L}(\varphi(x), \partial\varphi(x)/\partial x)\, d^m x, \tag{1.1}$$

where $\mathscr{L}$ is the Lagrangean and $\varphi(x)=(\varphi^1(x), \ldots, \varphi^n(x))$ a vector field defined in m-dimensional Euclidean space $\mathbb{R}^m$ (or in a region D of this space). Let us now consider the $(n+m)$-dimensional space $\mathbb{R}^{n+m}$ with coordinates $(\varphi, x) =(\varphi^1, \ldots, \varphi^n, x^1, \ldots, x^m)$. Each field $\varphi(x)$ defines an m-dimensional surface $\Omega(\varphi)$ in $\mathbb{R}^{n+m}$. The action functional S may be viewed as an additive functional given by the relation (1.1) on such m-dimensional surfaces in $\mathbb{R}^{n+m}$ which admit a one-to-one projection onto the subspace $\{\varphi^1=\ldots=\varphi^n=0\}$. Let us say that a functional Φ defined on m-dimensional surfaces in $\mathbb{R}^{n+m}$ is compatible with the action S if for any field φ holds

$$S(\varphi)=\Phi(\Omega(\varphi)). \tag{1.2}$$

Let us consider some concrete cases.

1. If the action has the form

$$S(\varphi)=\int A_i^\mu(\varphi, x)\, \partial_\mu \varphi^i\, d^m x, \tag{1.3}$$

then the following functional, defined for all m-dimensional surfaces in $\mathbb{R}^{n+m}$, is compatible with it:

$$\Phi(\Omega)=\int_\Omega \omega, \tag{1.4}$$

where ω is the m-form

$$\omega=\sum_i \sum_\mu A_i^\mu(\varphi, x)\, dx^1 \wedge \ldots \wedge dx^{\mu-1} \wedge d\varphi^i \wedge dx^{\mu+1} \wedge \ldots \wedge dx^m.$$

2. In the case of Yang-Mills fields one may consider the action

$$S(B, F)=-\tfrac{1}{8}\int \operatorname{tr}\{F^{\mu\nu} F_{\mu\nu}+\tfrac{1}{4} F^{\mu\nu}(\partial_\nu B_\mu-\partial_\mu B_\nu+[B_\mu, B_\nu])\}\, d^4 x,$$

where $\mu, \nu=1, \ldots, 4$ and the components of the vector field B_μ as well as those of the antisymmetric tensor $F_{\mu\nu}$ take their values in the Lie algebra $\mathscr{G}$ (of the gauge group G). Using the relations (1.3), (1.4) one can associate with this action an additive functional defined for all four dimensional surfaces in $(10q+4)$-dimensional space with coordinates $(B_\mu, F_{\mu\nu}, x^\mu)$, where q is the dimension of G. An analogous procedure can also be applied in the case of the gravitation field [43]. Below we show in which cases an additive functional compatible with the action functional may be defined for all surfaces of given dimension.

Let Ω be an m-dimensional surface in $\mathbb{R}^N$ and $y(\xi)$ a mapping from $\mathbb{R}^m$ into $\mathbb{R}^N$ giving a parametrization of the surface. (Here $y=(y^1, \ldots, y^N)$, $\xi=(\xi^1, \ldots, \xi^m)$.) An additive functional defined for m-dimensional surfaces may be written in the form

$$\Phi_A(\Omega)=\int A\left(y^a(\xi), \frac{\partial y^a(\xi)}{\partial \xi^{\mu_1}}, \ldots, \frac{\partial^r y^a(\xi)}{\partial \xi^{\mu_1} \ldots \partial \xi^{\mu_r}}\right) d^m \xi, \tag{1.5}$$

where conditions are imposed on the function A which guarantee the independence of the integral (1.5) from the choice of the parametrization $y(\xi)$ of Ω. We say that a function $A(p^a, p^a_{\mu_1}, \ldots, p^a_{\mu_1 \ldots \mu_r})$ satisfying these conditions is an *m-density* of rank r, where r is the maximal order of the derivatives entering in (1.5). The functional Φ_A, apparently, is additive in the following sense: if the surface Ω consists of two nonintersecting pieces Ω_1 and Ω_2 then

$$\Phi_A(\Omega)=\Phi_A(\Omega_1)+\Phi_A(\Omega_2).$$

Thus, in particular, an m-density of rank one is a function $A(y, K)$, where $y\in\mathbb{R}^N$ and K is an $N\times m$ matrix, subject to

$$A(y, KR)=A(y, K) \det R \tag{1.6}$$

for any $m\times m$ matrix R. This condition guarantees that the functional

$$\Phi_A(\Omega)=\int A\left(y(\xi), \frac{\partial y^a}{\partial \xi^\mu}\right) d^m \xi$$

is independent of the choice of the parametrization $y(\xi)$ of Ω.

Given the transformation $y=y(\xi)$ the collection of derivatives

$$\left\{\frac{\partial^K y^a}{\partial \xi^{\mu_1} \ldots \partial \xi^{\mu_K}}, a=1, \ldots, N; \mu_i=1, \ldots, m; K=0, 1, \ldots, r\right\}$$

at the point ξ_0 is termed the jet of order r of the map $y=y(\xi)$ at this point. A change of parametrization $\xi'=g(\xi)$ induces a transformation of the jets of order r of the map $y=y(\xi)$, which at each point is determined, of course, by the jet of the same order r of the map $\xi'=g(\xi)$. Under such transformations densities of rank r, considered as functions of jets of order r, have to be multiplied by the Jacobian of the reparametrization. For instance, for densities of rank one this leads to condition (1.6). Often it is natural to consider functionals defined only on surfaces without singular points. Then the function $A(p^a, p^a_{\mu_1}, \ldots, p^a_{\mu_1 \ldots \mu_r})$ defining the density only has to be defined under the condition that the matrix p^a_μ has rank m (the maximal rank possible).

Sometimes one also has to consider functionals defined in even narrower classes of surfaces. Then the domain of definition of A must be shrinken correspondingly.

The above definition of density is adjusted so that we can consider the corresponding functionals on surfaces in parametric form. Often one has also to deal with surfaces given with the aid of equations. Let us introduce the notion of D-density, which again is adjusted to integration over surfaces given by equations.

Let us call an m-D-density of rank one a function $B(y, M)$ (where $y \in \mathbb{R}^N$ and M is an $(N-m) \times N$ matrix) satisfying the condition

$$B(y, LM) = B(y, M) \det L$$

for any $(N-m) \times (N-m)$ matrix L. Let Ω be an m-dimensional surface in $\mathbb{R}^N$ given by the equations

$$F^i(y) = 0, \quad i = 1, \ldots, N-m.$$

Then the integral

$$\int B\left(y, \frac{\partial F^i}{\partial y^a}\right) \delta(F^i(y))\, d^N y$$

(where δ is the δ-function in $N-m$ variables) is defined only by the surface Ω itself and does not depend on the choice of the F^i giving the surface. In an analogous manner we define D-densities of arbitrary rank. D-densities and ordinary densities are in one-to-one correspondence [9].

A density is the most general object which may be integrated over a surface. To the usual objects of integration over an m-dimensional surface – the m-forms – there correspond then those m-densities of rank one, $A(y, K)$, which depend linearly on the matrix K.

In the field theoretic applications of functionals of the type $\Phi_A(\Omega)$ in place of the usual action S there arises the question when a given action S may be extended in a continuous manner to a functional $\Phi_A(\Omega)$ defined for all surfaces of given dimension (i.e., not only those of the type $\Omega(\varphi)$) and compatible with the original action S, as stated in (1.2). One may show that in the case when the Lagrangean depends in a polynomial way on the derivatives of the field at most of order one then the functional Φ_A satisfying the compatibility condition (1.2) may be extended in a continuous manner to all m-dimensional surfaces iff the density A corresponds to an m-form [7] (cf. [19]).

[7] The topology in the space of surfaces is defined in the standard way; a sequence of surfaces Ω_n, $n = 1, 2, \ldots$, with the parametrizations $y_n(\xi)$ converges to a surface Ω ($\Omega_n \to \Omega$) if for the parametrization $y(\xi)$ of Ω holds $y_n(\xi) \to y(\xi)$ and

$$\frac{\partial^k y_n(\xi)}{\partial \xi^{\mu_1} \ldots \partial \xi^{\mu_k}} \to \frac{\partial^k y(\xi)}{\partial \xi^{\mu_1} \ldots \partial \xi^{\mu_k}}$$

for $k = 1, \ldots, r$ and some fixed r. Functionals corresponding to densities of rank r are continuous in this topology. Let us remark that the functionals corresponding to ordinary differential forms are, in view of Stokes's theorem, distinguished among general functionals also by the condition of continuity in a much weaker sense (cf. [21]).

In a field theoretic formalism where field and space variables are treated equally [38] the rôle of field is played by surfaces and the rôle of Lagrangean by a density defining an additive functional on the space of surfaces. In this case the rôle of the symmetry group in the physical theory is played by the symmetry group of the functional in question. Frequently one has a group G for which the action on points of the space $\mathbb{R}^N$ is given. Then an action of G on the space of surfaces may be defined in such a manner that for $g \in G$ we have a map of $\mathbb{R}^N$ sending the surface Ω into Ω^g. The functional Φ is invariant with respect to G if $\Phi(\Omega^g) = \Phi(\Omega)$ for all $g \in G$ and all surfaces Ω. In an analogous way, a density A is G-invariant if $A^g = A$ for all $g \in G$, where A^g is defined by the condition $\Phi_{A^g}(\Omega) = \Phi_A(\Omega^g)$ for all Ω. The group G is selected because of physical considerations and the requirement of invariance with respect to it diminishes the supply of possible Lagrangeans (densities). In some cases such a symmetry requirement dictates uniquely the choice of the invariant Lagrangean (the G-invariant density). If among the m-densities of a given rank there exists a unique (up to a constant factor) G-invariant density, then the corresponding invariant functional is termed ideal with respect to the group G [12]. Let us consider a few examples.

1. Let G be the group of motions of the Euclidean space $\mathbb{R}^N$. Then the G-invariant functionals (and the corresponding densities) defined on m-dimensional surfaces ($m < N$) in $\mathbb{R}^N$ are exhausted by the functionals $\Phi_A(\Omega) =$ [the m-dimensional volume of the m-dimensional surface Ω] and the corresponding densities. Thus, the m-dimensional volume element in N-dimensional Euclidean space defines an ideal density of rank one.

2. Let G be the group of canonical transformations of the space $\mathbb{R}^{2n}$, that is, transformations preserving a nondegenerate closed 2-form ω given on $\mathbb{R}^{2n}$. In this case the G-invariant functionals on $2m$-dimensional surfaces ($m \leq N$) in $\mathbb{R}^{2N}$ are exhausted [22] by the well-known *integral invariants of Poincaré-Cartan:*

$$\Phi(\Omega) = \int_{\Omega} \underbrace{\omega \wedge \ldots \wedge \omega}_{m \text{ times}}, \tag{1.7}$$

that is, within the class of densities of arbitrary rank the $2m$-density $\omega \wedge \ldots \wedge \omega$ is ideal with respect to the group of canonical transformations in $\mathbb{R}^{2n}$.

In the following Section we consider more interesting examples.

Lately the formalism of field theory with field and space variables treated on equal footing has obtained an increasingly greater importance in the construction of superspace formulations of various supersymmetric theories and, especially, supergravity. In particular, one uses a generalization of the notion of density in the case of superspace, which allows one to consider integration over supersurfaces. (Let us remark that the problem of constructing an object for integration over supersurfaces also has an independent interest.) The basic notions about superspaces will be outlined in the following section.

§ 2. Superspace

Let us denote by Λ^q the *Grassmann algebra* with q anticommuting free generators $\varepsilon^1, \ldots, \varepsilon^q$. A general element λ of Λ^q may, in view of the anticommutativity of the generators: $\varepsilon^\alpha \varepsilon^\beta = -\varepsilon^\beta \varepsilon^\alpha$, be written in the form:

$$\lambda = \sum_{0 \leqq k \leqq q} a_{\alpha_1 \ldots \alpha_k} \varepsilon^{\alpha_1} \ldots \varepsilon^{\alpha_k}, \tag{2.1}$$

where the coefficients $a_{\alpha_1 \ldots \alpha_k}$ are antisymmetric under permutation of the indices.

The Grassmann algebra may be considered over the field of real as well as over the field of complex numbers and, correspondingly, the coefficients in (2.1) have to be taken real or complex.

Each element of the Grassmann algebra Λ^q may be written in a unique way as the sum of even and odd elements in the algebra (the parity of a monomial $\varepsilon^{\alpha_1} \ldots \varepsilon^{\alpha_k}$ is defined as the parity of the number k). An even element commutes with every element of the algebra. Two odd elements anticommute.

Each element λ of the Grassmann algebra Λ^q may be presented in the form

$$\lambda = m(\lambda) + n(\lambda), \tag{2.2}$$

where $n(\lambda)$ is a nilpotent[8] element in Λ^q and $m(\lambda)$ an ordinary number. The number $m(\lambda)$ is called the numerical part of λ.

The expression (2.1) reminds in form of a polynomial in the ε^α and, in fact, it is convenient to think of it as a function of the "anticommuting variables" ε^α. One can define the derivative $d\lambda/d\varepsilon^\alpha$ with respect to ε^α and the integral of λ with respect to the variables $\varepsilon^1, \ldots, \varepsilon^q$. These operations are uniquely determined by the relations

$$\frac{d\varepsilon^\alpha}{d\varepsilon^\beta} = \delta^\alpha_\beta, \qquad \int d\varepsilon^\alpha = 0; \qquad \int \varepsilon^\alpha d\varepsilon^\alpha = 1.$$

It is understood that both operations are linear, that a multiple integral may be reduced to a repeated one and that differentiation has to satisfy Leibniz's rule: if λ_1 is an element of arbitrary parity then

$$\frac{d}{d\varepsilon}(\lambda_1 \lambda_2) = \left(\frac{d\lambda_1}{d\varepsilon}\right) \lambda_2 \pm \lambda_1 \left(\frac{d\lambda_2}{d\varepsilon}\right),$$

where the sign is determined by the parity of λ_1.

Besides the Grassmann algebra Λ^q, an algebra with q anticommuting variables $\varepsilon^1, \ldots, \varepsilon^q$, we consider likewise the algebra $\Lambda^{p|q}$, which may be interpreted as an algebra with p commuting variables x^i and q anticommuting variables ε^α. An element of the algebra $\Lambda^{p|q}$ (the *Berezin algebra*, as we will refer to it) may be written in the form (2.1), where the coefficients $a_{\alpha_1 \ldots \alpha_k}$ are smooth

[8] That is, $(n(\lambda))^k = 0$ for some natural number k.

functions in the real variables $x^1, \ldots, x^p$, that is, for $\omega \in \Lambda^{p|q}$ holds

$$\omega = \sum a_{\alpha_1 \ldots \alpha_k}(x^1, \ldots, x^p)\, \varepsilon^{\alpha_1} \ldots \varepsilon^{\alpha_k}. \tag{2.3}$$

In the algebra $\Lambda^{p|q}$, as well as the Grassmann algebra Λ^q, one defines in the natural way even and odd elements. More formally, one can also define $\Lambda^{p|q}$ as the algebra of smooth functions on $\mathbb{R}^p$ with values in Λ^q.

There is further a natural notion of derivative of elements of the Berezin algebra with respect to x^i and ε^α and of the integral with respect to these variables. For example, if ω is given by (2.3) then

$$\int \omega\, d^p x\, d^q \xi = q! \int a_{1 \ldots q}(x^1, \ldots, x^p)\, d^p x.$$

It is of importance to notice that an expression such as (2.3) remains meaningful if we replace the real numbers x_i by even elements of an arbitrary Berezin algebra Λ and ε^α by odd elements of the same algebra. (If $a(x)$ is a smooth function of a real variable x, then in order to give a meaning to the expression $a(\lambda)$, where λ is an even element of the Berezin algebra Λ, one has to write λ in the form (2.2) and expand $a(m(\lambda)+n(\lambda))$ formally in a Taylor series in $n(\lambda)$. Because $n(\lambda)$ is nilpotent, the series will terminate and, consequently, defines an element of Λ.)

Now we are ready to introduce the notion of *superspace*. If Λ is an Grassmann algebra over $\mathbb{R}$, then a Λ-point of the $(p|q)$-dimensional superspace $\mathbb{R}^{p|q}$ is a row $(u^1, \ldots, u^p, \theta^1, \ldots, \theta^q)$ where $u^1, \ldots, u^p$ are even elements of Λ and $\theta^1, \ldots, \theta^q$ odd elements. More pictorially, we may say that the coordinates of the superspace $\mathbb{R}^{p|q}$ comprise p commuting and q anticommuting elements[9].

Each even element of the Berezin algebra $\Lambda^{p|q}$ defines a map of the set of Λ-points of $\mathbb{R}^{p|q}$ into the set of even elements of Λ, that is, the set of Λ-points of $\mathbb{R}^{1|0}$. Analogously, the odd elements define a map from the Λ-points of $\mathbb{R}^{p|q}$ into the set of Λ-points of $\mathbb{R}^{0|1}$. In fact, in view of the above remark, we may in (2.3) put in place of x^i an even element of Λ and in place of ε^α an odd one. This gives a possibility to interpret even (respectively odd) elements of the algebra $\Lambda^{p|q}$ as even (respectively odd) functions on $\mathbb{R}^{p|q}$.

We define a *Λ-function* on $\mathbb{R}^{p|q}$ to be a linear combination with coefficients in Λ of functions on $\mathbb{R}^{p|q}$ corresponding to elements of the algebra $\Lambda^{p|q}$. (More precisely, an even Λ-function on $\mathbb{R}^{p|q}$ is defined as a formal linear combination $\Sigma \lambda_i \omega_i$, where $\lambda_i \in \Lambda$, $\omega_i \in \Lambda^{p|q}$ and λ_i and ω_i have the same parity. An even Λ-function gives a map of the set of Λ-points of $\mathbb{R}^{p|q}$ onto the set of Λ-points of $\mathbb{R}^{1|0}$. An odd Λ-function is defined in exactly the same way but the elements λ_i and ω_i now have to have opposite parity.)

---

[9] With each Grassman algebra we associate the set $\mathbb{R}^{p|q}_\Lambda$ of *Λ-points* of the superspace $\mathbb{R}^{p|q}$. Strictly speaking, one has to consider superspace as a functor from the category of Grassman algebras into the category of sets (for details see [39]). However, in the sequel we will often assume that the algebra Λ is fixed; this allows us to say simply "point of superspace" in place of "Λ-point".

By definition a Λ-map of the superspace $\mathbb{R}^{p|q}$ into the superspace $\mathbb{R}^{p'|q'}$ is given by a row of p' even Λ-functions and q' odd Λ-functions on $\mathbb{R}^{p|q}$. Thus, every Λ-map gives a map of the set $\mathbb{R}^{p|q}_\Lambda$ of all Λ-points of $\mathbb{R}^{p|q}$ into $\mathbb{R}^{p'|q'}_\Lambda$, the set of Λ-points of $\mathbb{R}^{p'|q'}$. More pictorially, one can say that under a Λ-map the coordinates of points in $\mathbb{R}^{p'|q'}$ are expressed by the coordinates of points in $\mathbb{R}^{p|q}$ in the following manner:

$$\begin{aligned} u^{a'} &= f^{a'}(u, \theta), \\ \theta^{\alpha'} &= \varphi^{\alpha'}(u, \theta). \end{aligned} \tag{2.4}$$

where $f^1, \ldots, f^{p'}$ are even λ-functions and $\varphi^1, \ldots, \varphi^{q'}$ odd ones. In particular, a linear Λ-map is given by the formula

$$\begin{aligned} u^{a'} &= A^{a'}_a u^a + B^{a'}_\alpha \theta^\alpha, \\ \theta^{\alpha'} &= C^{\alpha'}_a u^a + D^{\alpha'}_\alpha \theta^\alpha, \end{aligned} \tag{2.5}$$

where the matrices A, D consist of even elements of Λ and B, C of odd ones.

For simplicity we will sometimes combine the even and odd coordinates u^a and θ^α into one symbol z^A. Then the superspace map (2.4) may be written as

$$z^{A'} = z^{A'}(z),$$

and the linear map (2.5) as

$$z^{A'} = I^{A'}_A z^A,$$

where

$$I^{A'}_A = \begin{Bmatrix} A^{a'}_a & B^{a'}_\alpha \\ C^{\alpha'}_a & D^{\alpha'}_\alpha \end{Bmatrix}$$

is the matrix of the linear map from $\mathbb{R}^{p|q}$ into $\mathbb{R}^{p'|q'}$. The matrix $I^{A'}_A$ is said to be regular if the rank of the numerical part of the matrix A equals $\min(p, p')$ while the rank of the numerical part of the matrix D equals $\min(q, q')$. A linear map is regular when its matrix is regular.

The *Berezinian*, or the superdeterminant, of a linear map in the case $p' = p$, $q' = q$ is an expression of the type

$$\operatorname{Ber} I = \det(A - BD^{-1} C) \det D^{-1}.$$

The Berezinian of a regular map is an invertible element in Λ. These maps are precisely the only invertible linear maps.

The definition of integral of elements of the algebra $\Lambda^{p|q}$, as displayed above, gives us a possibility to define the integral of Λ-functions with respect to the superspace $\mathbb{R}^{p|q}$. By linearity this defines the integral of Λ-functions with respect to $\mathbb{R}^{p|q}$ as an element of Λ. One can show that the rule for changing variables in an integral of Λ-functions under superspace transformations is the same as in ordinary analysis: one just has to substitute the determinant for the Berezinian. By transformation we understand here, as always, invertible transformation.

Moreover, by abuse of language, we may will use the term "transformation" in place of "Λ-transformation", "point" in place of "Λ-point" etc.

Let

$$z^A = z^A(\zeta^B),$$

be a map from the superspace $\mathbb{R}^{m|n}$ into the superspace $\mathbb{R}^{M|N}$ ($m \leqq M$, $n \leqq N$), where $z^A = (x^1, \ldots, x^M, \theta^1, \ldots, \theta^N)$ and $\zeta^B = (\xi^1, \ldots, \xi^m, \nu^1, \ldots, \nu^n)$ are the coordinates in $\mathbb{R}^{M|N}$ and $\mathbb{R}^{m|n}$ respectively. The differential $Dz(\zeta)$ of this map at the point ζ may be viewed as a linear map (2.5) whose matrix is the Jacobian matrix

$$I_{AB} = \frac{\partial z^A}{\partial \zeta^B}, \qquad I_{AB} = \begin{Bmatrix} \dfrac{\partial x^a}{\partial \xi^b}, & \dfrac{\partial x^a}{\partial \nu^\beta} \\ \dfrac{\partial \theta^\alpha}{\partial \xi^b}, & \dfrac{\partial \theta^\alpha}{\partial \nu^\beta} \end{Bmatrix}.$$

We say that the map under considerations defines an $(m|n)$-dimensional supersurface in $\mathbb{R}^{M|N}$ if its Jacobian is regular at each point. Two maps $z(\zeta)$ and $z'(\zeta)$ define the same supersurface if there exist an invertible map g of the superspace $\mathbb{R}^{m|n}$ into itself such that $z' = z \circ g$. The map g is called a reparametrization of the surface. (This definition of supersurface is a local one; this remark applies to all the following considerations. The passage to the study of supersurfaces in the large is performed in the standard way and causes absolutely no difficulties.)

The notion of density as introduced in § 1 has a natural generalization to the super case. Let us consider a function $A(z, K)$ where z is a point of $\mathbb{R}^{M|N}$ and K the Jacobian matrix of the linear map $z^A = K^A_B \zeta^B$ from $\mathbb{R}^{m|n}$ into $\mathbb{R}^{M|N}$.

The function $A(z, K)$ will be termed an *$(m|n)$-density* of rank one or simply a density if

$$A(z, KL) = A(z, K) \operatorname{Ber} L, \tag{2.6}$$

where L is the Jacobian matrix of a linear map from $\mathbb{R}^{m|n}$ into itself.

As in ordinary space, to each such density one may assign a functional defined over supersurfaces. Let Ω be an $(m|n)$-dimensional supersurface in $\mathbb{R}^{M|N}$ given by the map $z(\zeta)$. Then the functional

$$\Phi_A(\Omega) = \int A\left(z(\zeta), \frac{\partial z^A}{\partial \zeta^B}\right) d^{m|n}\zeta,$$

$$d^{m|n}\zeta = d^m \xi\, d^n \nu = d\xi^1 \ldots d\xi^m\, d\nu^1 \ldots d\nu^n,$$

corresponding to the density A, does not depend on the choice of the parametrization $z(\zeta)$ of Ω, this in view of (2.6).

In an analogous manner one can define an $(m|n)$-density of rank r by considering integrals of the form

$$\int A\left(z(\zeta), \frac{\partial z^A}{\partial \zeta^B}, \ldots, \frac{\partial^r z^A}{\partial \zeta^{B_1} \ldots \partial \zeta^{B_r}}\right) d^{m|n}\zeta, \tag{2.7}$$

where the map $z(\zeta)$ defines an $(m|n)$-dimensional supersurface. If (2.7) does not depend on the choice of the parametrization $z(\zeta)$, then A is called an $(m|n)$-density of rank r.

Similarly, one may generalize to the super case the notion of D-density, an object which is integrated over surfaces defined by equations ([9]).

The description of the general object of integration over a surface with the aid of the notion of density admits a transfer to the case of superspace.

In the usual case one has most often to integrate differential forms; however, in superspace differential forms are not objects for integration. In superspace a somewhat distinguished rôle among objects for integration is taken up by the integral forms of Bernshteĭn-Leĭtes [4]. These forms have a simple interpretation in terms of D-densities.

As in ordinary space, one defines in superspace the notion of G-invariant density and likewise that of ideal density.

Let us give an example of an ideal density which is a nontrivial generalization of the density (1.7), corresponding to the integral invariants of canonical transformations. Let G be the group of transformations of the superspace $\mathbb{R}^{2m|n}$ which leaves invariant the nondegenerate closed even 2-form $\omega = dz^A I_{AA'} dz^{A'}$ (i.e. the group of supercanonical transformation of $\mathbb{R}^{2m|n}$). Then the $(2p|q)$-density $(p \leqq m,\ q \leqq n)$

$$A\left(z(\zeta), \frac{\partial z^A}{\partial \zeta^B}\right) = \sqrt{\left|\operatorname{Ber} \frac{\partial z^A}{\partial \zeta^B} I_{AA'} \frac{\partial^{A'}}{\partial \zeta^{B'}}\right|}$$

is a G-invariant ideal density in the class of all densities of rank k. The functional corresponding to this density $\Phi = \int_\Omega A$, $\dim \Omega = (2p|q)$, an integral invariant for supercanonical transformations, provides a generalization to the super case of the *integral invariant of Poincaré-Cartan* [19].

In conclusion let us say a few words about the complexification of real structures. We arrive at the notion of $(p|q)$-dimensional complex superspace $\mathbb{C}^{p|q}$ by considering the real superspace $\mathbb{R}^{2p|2q}$ and the automorphism J which sends the Λ-point $(x^1, \ldots, x^{2p}, \theta^1, \ldots, \theta^{2q})$ of this space into the Λ-point $(-x^2, x_1, \ldots, -\theta^2, \theta^1, \ldots)$. Instead of the coordinates x^a, θ^α, pertaining to the real Grassmann algebra, we may consider the coordinates $w^1 = x^1 + ix^2, \ldots, v^1 = \theta^1 + i\theta^2, \ldots$, taking their values in the complexified Grassmann algebra. Then J sends $(w^1, \ldots, w^p, v^1, \ldots, v^q)$ into $(iw^1, \ldots, iw^q, iv^1, \ldots, iv^q)$. The superspace $\mathbb{R}^{2p|2q}$ equipped with the automorphism J may be interpreted as the complex superspace $\mathbb{C}^{p|q}$.

The complex Grassmann algebra differs from the real one only in the respect that the coefficient in the expression (2.1) take complex values. We may always assume that in the complex Grassmann algebra there is a complex conjugation (i.e., an antilinear map $A \to \bar{A}$) satisfying the condition

$$\overline{AB} = \bar{B}\bar{A}. \tag{2.8}$$

We may also assume that complex conjugation maps the generators $\varepsilon^1, \ldots, \varepsilon^q$ into themselves. Then the conjugation is uniquely determined by the requirement (2.8).

A linear map of the superspace $\mathbb{R}^{2p|2q}$ into $\mathbb{R}^{2p'|2q'}$ is termed a complex linear map of the complex superspace $\mathbb{C}^{p|q}$ into $\mathbb{C}^{p'|q'}$ if it commutes with J. A map from $\mathbb{R}^{2p|2q}$ into $\mathbb{R}^{2p'|2q'}$ defines an analytic map from the complex superspace $\mathbb{C}^{p|q}$ into $\mathbb{C}^{p'|q'}$ if its differential is a complex linear map at each point.

Using the conjugation in the Grassmann algebra, one can define a complex conjugation in $\mathbb{C}^{p|q}$ mapping the point $(w^1, \ldots)$ into the point $(\bar{w}^1, \ldots)$.

We see that in superspace all definitions to a high degree resemble the corresponding definitions in ordinary space. This allows us often to give exact definitions of notions in the super case, generalizing the classical ones.

In the sequel we will sometimes omit the prefix "super" where this does not cause any confusion. For instance, we will say space instead of superspace, surface instead of supersurface and so forth.

§ 3. Supersymmetry

As we told above, almost all mathematical notions can be carried over to the case of superspace. In particular, this applies to the group concept. A *supergroup* is defined as a group with parameters belonging to a Grassmann algebra (similarly as in the case of superspace the coordinates belong to a Grassmann algebra). More presicely, in order to define a supergroup we have to associate with each Grassmann algebra Λ a group G_Λ, the group of Λ-elements of the group; for different algebras Λ the groups G_Λ have to agree with each other. (To each parity preserving homomorphism ρ of the Grassmann algebras there has to correspond a homomorphism $\tilde{\rho}$ of the corresponding groups and it is thereby required that $\widetilde{\rho_1 \circ \rho_2} = \tilde{\rho}_1 \circ \tilde{\rho}_2$. Thus, a supergroup defines a functor from the category of Grassmann algebras into the category of groups.) The simplest example of a supergroup is the supergroup $GL(p|q; \mathbb{R})$ of linear transformations of the space $\mathbb{R}^{p|q}$. (For this group the Λ-elements are all invertible linear Λ-maps from $\mathbb{R}^{p|q}$ into itself.) An example of an infinite dimensional supergroup is the supergroup of all selfmaps of $\mathbb{R}^{p|q}$. (Its Λ-elements are all invertible Λ-transformations.)

Let us give the construction of the *Poincaré supergroup*, which in some sense is the smallest supergroup containing the usual Poincaré group. Let us first notice that the Poincaré group may be realized as a subgroup of the group of affine analytic transformations of the space $\mathbb{C}^{4|2}$ with coordinates z^a, $a=0, 1, 2, 3$; θ^α, $\alpha=1, 2$. (The even coordinates z^a transform as a vector and the odd ones θ^α as a two dimensional spinor.) Apparently, the transformations of the Poincaré group map into itself the surface defined by the equations:

$$z^a - \bar{z}^a = 2i\theta^\alpha \sigma^a_{\alpha\dot{\beta}} \bar{\theta}^{\dot{\beta}}. \tag{3.1}$$

Here $\bar{z}^a$ and $\bar{\theta}^\alpha$ are the complex conjugates of z^a and θ^α respectively. (The matrices σ^a are defined by the formulae (0.2).) Besides this, there exist transformations containing anticommuting parameters and mapping the surface (3.1) into itself. Among these one has transformations of the form

$$z^a \to z^a + 2i\theta^\alpha \sigma_{\alpha\dot\beta} \bar\varepsilon^{\dot\beta} + i\sigma_{\alpha\dot\beta} \varepsilon^\alpha \bar\varepsilon^{\dot\beta}, \qquad (3.2)$$
$$\theta^\alpha \to \theta^\alpha + \varepsilon^\alpha,$$

where ε^α are two anticommuting complex parameters. The Poincaré supergroup is now generated by the transformations (3.2) together with the transformations of the ordinary Poincaré group.[10] The Poincaré supergroup induces, in the usual way, transformations on the $(4|4)$-dimensional real surface (3.1). In the coordinates $x^a = 1/2(z^a + \bar z^a)$, θ^α, $\bar\theta^{\dot\alpha}$ on this surface the transformations (3.2) take the form (0.1).

One can also introduce an N-extension of the Poincaré supergroup, by replacing the space $\mathbb{C}^{4|2}$ by $\mathbb{C}^{4|2N}$ and the surface (3.1) by the $(4|4N)$-dimensional surface

$$z^a - \bar z^a = 2i\theta^{\alpha j} \sigma^a_{\alpha\dot\beta} \bar\theta^{\dot\beta}_j, \qquad (3.3)$$

where z^a and $\theta^{\alpha j}$ $(j=1, \ldots, N)$ are the coordinates in $\mathbb{C}^{4|2N}$.

Then the N-extension of the Poincaré supergroup may be defined as a group of affine analytic transformations of $\mathbb{C}^{4|4N}$ mapping the surface (3.3) into itself. Considering the action of the N-extension of the Poincaré supergroup on the surface (3.3) we get a realization of that group as mappings of the space $\mathbb{R}^{4|4N}$ and the definition of the (extended) Minkowski space for any N.

By a supersymmetric theory we mean a theory which is invariant with respect to Poincaré supergroup. It is convenient to construct such theories from fields defined on $\mathbb{R}^{4|4}$ or from analytic fields defined on $\mathbb{C}^{4|2}$. Let us remark that to each analytic field on $\mathbb{C}^{4|2}$ there corresponds a field on $\mathbb{R}^{4|4}$, which is the restriction of an analytic field to the surfaces (3.1). The fields Φ on $\mathbb{R}^{4|4}$ which arise in this way are termed chiral fields. It is easy to see that they satisfy the condition

$$D_{\dot\beta} \Phi = 0,$$

where

$$D_{\dot\beta} = \frac{\partial}{\partial \bar\theta^{\dot\beta}} + i\theta^\alpha \sigma^a_{\alpha\dot\beta} \frac{\partial}{\partial x^a}.$$

The simplest supersymmetric model, the Wess-Zumino model, contains one chiral superfield. Considering this field as an analytic field on $\mathbb{C}^{4|2}$, one may present it in the form

$$\Phi(z^a, \theta^\alpha) = \varphi(z) + \Psi_\alpha(z)\theta^\alpha + F(z)\theta\theta,$$

where $\theta\theta \equiv \varepsilon_{\alpha\beta}\theta^\alpha\theta^\beta \equiv 2\theta^1\theta^2$. The field φ corresponds to a scalar particle and the field Ψ_α to a fermion of spin 1/2.

[10] The analytic selfmaps of $\mathbb{C}^{4|2}$ which map the surface (3.1) into itself generate the socalled superconformal group. The Poincaré supergroup consists of those superconformal maps which are supervolume preserving, i.e. have a Jacobian matrix with unit Berezinian.

§ 4. Supergravity

Let us consider the complex superspace $\mathbb{C}^{4|2}$. Let us denote the coordinates in $\mathbb{C}^{4|2}$ by z^a ($a=0, 1, 2, 3$) and θ^α ($\alpha=1, 2$). We will also employ the unified notation $z^A=(z^a, \theta^\alpha)$. As before, $\mathscr{L}$ denotes the supergroup of analytic transformations of $\mathbb{C}^{4|2}$ whose Jacobian matrix has unit Berezinian. (Geometrically this means that supervolume in $\mathbb{C}^{4|2}$ is preserved.)

As mentioned in the Introduction, a supergravitational field will be considered as a real $(4|4)$-dimensional surface in $\mathbb{C}^{4|2}$. The group $\mathscr{L}$ acts in a natural way in the superspace made up by the $(4|4)$-dimensional surfaces. The action functional of supergravity is defined on the superspace of $(4|4)$-dimensional surfaces; it is defined by a $(4|4)$-density of rank 2. The following proposition holds true: there exists a unique (up to a facton) $(4|4)$-density of rank 2 which is invariant with respect to $\mathscr{L}$ [12]. This density defines the action functional of supergravity.

The proposition just stated means that the density of supergravitational action is uniquely given by its symmetry properties, that is, it is ideal in the sense of the definition in § 1. The proof of this proposition again is based on the remark that a $(4|4)$-dimensional surface in general position in $\mathbb{C}^{4|2}$ in the neighborhood of any point can with the aid of transformations in $\mathscr{L}$ and reparametrizations be reduced, up to infinitesimals of order higher than two, to the standard form

$$\begin{aligned} z^a &= x^a + i\sigma^a_{\alpha\dot\beta}\, v^\alpha \bar v^{\dot\beta}, \\ \theta^\alpha &= v^\alpha, \\ \theta^{\dot\alpha} &= \bar v^{\dot\alpha}. \end{aligned} \tag{4.1}$$

Here x^α, v^α, $\bar v^\alpha$ are the coordinates of real $(4|4)$-dimensional superspace $\mathbb{R}^{4|4}$. (Otherwise put, this means that almost each jet of the second order of a map from $\mathbb{R}^{4|4}$ into $\mathbb{C}^{4|2}$ can be obtained from the standard jet corresponding to (4.1) with the aid of transformations belonging to the group $\mathscr{L}$ and reparametrization maps.) Let us remark that the surface (4.1) coincides with the surface (3.1), which was used in the definition of the Poincaré supergroup. Therefore one can say that it corresponds to a flat superspace. Since the procedure for reducing a surface into the form (4.1) (modulo infinitesimals up to the third order) is completely constructive, the proposition just formulated may be utilized to get an explicit form of the action functional of supergravity [12]. However, a more elegant and also simpler expression for the action functional may be obtained using the Levi form of a $(4|4)$-dimensional surface in $\mathbb{C}^{4|2}$. We will see that with the aid of the Levi form one can construct a density of rank two satisfying the desired invariance requirements [40]. It follows from the above proposition that this density is the one we were looking for.

Let Ω be a $(4|4)$-dimensional surface in general position in $\mathbb{C}^{4|2}$ given by the equations

$$f^b(z, \bar z) = 0 \tag{4.2}$$

where f^b, $b=0, 1, 2, 3$, are real functions. Then the real tangent plane $T_z(\Omega)$ (a real subspace) of Ω at the point z is given by the equations

$$dz^A\,\partial_A f^b + d\bar{z}^A\,\bar{\partial}_A f^b = 0. \tag{4.3}$$

Here $\partial_A = (\partial_a, \partial_\alpha)$, $\partial_a = \partial/\partial z^a$, $\partial_\alpha = \partial/\partial\theta^\alpha$ and analogously for $\bar{\partial}_A$. The maximal complex subspace H_z contained in the real subspace $T_z(\Omega) \subset \mathbb{C}^{4|2}$ is defined by the equations

$$dz^A\,\partial_A f^b = 0. \tag{4.4}$$

The subspace H_z is termed the complex tangent plane and in the case at hand, for surfaces in general position, has complex dimension $(0|2)$. The complex tangent planes are mapped into each other under transformations of the surfaces generated by analytic transformation of the ambient complex space. Let us consider the expression

$$\omega^b = dz^A\,\partial_A\,\overline{dz^B}\,\bar{\partial}_B f^b, \tag{4.5}$$

which defines a vector valued Hermitian form, invariant for analytic transformations $z \to \Phi(z)$. The vector valued form ω^b, considered on the surface Ω and restricted to the complex tangent plane H_z, $z \in \Omega$, is called the *Levi form* of the surface in question. It is not hard to see that the Levi form is invariant with respect to analytic maps. If we replace the f^b by the functions

$$\hat{f}^b(z, \bar{z}) = \eta^b_c(z, \bar{z})\, f^c(z, \bar{z}),$$

then the Levi form transforms in the following manner: $\hat{\omega}^b = \eta^b_c\,\omega^c$. Here the expression (4.5) for ω^b may be viewed as defining the Levi form in the hypothesis that (4.2) and (4.4) are fulfilled. This means, in particular, that on the differentials in (4.5) we have imposed the relation $dz^A\,\partial_A f^b \equiv dx^a\,\partial_a f^b + d\theta^\alpha\,\partial_\alpha f^b = 0$. With the aid of this relation one can express dx^a in terms of $d\theta^\alpha$, that is, one can consider the $d\theta^\alpha$ as coordinates on the $(0|2)$-dimensional complex tangent plane H_z. Let v^α be any other coordinates on H_z. Then $d\theta^\alpha = R^\alpha_\beta\, v^\beta$ and $dx^a = R^a_\beta\, v^\beta$, where R^α_β and R^a_β satisfy the relation

$$R^a_\beta\,\partial_a f^b + R^\alpha_\beta\,\partial_\alpha f^b = 0.$$

(The vectors $R^A_1 = (R^a_1, R^\alpha_1)$ and $R^A_2 = (R^a_2, R^\alpha_2)$ form a basis for the complex plane H_z.) In the coordinates v^α the Levi form takes the form

$$\omega^c = \bar{v}^{\dot{\beta}}\, v^\alpha\, \Gamma^c_{\alpha\beta},$$

where $\bar{v}^{\dot{\beta}} = (v^\beta)^*$ and

$$\Gamma^c_{\alpha\beta} = R^A_\alpha\,\partial_A\,\overline{R^B_\beta}\,\bar{\partial}_B f^c. \tag{4.6}$$

The expression (4.6) will be called the Levi matrix in the basis R^A_α.

The above rule for transforming the Levi form under change of functions defining the surface Ω by the equation (4.2) means that the Levi form characterizes the surface itself and, essentially, does not depend on the choice of the defining equations. It is therefore natural to use the Levi form to build up a functional on the space of surfaces. Let us display a functional, invariant under the group $\mathscr{L}$ and which, consequently (in view of the remarks in the beginning of this section), is the supergravitational action functional. Let us introduce the quantity $\Gamma = \det \Gamma_b^c$, where Γ_b^c is defined by the expansion of the Levi form in a basis for Hermitian matrices: $\Gamma_{\alpha\dot{\beta}}^c = \Gamma_b^c \sigma_{\alpha\dot{\beta}}^b$. If the Levi form is written with the aid of the coordinates $d\theta^\alpha$ on the complex tangent plane (that is, $R_\beta^\alpha = \delta_\beta^\alpha$) then the expression for the action functional takes the form [40]

$$S(\Omega) = C \int \left| \det \left(\frac{\partial f}{\partial x^a} \right)^b \right|^{4/3} |\Gamma|^{-1/3} \cdot \prod_b \delta(f^b(z, \bar{z}))\, dz\, d\bar{z}, \tag{4.7}$$

where C is an arbitrary constant and the δ-function converts the integral on $\mathbb{C}^{4|2}$ into an integral over the surface Ω.[11]

If the basis used in the definition of the Levi form is taken arbitrary, then the expression for the action is somewhat more complicated. Let us consider forms $\sigma^B = dz^A \sigma_A^B$, satisfying $\sigma_A^b = \partial_A f^b$ and the unimodularity condition Ber $\sigma_A^B = 1$. (In other words, $\sigma^B = dz^A \partial_A f^B$ and the forms σ^β complement the system of forms σ^b to a unimodular system $\sigma^B = (\sigma^b, \sigma^\beta)$.) In the basis R_β^A in H_z let us construct the matrix $W_\beta^\alpha = R_\beta^A \sigma_A^\alpha$. The determinant of this matrix will be written W. It is essential that W defines a density with the aid of the functions f^b and the basis R_β^A (that is, it does not depend on the choice of the forms σ^α). If the basis R_β^A in H_z is chosen in an arbitrary way, then the action can be written as

$$S(\Omega) = C \int |W|^{4/3} |\Gamma|^{-1/3} \prod_b \delta(f^b(z, \bar{z}))\, dz\, d\bar{z}. \tag{4.8}$$

Notice that (4.7) arises from (4.8) if we put $\sigma^\alpha = (\det \partial_a f^b)^{1/2} d\theta^\alpha$. One can show (cf. [40]) that the functional S is invariant for the group $\mathscr{L}$ of the supervolume preserving analytic transformations of $\mathbb{C}^{4|2}$ and does not depend on the choice of the functions f^b determining the surface Ω by the equations $f^b = 0$.

In conclusion, let us point out that functionals invariant for supervolume preserving analytic transformations of $\mathbb{C}^{M|N}$ can in an analogous way be obtained with the Levi form of an $(m|n)$-dimensional surface in $\mathbb{C}^{M|N}$ also in the case of some other dimensions $(m|n)$ and $(M|N)$ [40].

§ 5. Transformation of Surfaces into Normal Form

In this section we consider in some greater detail the (4|4)-dimensional surfaces in the superspace $\mathbb{C}^{4|2}$, which, as we have seen, play the rôle of supergravitational fields.

[11] Let us remark that if we write the action in the form (4.7) we can consider the Lagrangean as a (4|4)-D-density of rank 2.

Often it turns out to be useful that a (4|4)-dimensional surface in general position can be reduced with the aid of transformations in the group $\mathscr{L}$ to the standard form (4.1) modulo infinitesimals up to the third order. One application of this proposition we saw in § 4. Now we study the question of putting a (4|4)-dimensional surface in general position in $\mathbb{C}^{4|2}$ in a suitable standard form modulo infinitesimals up to any order. In supergravity the problem of transforming to standard form modulo infinitesimals of the fifth order was studied by V.I. Ogievetskiĭ and E.S. Sokachev [29].

For a general surface we may assume that it is given by the parametric equations

$$\begin{aligned} z^a(x, v, \bar{v}) &= x^a + iH^a(x, v, \bar{v}), \\ \theta^\alpha &= v^\alpha, \end{aligned} \tag{5.1}$$

that is, corresponds to a (real) vector superfield H^a (cf. Introduction). Let us denote the surface (5.1), thus corresponding to the superfield $H^a(x, v, \bar{v})$, by the symbol Ω. Under the action of a map g in $\mathscr{L}$ the field H corresponding to the surface Ω is transformed into a field H^g such that $\Omega(H^g) = \Omega^g(H)$ (g sends Ω into a surface Ω^g).

Let us notice that it is natural to distinguish in $\mathscr{L}$ the subgroup L consisting of those maps $z'(z)$ for which

$$\begin{aligned} z'^a &= \sigma^a_{\alpha\dot{\beta}}\, g^\alpha_\pi\, \bar{g}^{\dot{\beta}}_{\dot{\rho}}\, z^{\pi\dot{\rho}}, \\ \theta'^\alpha &= g^\alpha_\pi\, \theta^\pi, \end{aligned} \tag{5.2}$$

where $\det g^\alpha_\pi = 1$, $\sigma^a_{\pi\dot{\rho}}\, z^{\pi\dot{\rho}} = z^a$. This subgroup L is isomorphic to $SL(2, \mathbb{C})$ and locally isomorphic to the Lorentz group. The coefficients of the expansion of $H^a(z, v, \bar{v})$ in a series in $v, \bar{v}$,

$$\begin{aligned} H^a(x, v, \bar{v}) = S^a(x) &+ [\Gamma^a_\alpha(x)\, \theta^\alpha + \text{complex conjugate}] + \\ &+ [D^a(x)\, v v + \text{complex conjugate}] + E^a_{\alpha\dot{\beta}}(x)\, v^\alpha \bar{v}^{\dot{\beta}} + \\ &+ [\Psi^a_\alpha(x)\, v^\alpha \bar{v}\bar{v} + \text{complex conjugate}] + A^a(x)\, v v \bar{v} \bar{v}, \end{aligned} \tag{5.3}$$

are fields admitting an immediate physical interpretation: $E^a_{\alpha\dot{\beta}}(x)$ is connected with the field of the graviton and Ψ^a_α with the field of the gravitino. (This identification is suggested by the rule under which the fields change under the action of the group L, cf. [27], [28].)

In order to transform (4|4)-dimensional surfaces into standard form we have to analyze the action of $\mathscr{L}$ on a surface Ω or on the field H corresponding to it. It will be convenient to modify somewhat the approach to the solution of this problem. We extend $\mathscr{L}$ to the group $\mathscr{A}$ of all analytic selfmaps of $\mathbb{C}^{4|2}$ subject to the requirement of supervolume preservation. Instead of the space $\mathscr{E}$ of surfaces Ω we consider the space $\mathscr{H}$ of pairs (V, Ω) where Ω is a surface and $V = V(z^A)$ is an even nowhere vanishing function on $\mathbb{C}^{4|2}$. By definition, under the action of the analytic map $z^g = z^{A'}(z^A)$ of the space $\mathbb{C}^{4|2}$ ($g \in \mathscr{A}$) the pair (V, Ω) is mapped onto the pair $(V, \Omega)^g = (V^g, \Omega^g)$, where

$$V^g(z^{A'}) = V(z^A(z^{A'}))\,\mathrm{Ber}\,\frac{\partial z^A}{\partial z^{A'}}. \tag{5.4}$$

In the sequel we refer to $V(z)$ as the volume form.

The solution of the problem of reducing surfaces Ω to a standard form under the action of the group $\mathscr{L}$ can now be replaced by the equivalent problem of reducing pairs (V, Ω) into standard form under the action of maps in $\mathscr{A}$. Indeed, the action of the group $\mathscr{A}$ on the set of pairs (V_0, Ω), where $V_0=1$, restricts to the action of $\mathscr{L}$ on surfaces Ω; moreover, under a suitable map $g\in\mathscr{A}$ any volume form $V(z^A)$ can be written as

$$V^g = V_0. \tag{5.5}$$

From this it follows that the two approaches are equivalent[12].

In particular, by associating with the functional $\Phi(\Omega)$, defined on the space $\mathscr{E}$ of $(4|4)$-dimensional surfaces, the functional $F[(V, \Omega)]$ on the space $\mathscr{H}$ of pairs (V, Ω) subject to

$$F[(V, \Omega)] = \Phi(\Omega^g),$$

where g is defined by the condition (5.5), we get a one-to-one correspondence between $\mathscr{L}$-invariant functionals, defined on $\mathscr{E}$, and $\mathscr{A}$-invariant ones, defined on $\mathscr{H}$. We emphasize that all these considerations are local.

In the sequel, it will be convenient to consider, instead of the space of pairs (V, Ω), where $\Omega=\Omega(H)$, the space of pairs (V, H), which we likewise denote by $\mathscr{H}$.

We define now the subspace $\mathscr{H}_0$ of $\mathscr{H}$ as the set of pairs (V, H) fulfilling the following conditions. The field $V(z^A)$, $z^A=(z^a, \theta^\alpha)$ is subject to the conditions

$$\begin{aligned} &V(z^A) = 1 + V_a z^a + \text{terms of order higher than two in } z^a, \theta^\alpha, \\ &\operatorname{Im} V_a = 0. \end{aligned} \tag{5.6}$$

The field $H(x, v, \bar v)$ is subject to the condition

$$H(x, v, \bar v)|_{x=v=\bar v=0} = 0, \tag{5.7}$$

and the coefficients of the expansion (5.3) of H in powers of v, $\bar v$ to the equations

$$S^a(x)=0, \quad B^a_\alpha(x)=0, \quad D^a(x)=0, \tag{5.8}$$

$$E^a_b(x) = E^b_a(x), \quad E^a_a(x)=4, \quad E^a_b(x)\, x^b = x^a, \tag{5.9}$$

$$\Psi^{\alpha\dot\beta}_\alpha(x)=0, \quad \Psi^{\alpha\dot\beta}_\gamma(x)\, \sigma^\gamma_{\dot\beta b}\, x^b = 0, \tag{5.10}$$

$$A_b(x)\, x^b = 0, \tag{5.11}$$

where $E^a_b\, \sigma^b_{\alpha\dot\beta} = E^a_{\alpha\dot\beta}$, $\Psi^{\alpha\dot\beta}_\gamma\, \sigma^a_{\alpha\dot\beta} = \Psi^a_\gamma$.

The conditions (5.6)–(5.11) defining the subspace $\mathscr{H}_0$ will be called normality conditions for the pair $(V, H)\in\mathscr{H}_0$. (Let us remark that these conditions are Lorentz-invariant, that is, invariant for the subgroup L of $\mathscr{A}$.)

The following proposition holds true [21]. Every pair (V, H) can modulo infinitesimals up to arbitrary order (in z^a, θ^α, x^a, v^α) be transformed into nor-

---

[12] The field V introduced in this way is in physics called a compensating field [11]. The introduction of the field V compensates the extension of the group $\mathscr{L}$ to the group $\mathscr{A}$.

mal form using an analytic map, uniquely determined up to a transformation in the Lorentz group L. More formally, for any pair $(V, H) \in \mathscr{H}$ and each given integer r there exists a map $g \in A$ such that up to infinitesimals of order r the pair $(V, H)^g$ satisfies the normality conditions (5.6)–(5.11). If g and $\tilde{g}$ are two such maps then $\tilde{g} g^{-1} = l(1+\lambda)$, where $l \in L$ and λ is an infinitesimal of order r. (In the proof [21] one can use reasonings analogous to those employed by Chern and Moser [5] in the study of surfaces of real codimension one in $\mathbb{C}^n$. Cf. Chap. V.)

A consequence of this proposition is the fact that the $\mathscr{A}$-invariant functionals on $\mathscr{H}$ are in one-to-one correspondence with the Lorentz-invariant functionals on $\mathscr{H}_0$. Thus, if $I[(V, H)]$ is a Lorentz-invariant functional, defined for $(V, H) \in \mathscr{H}_0$, then it can be continued to an $\mathscr{A}$-invariant functional $F[(V, H)]$ on the entire space $\mathscr{H}$ by requiring that

$$F[(V, H)] = I[(V, H)^g],$$

where $g \in A$ is a map such that $(V, H)^g \in \mathscr{H}_0$. Let us remark that the reduction of fields (V, H) to normal form can be done in a more constructive way (cf. [21]). This can be exploited for an explicit construction of $\mathscr{L}$-invariant functionals corresponding to invariant $(4|4)$-densities of arbitrary rank in $\mathbb{C}^{4|2}$. (We consider only functionals corresponding to densities. Therefore in their study one requires only the above theorem on the reduction to normal form up to infinitesimals of arbitrary order.)

§ 6. Supergravity and G-Structures

The language of G-structures is very useful in the study of supergravity. Below we formulate the notions of the theory of G-structures required [42]. In order to simplify the notation we restrict ourselves to the case of ordinary space (the transition to the case of superspace goes without any trouble). Let G be a subgroup of $GL(m, \mathbb{R})$. We say that an m-dimensional smooth manifold M carries a *G-structure* if at each point of the manifold there is given a distinguished family of tangent frames (depending smoothly on the point) connected with each other by transformations of the group G. Otherwise put, two frames $(e_1, \ldots, e_m)$ and $(e'_1, \ldots, e'_m)$ in the family must be connected by a relation $e'_a = g_a^b e_b$, where g_a^b is a matrix in G. Frames, belonging to the distinguished family, are said to be admissible. Thus for instance, a Riemannian manifold may be regarded as a manifold equipped with an $O(m)$-structure generated by orthonormal frames. A complex manifold may be viewed as a manifold with a $GL(n, \mathbb{C})$-structure, with $m = 2n$. A *G-structure* in an m-dimensional manifold will be termed *trivial* if each point of M has a neighborhood U such that for a suitable system of coordinates $(x^1, \ldots, x^m)$ in U the corresponding coordinate frame $(\partial/\partial x^1, \ldots, \partial/\partial x^m)$ at all points of U is admissible for the G-structure in question. For instance, the $GL(n, \mathbb{C})$-

structure mentioned above on a complex manifold is trivial. (It is generated by holomorphic coordinates.)

A G-structure given on an m-dimensional manifold M induces a certain geometry on an n-dimensional surface $N \subset M$. We say that a tangent frame on N is admissible if it comes from a frame which is admissible for the G-structure on M. It is not hard to write down the restrictions on a surface N under which the family of admissible frames on N just described defines a G'-structure on N – the induced structure (here G' is a subgroup of $GL(n, \mathbb{R})$). Indeed, one has to require that the tangent spaces of N at different points are connected with each other with the aid of maps in G'. (A tangent space of M is considered to be identified with $\mathbb{R}^m$ using any of the admissible frames. This identification is given only up to a map in G.) If admissible frames exist at each point of the surface N, then the surface is said to be regular. On a regular surface two admissible frames are connected by a map belonging to the group G' consisting of those n-dimensional matrices g' which can be completed to a block-triangular matrix

$$g=\begin{pmatrix} g' & h \\ 0 & r \end{pmatrix},$$

belonging to G. Thus, the admissible frames on N define a *G'-structure, induced* on the surface by the G-structure on the ambient manifold.

In particular, if M is a k-dimensional complex manifold defined, as before, by a trivial $GL(k, \mathbb{C})$-structure and N an n-dimensional real surface in general position in M, then for $n>k$ a G'-structure is induced on N, where G' can be described as the group consistings of selfmaps of $\mathbb{R}^n$ leaving a certain fixed $2(n-k)$-dimensional subspace of $\mathbb{R}^n$ invariant and inducing on this subspace a map, complex-linear with respect a certain fixed complex structure on it. This group G' will be denoted by the symbol CR and the corresponding structure will be termed a CR-*structure (Cauchy-Riemann structure)*, cf. chap. VI. This appelation will be employed also in the supercase.

One can prove a general theorem characterizing an induced structure in terms of restrictions on the so-called structural functions of the G'-structure [33]. (The first two structure functions are closely connected with the torsion and the curvature of a connection on the G'-structure.) These restrictions constitute a generalization of the well-known Gauss-Codazzi equations in Riemannian geometry [42]. It turns out that it suffices to consider these restrictions only on a finite number of structural functions; this number is defined with the aid of the Spencer cohomology of a certain linear space of matrices (cf. [33]). In the most interesting cases the restrictions are formulated in terms of only the torsion of a connection of the G'-structure or only the torsion and the curvature [39], [33].

In supergravity, as we have already told, the main object of study is the $(4|4)$-dimensional real surface in $\mathbb{C}^{4|2}$. Let us introduce a trivial $SL(4|2, \mathbb{C})$-structure in $\mathbb{C}^{4|2}$. (Here $SL(4|2, \mathbb{C})$ denotes the group of complex linear selfmaps of $\mathbb{C}^{4|2}$ whose Berezinian is unity.) The Ogievetskiĭ-Sokachev group

$\mathscr{L}$ may be characterized as the automorphism group of this $SL(4|2,\mathbb{C})$-structure (that is, the group of selfmaps of $\mathbb{C}^{4|2}$ mapping admissible frames into admissible ones). On a $(4|4)$-dimensional surface in general position in $\mathbb{C}^{4|2}$ there is induced a SCR-structure, where SCR is the group obtained from the above group CR by imposing a certain determinantal condition [39]. The Levi form of a surface can be viewed as a component of the first structural function of this SCR-structure. This allows us to utilize the results of the theory of G-structures in order to arrive at the expression for the supergravitational action mentioned in § 4. The above mentioned general theorem on induced structures allows us to come from the formulation of supergravity in § 4 to a formulation where the basic object is a SCR-structure on a $(4|4)$-dimensional manifold. On this SCR-manifold we must superimpose restrictions originating from the theorem on induced structures. From the formalism where the basic object is a SCR-structure one can, using a well-known procedure of reduction [42], pass to a formulation of supergravity where the rôle of the structure group is played by the Lorentz group (a formalism of Wess-Zumino type).

Until now, when speaking of supergravity we had only in mind the minimal $N=1$ supergravity. Statements analogous to the ones above can be proved also for so-called non-minimal and alternative minimal supergravities [35], [32], [33]. At present, for more extended supergravities with $N\geqq 3$ one has in superspace only formulations of Wess-Zumino type (cf., for example, [17]). For $N=2$ supergravities essential progress has been made in [10] on the basis of methods analogous to the ones set forth in the next Section.

§ 7. Super Gauge Fields

The notion of *gauge field* on a manifold M can be identified with the notion of connection on a principal G-bundle on the same manifold. Locally such a connection can be defined with the aid of a 1-form A on M taking its values in the Lie algebra $\mathscr{G}$ of the Lie group G (the gauge group). Two fields A and $\tilde{A}$ are termed gauge equivalent if they are related by the formula

$$\tilde{A}=g^{-1}Ag+g^{-1}dg$$

for some G-valued function $g(x)$ on M (the corresponding connections are mapped into each other by a bundle automorphism). The strength of a gauge field (curvature of the connection) is defined by a 2-form F,

$$F=dA+1/2[A\wedge A].$$

The condition $F=0$ is equivalent to the field A being gauge equivalent to zero. The definition of gauge field and strength carries over without any trouble to the case when M is a superspace (or a supermanifold).

In contemporary physics a major rôle is played by theories based on the use of gauge fields in $\mathbb{R}^4$. Lately supersymmetric analogues of these theories have gained a great popularity. It would be natural to think that such analogues could be obtained by considering gauge fields on $\mathbb{R}^{4|4}$ (or on $\mathbb{R}^{4|4N}$, if we are interested in N-extended supersymmetry, in which case for supersymmetric gauge theories N has to equal 1, 2, 3 or 4). It turns out that it is in fact so. However, in the construction of supersymmetric gauge theories there arise not arbitrary gauge fields on $\mathbb{R}^{4|4N}$ but only fields satisfying certain restrictions on their strenghts *(constraints)* [48], [15], [51], [42].

The restrictions on a gauge field in $\mathbb{R}^{4|4N}$ are conveniently written in a basis generated by the vectors $D_a = \partial/\partial x^a$ and

$$D_{\alpha i} = \frac{\partial}{\partial \theta^{\alpha i}} + i\sigma^a_{\alpha\dot\beta}\,\bar\theta^{\dot\beta}_i \frac{\partial}{\partial x^a},$$
$$D^j_{\dot\beta} = \frac{\partial}{\partial \bar\theta^{\dot\beta}_j} + i\theta^{\alpha j}\sigma^a_{\alpha\dot\beta}\frac{\partial}{\partial x^a}. \tag{7.1}$$

(Same notation as in § 3.) The components of the 1-form A of the gauge field in this basis will be denoted by A_a, $A_{\alpha i}$, $A_{\dot\alpha}^{\ i}$ and analogously for the strength. (For $N=1$ the index $i=1, \dots, N$ is not needed.) For each N it is necessary to impose, in particular, the condition $F_{\alpha i\dot\beta}^{\ \ \ i}=0$, which allows us to express the components A_a of the gauge field in terms of the components $A_{\alpha i}$, $A_{\dot\beta}^{\ j}$. Therefore in what follows we need not consider the A_a at all. The remaining conditions on the strength are most conveniently stated as conditions for the strength to vanish on certain subspaces of the complexification of the tangent space to $\mathbb{R}^{4|4N}$.

For $N=1$ one has to require that the strength vanishes on the $(0|2)$-dimensional subspaces generated by the vectors D_α at each point of $\mathbb{R}^{4|4}$. For $N=2$ and $N=3$ one has to impose the vanishing of strength on all $(0|4)$-dimensional complex subspaces generated by vectors of the form

$$\tilde D_\alpha = p^i D_{\alpha i}, \qquad \tilde D_{\dot\beta} = u_j D_{\dot\beta}^{\ j}, \qquad p^i u_i = 0, \tag{7.2}$$

where p^i, u_j are complex numbers. (This formulation of the conditions of the strength is given in [34]. A formulation of the conditions on the strength for $N=1, 2, 3$ using another family of subspaces can be found in [51]; cf. § 8). If $N=4$ the strength has to vanish on all $(0|4)$-dimensional complex subspaces generated by the vectors (cf. [34])

$$\tilde D_1 = p^i\rho^\alpha D_{\alpha i} + u_i\rho^{\dot\beta} D_{\dot\beta}^{\ i},$$
$$\tilde D_2 = q^i\rho^\alpha D_{\alpha i} + v_i\rho^{\dot\beta} D_{\dot\beta}^{\ i},$$
$$\tilde D_3 = p^i\omega^\alpha D_{\alpha i} + u_i\omega^{\dot\beta} D_{\dot\beta}^{\ i},$$
$$\tilde D_4 = q^i\omega^\alpha D_{\alpha i} + v_i\omega^{\dot\beta} D_{\dot\beta}^{\ i},$$

where the complex parameters p^i, q^i, u_i, v_i, ρ^α, $\rho^{\dot\beta}$, ω^α, $\omega^{\dot\beta}$ satisfy the relations

$$p^i u_i = p^i v_i = q^i u_i = q^i v_i = 0,$$
$$(u_i v_j - u_j v_i)\,\rho^{\dot\alpha}\omega^{\dot\beta}\varepsilon_{\dot\alpha\dot\beta} = \varepsilon_{ijkl}\,q^k p^l \rho^\alpha\omega^\beta\varepsilon_{\alpha\beta}.$$

(Here $\varepsilon_{\alpha\beta}$, $\varepsilon_{\dot\alpha\dot\beta}$ and ε_{ijkl} are absolutely antisymmetric symbols.) Let us remark that for $N=3$ and $N=4$ the restrictions on the strength of gauge fields in superspace just described mean, in fact, the equations of motion for ordinary fields on $\mathbb{R}^4$ – the components of the expansion of the superfields in a series in the odd variables. (Moreover $N=3$ and $N=4$ give equivalent equations of motion.) On the contrary, the conditions on the superfield for $N=1$ and $N=2$ are kinematic and the equations of motion constitute additional equations obtained from the Lagrangian, written down in terms of superfields satisfying the above kinematic restrictions.

It is important to notice that the formulations of supersymmetric theories in terms of superfields, on which there are imposed constraints in the form of nontrivial differential equations in superspace, are not sufficiently adequate. It is necessary to pass to formulations with no restrictions on the superfield. It turns out that in the construction of such formalisms (that is, solution of constraints) complex geometry is a most convenient tool. In the case of super gauge theories the point of departure for application of complex geometry is provided by the above interpretation of the restrictions on the strength.

As in the description of supergravity (cf. §§ 4–6), it is in the case of super gauge theories convenient to employ the notion of CR-structure. We say that there is given a CR-*structure* on the manifold M if in the tangent space $T_z(M)$ to each point $z \in M$ there is a distinguished subspace H_z, smoothly depending on z, and if in each of these subspaces H_z there is defined a complex structure, which likewise varies smoothly with z. (Let us recall that if there is defined a complex structure in a real vector space then the complexification of the space decomposes into a sum of two subspaces corresponding of vectors of type (1, 0) and type (0, 1) respectively. Conversely, the presence of such a decomposition is equivalent to a complex structure being introduced. This is often used in the definition of CR-structures.) Let D_α ($\alpha=1, \ldots, h$, where h is the complex dimension of H_z) be vector fields on M spanning at each point $z \in M$ a basis for vectors of type (1, 0) for the subspace H_z. A CR-structure on M is termed integrable if $[D_\alpha, D_\beta]=c^\gamma_{\alpha\beta} D_\gamma$. (If M is a supermanifold one has to replace the commutator by the anticommutator $\{X, Y\}$, provided both vector fields X and Y are odd.) A CR-structure arises in a natural way on a real surface in general position in complex space. In this case $T_z(M)$ is a real subspace of a complex space. We define H_z as the maximal complex subspace contained in $T_z(M)$. This gives a CR-structure on the surface under consideration, called the induced one. It is not hard to convince oneself that such CR-structures (and, in the real-analytic case, only such) are integrable. (It is clear that the definition of CR-structure and induced CR-structure coincides with the one given in § 6 in terms of G-structures. Moreover, the statement on integrable CR-structures is a special case of a general theorem on induced structure [33] mentioned in § 6.)

The superspace $\mathbb{R}^{4|4}$, which serves for the construction of supersymmetric theories for $N=1$, has a natural CR-structure, which is defined by the $(0|2)$-dimensional complex subspaces spanned by the vectors D_α ($\alpha=1, 2$) (cf. (7.1)).

This CR-structure is integrable, as $\{D_\alpha, D_\beta\}=0$. It is easy to see that this CR-structure on $\mathbb{R}^{4|4}$ arises as the induced CR-structure on the surface in $\mathbb{C}^{4|2}$ given by the equations (3.1).

Manifolds equipped with a CR-structure will be termed CR-*manifolds*. A CR-*bundle* on a CR-manifold is bundle such that a typical fiber is a complex manifold whereas the bundle space carries an integrable CR-structure compatible with the complex structure on the fibers and the CR-structure on the basis.

In the case of the $N=1$ super-gauge theory with gauge group G the following statement holds true [30]. There exists a one-to-one correspondence between gauge fields A_α, $A_{\dot\alpha}$ on the CR-manifold $\mathbb{R}^{4|4}$, satisfying the above restrictions on the strength, and CR-fibrations with structure group G^c and a given reduction to the group G. (Here G^c is the complexification of G; it is assumed that $G\subset G^c$.) This statement is the immediate CR-analogue of the known fact on the existence of a canonical connection in a holomorphic Hermitian bundle (as an Hermitian structure on a complex vector bundle may be regarded as a reduction of a $GL(n, \mathbb{C})$-bundle to a $U(n)$-bundle.) Let us remark that such objects admit an interpretation in terms of induced structures: as the ambient space we take a principal holomorphic (or CR-) bundle with group G^c and for the surface we take the subbundle obtained by reducing to the subgroup G (for details see [30]).

This connection between gauge fields and CR-fibrations over $\mathbb{R}^{4|4}$ leads to the known superfield formulation [8], [37] of a supergauge theory with $N=1$ without any constraints on the superfield. The description in terms of CR-bundles admits a generalization also in the case $N>1$, provided we use a trick similar to the Ward transformation for selfdual gauge fields in ordinary four dimensional space [47]. For $N=2$ it is question of gauge fields $A_{\alpha i}$, $A_{\dot\alpha}^{\ i}$ on $\mathbb{R}^{4|8}$ satisfying the above restrictions (constraints). Let us remark that at each point of $\mathbb{R}^{4|8}$ the family of $(0|4)$-dimensional complex tangent spaces appearing in the formulation of these restrictions defines a complex manifold isomorphic to $\mathbb{CP}^1$ and the parameters p^i, $i=1, 2$ can be used as homogeneous coordinates on it (cf. (7.2)). Let us consider the $(6|8)$-dimensional manifold P_2 formed by all tangent subspaces at all points of $\mathbb{R}^{4|8}$ (that is, at each point the original superspace $\mathbb{R}^{4|8}$ is supplemented with a copy of $\mathbb{CP}^1$). The manifold P_2 has a natural CR-structure given by $(1|4)$-dimensional complex subspaces. These subspaces are obtained from the one dimensional complex tangent spaces to each $\mathbb{CP}^1$ in P_2 and $(0|4)$-dimensional complex subspaces in $\mathbb{R}^{4|8}$, each of which being lifted in an obvious way from $\mathbb{R}^{4|8}$ to the very point in P_2 to which it corresponds by the construction. In other words, if we take x^a, $\theta^{\alpha i}$, $z=p^1/p^2$ as coordinates on the $(6|8)$-dimensional manifold P_1 then the vector field (7.2), considered as a field on P_2, spans together with $\partial/\partial\bar z$ at each point the $(1|4)$-dimensional subspace of type $(0, 1)$ defining the CR-structure on P_2. This CR-structure is integrable, as is easy to see. It is likewise clear that if a gauge field is lifted (as a connection) to P_2 then its strength will vanish on the complex subspaces of type

(0, 1) corresponding to the CR-structure on P_2 if and only if the original field on $\mathbb{R}^{4|8}$ satisfies the above restrictions (constraints). We are thus now in a situation analogous to the case $N=1$ but on the auxiliary CR-manifold P_2.

In complete analogy with the Ward transformation we get the following statement [34]. There exists a one-to-one correspondence between complex gauge fields on $\mathbb{R}^{4|8}$, satisfying the said restrictions, and CR-bundles on P_2 which are holomorphically trivial on each $\mathbb{CP}^1$ lying over a point of $\mathbb{R}^{4|8}$. (If we consider real gauge fields, we have to impose specific reality conditions on the bundle obtained.)

In the case $N=3$ one has an analogous proposition, provided we pass to the CR-manifold P_3, consisting of all $(0|4)$-dimensional complex tangent subspaces generated by vectors of the type (7.2) at all points. This time the complex subspaces under consideration define at each point of $\mathbb{R}^{4|12}$ a manifold isomorphic to the flag manifold $F(1, 2, \mathbb{C}^3)$. (This manifold is given as a quadric in $\mathbb{CP}^2 \times \mathbb{CP}^2$, defined in the homogeneous coordinates p^i and u_i by the equations $p^i u_i=0$, cf. (7.2).)

Let us remark that analogous constructions are also possible for supergauge theories in curved $(4|4N)$-dimensional superspace, corresponding to N-extended supergravity. In this case the proposition on integrability of the CR-structures arising follows from conditions imposed on supergravity (analogously to super gauge theories) in the form of restrictions on the torsion and curvature of a connection in a frame bundle. (The structure group usually is the Lorentz group.)

Above we dealt with CR-manifolds and CR-bundles. We must say that in the real analytic case these objects can be obtained from corresponding holomorphic objects. Thus, a CR-bundle is gotten by restricting a holomorphic bundle to a real surface on the base. It is likewise convenient to give a formulation of the said propositions in complexified form, in a spirit close to the well-known twistor constructions [47], [51], [14], [16], [25]. In order to get the complex version it is convenient to use the fact that on the CR-manifold employed above the super conformal group acts transitively. This group may be realized as the group $SU(2, 2|N)$ and its complexification is isomorphic to $SL(4|N; \mathbb{C})$. Therefore the complex analogues $\mathscr{P}_N$ $(N=2, 3)$ of the CR-manifolds P_N arise in a natural way as factor spaces of $SL(4|N; \mathbb{C})$. It turns out that for $\mathscr{P}_2$ one can take the manifold of $(2|1)$-dimensional subspaces in $\mathbb{C}^{4|2}$ (twistor space), that is, $\mathscr{P}_2=G(2|1; \mathbb{C}^{4|2})$, and that for $\mathscr{P}_3$ one can take the flag manifold $\mathscr{P}_3=F(2|1; 2|2; \mathbb{C}^{4|3})$ (for details see [31]). Using these manifolds one can interpret the gauge fields satisfying the above conditions on strength and the requirement of complex analyticity in a certain domain as holomorphic bundles over the corresponding domain in $\mathscr{P}_N$.

The reduction given in this Section of the constraints on the gauge fields to problems in complex geometry makes it possible to construct a formalism of super-symmetric gauge theory (for $N=2$ and $N=3$) in terms of superfields on which no restrictions are imposed (cf. [10]).

§ 8. Geometry of Super Light Like Geodesics

The conditions on the strength of super gauge fields, which we spoke of in the preceding Section, for $N=1, 2, 3$, can also be formulated as vanishing conditions for the strength on so-called super light like planes. (Such a formulation was given by Witten and used in twistor constructions [51].) In order to realize such a formulation of the conditions on the strength one requires a certain family of surfaces in superspace, the tangent spaces of which at each point are super light like planes. In flat superspace (in the sense of § 3) there exists a natural definition of such planes, termed super light like geodesic planes (for details cf. below). It turns out to be useful to consider likewise the analogue of the family of these planes in a more general situation. (Cf. [23], [24], [16], [25], in which papers the space of super light like geodesics in a flat as well as in a curved superspace is used in the study of the solutions of supersymmetric Yang-Mills equations.)

Here we study super light like geodesics in a curved superspace corresponding to supergravity. In an approach of "Wess-Zumino type" in N-extended supergravity the basic object is a $(4|4N)$-dimensional superspace M such that at each point there is chosen a tangent frame (super "Vierbein") defined up to a map belonging to the Lorentz group L (that is, an L-structure on $\mathbb{R}^{4|4N}$). Moreover, there are imposed certain restrictions (constraints) on the torsion and curvature of this L-structure (cf. [17]). It turns out that from the geometry of the superspace arising in supergravity one can draw specific conclusions about the geometry of super light like geodesic surfaces in it [1]. This means also that the constraints in a Wess-Zumino type approach (at least to some extent) gain a definite geometric meaning. (We discussed the geometric meaning of the constraints from a different point of view already in § 6.)

In the flat superspace (Minkowski superspace) $\mathbb{R}^{4|4N}$ [13], with coordinates denoted, as before (cf. § 3), by x^a, $\theta^{\alpha i}$, $\bar{\theta}^{\dot{\alpha}}{}_i$, a *super light like geodesic surface* is defined [7], [51] as a $(1|2N)$-dimensional surface which can be written in the parametric form

$$\begin{aligned} x^a &= c^a + t\rho^\alpha \sigma^a_{\alpha\dot{\beta}} \bar{\rho}^{\dot{\beta}} + i\rho^\alpha \varepsilon^j \sigma^a_{\alpha\dot{\beta}} \bar{\theta}^{\dot{\beta}}_j - i\theta^{\alpha j} \sigma^a_{\alpha\dot{\beta}} \bar{\rho}^{\dot{\beta}} \bar{\varepsilon}_j, \\ \theta^{\alpha i} &= \eta^{\alpha i} + \rho^\alpha \varepsilon^i, \\ \bar{\theta}^{\dot{\alpha}}_i &= \bar{\eta}^{\dot{\alpha}}_i + \bar{\rho}^{\dot{\alpha}} \varepsilon_i, \end{aligned} \tag{8.1}$$

where ρ^α is a numerical (commuting) complex spinor, defining a super-light like geodesic issuing from the point $(c^a, \eta^{\alpha i}, \bar{\eta}^{\dot{\alpha}}_i)$; $t, \varepsilon^i, \bar{\varepsilon}_i$ parametrize the surface (here $\bar{\varepsilon}_i$ is the complex conjugate of ε^i and t real).

---

[13] We remark that in twistor constructions [7], [51], [23], [24], [16] one uses complex super light like geodesics in complexified superspace. We shall deal with the superspace $\mathbb{R}^{4|4N}$. However, the passage to the complex case turns out to be trivial.

It is clear that if $N=0$, that is, in ordinary Minkowski space, we get ordinary one dimensional light like geodesics.

The space of all super light like geodesics in the complexified Minkowski superspace $\mathcal{M}$ can be realized as a $(5|2N)$-dimensional quadric L in $\mathbb{CP}^{3|N} \times \mathbb{CP}^{3|N}$. Witten [51] proved that for $N=0, 1, 2, 3$ there exists a one-to-one correspondence between super gauge fields holomorphic in a suitable domain in $\mathcal{M}$ and certain holomorphic bundles on the corresponding domain in L. (The case $N=0$ is also considered in [14].) For this one uses the fact that a super gauge field is flat on each super light like geodesic. For $N=0$ this follows from the fact that light like geodesics are one dimensional but for $N=1, 2, 3$ one requires the restrictions which, as we have told, must be imposed on the strength of super gauge fields. A most beautiful interpretation can be given also for the equations of motion. It turns out [51], [14] that for $N=0$ gauge fields, satisfying the Yang-Mills equations, correspond to those of the bundles considered which can be continued to a third order infinitesimal neighborhood of the quadric L in $\mathbb{CP}^3 \times \mathbb{CP}^3$. On the other hand, if $N=3$ a bundle on the $(5|6)$-dimensional manifold L describes a field which already satisfies the field equations [51], because in this case the restrictions imposed on the strength are equivalent to these equations. In the general case, for $N=0, 1, 2, 3$, we get the proposition [16] that a field satisfying the (supersymmetric) Yang-Mills equations can be distinguished by the condition that the bundle corresponding to it can be extended to a $(3-N)$-th order infinitesimal neighborhood of L of $\mathbb{CP}^{3|N} \times \mathbb{CP}^{3|N}$.

In a curved superspace M we define a *super light like geodesic* as a $(1|2N)$-dimensional surface such that at each point the tangent plane is arranged similarly as the tangent plane to a super light like geodesic in a flat space. More exactly, we require that at each point of this $(1|2N)$-dimensional surface there exist complex numbers ρ^α, $\alpha=1, 2$, such that the vectors

$$Q_i=\rho^\alpha E_{\alpha i}, \quad \bar{Q}^i=\bar{\rho}^{\dot\alpha} \bar{E}^i_{\dot\alpha}, \quad D=(\rho^\alpha \sigma^a_{\alpha\dot\beta} \bar{\rho}^{\dot\beta}) E_a \tag{8.2}$$

span a basis of the tangent space at this point. (Here $E_A=(E_a, E_{\alpha i}, \bar{E}_{\dot\alpha}^{\,i})$ are the vectors of the super Vierbein corresponding to the L-structure on M defined by the supergravity field.) A super light like surface is said to be *super light like geodesic* if it is autoparallel (that is, if under parallel transport of a vector belonging to the tangent plane along a curve lying in the surface we again get a vector tangent to the surface). The existence of super light like surfaces (and, even more, super light like geodesics) is not *a priori* evident. However, one can show [1] that the accepted constraints in supergravity on the torsion and curvature imply the existence of super light like geodesics for supergravity with any N. Moreover, for each point of superspace and vectors at this point defined by the formulae (8.2) one can find a (unique) super light like geodesic passing trough this point and tangent to the given vector.

For $N=1$ and $N=2$ one can divide the constraints into kinematical ones and dynamical ones. (Strictly speaking, in these cases by constraints one usually understands only kinematical constraints.) It turns out that in these cases for the above proposition on the existence of super light like geodesics to hold it suffices to require only that the kinematic constraints are satisfied. Moreover, if in $N=1$ minimal supergravity we take the kinematic constraints in the Ogievetskiĭ-Sokachev form [28] then the inner geometry of super light like geodesics turns out to be flat. In other words, under parallel transport of a tangent vector of a super-light like geodesic surface along a closed curve on this surface this vector remains unchanged. Exactly the same result holds for nonminimal supergravity and for $N=2$ supergravity.

The equations of motion (the dynamical constraints) imply new restrictions on the geometry of super light like geodesics. Namely, if the equations of motions in Wess-Zumino form hold for minimal $N=1$ supergravity then the super light like geodesics turn out to be not only flat but also flatly imbedded in the ambient space. Otherwise put, under parallel transport on any vector along a closed curve lying on a super-light like geodesic we get back the original vector (we do not assume that the vector is tangent to the surface). The same statement is true in the case when the equations of motion are fulfilled in nonminimal $N=1$ supergravity or else the equations of motion of $N=2$ supergravity.

In conclusion, let us point out that in the proof of the propositions stated one uses an auxiliary space $\hat{P}$, the manifold of all tangent super light like subspaces at all points of the original $(4|4N)$-dimensional superspace M. Then the L-structure and the connection on it over M define a distribution of $(1|2N)$-dimensional subspaces on $\hat{P}$. Besides this, for instance, the proof of the existence of super-light like geodesics on M requires that one checks that the distribution on $\hat{P}$ mentioned is involutive.

References

For the convenience of the reader, references to reviews in Zentralblatt für Mathematik (Zbl.), compiled using the MATH database, have been included as far as possible.

1. Baranov, M.A., Roslyĭ, A.A., Schwarz, A.S.: Super light like geodesics in supergravitation. Yad. Fiz. *41*, 285–287 (1985) [Russian]. English transl.: Sov. J. Nucl. Phys. *41*, 180–181 (1985). Zbl. 592.58013
2. Berezin, F.A.: Introduction to the algebra and analysis of anticommuting variables. Moscow: MGU 1983 [Russian]. Zbl. 527.15020. English transl.: Introduction to superanalysis (Part I). Dordrecht: Reidel 1987
3. Berezin, F.A., Kats, G.I.: Lie groups with commuting and anticommuting variables. Mat. Sb. Nov. Ser. *82* (124), 343–359 (1970) [Russian]. English transl.: Math. USSR, Sb. *11* (1970), 311–325 (1971)
4. Bernshteĭn, I.N., Leĭtes, D.A.: How to integrate differential forms on supermanifolds. Funkts. Anal. Prilozh. *11*, No. 3, 70–71 (1976) [Russian]. English transl.: Funct. Anal. Appl. *11*, 219–221 (1978). Zbl. 364.58005
5. Chern, S.S., Moser, J.K.: Real hypersurfaces in complex manifolds. Acta Math. *113* (1974) 219–271 (1975). Zbl. 302.32015. Russian transl.: Usp. Mat. Nauk *38*, No. 2 (230) 149–193 (1983)

6. Fayet, P., Ferrara, S.: Supersymmetry. Phys. Rep. *32*, 250–334 (1977)
7. Ferber, A.: Supertwistors and conformal supersymmetry. Nucl. Phys. B *132*, 55–64 (1978)
8. Ferrara, S., Zumino, B.: Supergauge invariant Yang-Mills theories. Nucl. Phys. *B 79*, 413–421 (1974)
9. Gaĭduk, A.V., Khudaverdyan, O.M., Schwarz, A.S.: Integration over surfaces in superspace. Teor. Mat. Fiz. *52*, 375–383 (1982) [Russian]. English transl.: Theor. Math. Phys. *52*, 862–868 (1983). Zbl. 513.58015
10. Galperin, A., Ivanov, E., Kalitzin, S., Ogievetsky, V., Sokachev, E.: Unconstrained N=2 matter, Yang-Mills and supergravity theories in harmonic superspace. Classical Quantum Gravity *1*, 469–498 (1984)
11. Gates, S.J., Siegel, W.: Understanding constraints in superspace formulation of supergravity. Nucl Phys. B *163*, 519–545 (1980)
12. Gayduk, A.V. (=Gaĭduk, A.V.), Romanov, V.N., Schwarz, A.S.: Supergravity and field space democracy. Commun. Math. Phys. *79*, 507–528 (1981)
13. Gol'fand, Yu.A., Likhtman, E.P.: Extension of the algebra of generators of the Poincaré group and violation of P-invariance. Zh. Eksper. Teor. Fiz. *13*, 452–457 (1971) [Russian]
14. Green, P.S., Isenberg, J., Yasskin, P.B.: Non-self-dual gauge field theories. Phys. Lett. B *78*, 462–464 (1978)
15. Grimm, R., Sohnius, M, Wess, J.: Extended supersymmetry and gauge theories. Nucl. Phys. B *133*, 275–284 (1978)
16. Henkin, G.M. (=Khenkin, G.M.): Tangent Cauchy-Riemann equations and Yang-Mills, Higgs and Dirac fields. In: Proc. of the ICM, Warszawa, Aug. 16–24, 1983, 809–827. Amsterdam etc.: North Holland 1984. Zbl. 584.58050
17. Howe, P.: Supergravity in superspace. Nucl. Phys. B *199*, 309–364 (1982)
18. Khudaverdian, O.M. (=Khudaverdyan, O.M.), Schwarz, A.S. (=Shvarts, A.S.), Tyupkin, Yu.S.: Integral invariants for supercanonical transformations. Lett. Math. Phys. *5*, 517–522 (1981). Zbl. 521.58054
19. Khudaverdyan, O.M., Schwarz, A.S.: Additive and multiplicative functionals. Preprint ITEF-3. Moscow: 1980 [Russian]
20. Khudaverdyan, O.M., Schwarz, A.S.: Multiplicative functionals and gauge fields. Teor. Mat. Fiz. *46*, 187–198 (1981) [Russian]
21. Khudaverdyan, O.M., Schwarz, A.S.: Normal gauging in supergravity. Teor. Mat. Fiz. *57*, 354–362 (1983) [Russian]
22. Lee, H.-C.: The universal integral invariants of Hamiltonian systems and application to the theory of canonical transformations. Proc. R. Soc. Edinb., Sect. A *62*, 237–246 (1947). Zbl. 30, 55
23. Manin, Yu.I.: Flag superspaces and the supersymmetric Yang-Mills equations. In: Problems of high energy physics and quantum theory of fields. Tr. Mezhdunarodnogo Seminara, Protvino, July 1982, 46–73. Serdukhov: IFVE 1982 [Russian]
24. Manin, Yu.I.: Supersymmetry and supergravity in the space of null supergeodesics. In: Group-theoretic methods in physics. Tr. Mezdunarodnogo Seminara, Zvenigorod, November 1982, Vol. 1, 203–208. Moscow: Nauka 1983 [Russian]. Zbl. 599.58004
25. Manin, Yu.I.: Gauge fields and complex geometry. Moscow: Nauka 1984 [Russian]. Zbl. 576.53002. English transl.: Berlin etc.: Springer-Verlag 1988
26. Ogievetskiĭ, V.I., Mezinchesku, L.: The symmetry between bosons and fermions and superfields. Usp. Fiz. Nauk *117*, No. 4, 637–683 (1975) [Russian]
27. Ogievetskiĭ, V.I., Sokachev, E.S.: The simplest group for Einstein supergravity. Yad. Fiz. *31*, 264–279 (1980) [Russian]
28. Ogievetskiĭ, V.I., Sokachev, E.S.: Axial gravitational superfield and the formalism of differential geometry. Yad. Fiz. *31*, 821–840 (1980) [Russian]. Zbl. 569.35039. English transl.: Sov. J. Nucl. Phys. *31*, 424–433 (1980)
29. Ogievetskiĭ, V.I., Sokachev, E.S.: Normal gauge in supergravity. Yad. Fiz. *32*, 862–869 (1980) [Russian]
30. Rosly, A.A. (=Roslyĭ, A.A.): Geometry of N=1 Yang-Mills theory in curved superspace. J. Phys. A *15*, 1663–1667 (1982)

31. Rosly, A.A.: Gauge fields in superspace and twistors. Classical Quantum Gravity *2*, 693–699 (1985). Zbl. 576.35075
32. Rosly, A.A., Schwarz, A.S.: Geometry of N = 1 supergravity. Commun. Math. Phys. *95*, 161–184 (1984)
33. Rosly, A.A., Schwarz, A.S.: Geometry of N = 1 supergravity, II. Commun. Math. Phys. *96*, 285–309 (1984)
34. Roslyĭ, A.A.: Constraints in supersymmetric Yang-Mills theory as integrability conditions. In: Group-theoretic methods in physics. Tr. Mezhdunarodnogo Seminara, Zvenigorod, November, 1982, pp. 263–268. Moscow: Nauka 1982 [Russian]. Zbl. 599.58056
35. Roslyĭ, A.A., Schwarz, A.A.: The geometry of nonminimal and alternative minimal supergravity. Yad. Fiz. *37*, 786–794 (1983) [Russian]. Zbl. 592.53067. English transl.: Sov. J. Nucl. Phys. *37*, 466–471 (1983)
36. Salam, A., Strathdee, J.: Super-gauge transformations. Nuclear Phys. B *76*, 477–482 (1974)
37. Salam, A., Strathdee, J.: Super-symmetry and non-Abelian gauges. Phys. Lett. B *51*, 353–355 (1974)
38. Schwarz, A.S.: Are the field and space variables on equal footing? Nucl. Phys. B *171*, 154–166 (1980)
39. Schwarz, A.S.: Supergravity, complex geometry and G-structures. Commun. Math. Phys. *87*, 37–63 (1982)
40. Schwarz, A.S.: Supergravity and complex geometry. Yad. Fiz. *34*, 1114–1149 (1981) [Russian]
41. Sohnius, M.F.: Bianchi identities for supersymmetric gauge theories. Nucl. Phys. B *136*, 461–474 (1978)
42. Sternberg, S.: Lectures on differential geometry. Englewood Cliffs: Prentice Hall 1964. Zbl. 129, 131
43. Szczyrba, W.: A symplectic structure for the Einstein-Maxwell field. Rep. Math. Phys. *12*, 169–191 (1977). Zbl. 396.53020
44. Van Nieuwenhuizen, P.: Supergravity. Phys. Rep. *68*, 189–398 (1981)
45. Volkov, D.V.: Phenomenological Langrangeans. In: The Physics of elementary particles and the atomic nucleus, 3–41. Moscow: Atomizdat 1973 [Russian]
46. Volkov, D.V., Akulov, V.P.: The Goldstine field with one half spin. Teor. Mat. Fiz. *18*, 39–50 (1974) [Russian]
47. Ward, R.S.: On self-dual fields. Phys. Lett. A *61*, 81–82 (1977)
48. Wess, J.: Supersymmetry – supergravity. In: Topics in quantum field theory and gauge theories. Proc. VIII Internat. Seminar on Teor. Phys., Salamanca, June 1977. Lect. Notes Phys. 77, 81–125. Berlin etc.: Springer 1978
49. Wess, J., Zumino, B.: Supergauge transformations in four dimensions. Nucl. Phys. B *70*, 39–50 (1974)
50. Wess, J., Zumino, B.: Superspace formulation of supergravity. Phys. Lett. B *66*, 361–364 (1977)
51. Witten, E.: An interpretation of classical Yang-Mills theory. Phys. Lett. B *77*, 394–398 (1978)

Author Index

Subject Index